The Magellanic Clouds – a pair of nearby, satellite galaxies – are caught in a dynamic struggle internally and with our Milky Way. Given their close proximity, they offer a unique opportunity to study in detail the dynamics and composition of other galaxies. They have a long history of study, but interest in them has blossomed over the past four decades. This is the first book to provide a synthesized and comprehensive account of the Magellanic Clouds.

This authoritative volume presents the latest understanding of the structure, evolution and dynamics of these satellite galaxies. It draws together wide-ranging observations in the X-ray, far-ultraviolet, infrared and millimetre wavelengths, including results from the Hubble Space Telescope.

For graduate students and researchers, this timely edition provides a definitive reference on the Magellanic Clouds; it also gives useful supplementary reading for graduate courses on galaxies, the interstellar medium, stellar evolution and the chemical composition of galaxies.

THE MAGELLANIC CLOUDS

Cambridge astrophysics series

Series editors

Andrew King, Douglas Lin, Stephen Maran, Jim Pringle and Martin Ward

Titles available in this series

5. The Solar Granulation
   by R. J. Bray, R. E. Loughhead and C. J. Durrant
7. Spectroscopy of Astrophysical Plasmas
   by A. Dalgarno and D. Layzer
10. Quasar Astronomy
    by D. W. Weedman
13. High Speed Astronomical Photometry
    by B. Warner
14. The Physics of Solar Flares
    by E. Tandberg-Hanssen and A. G. Emslie
15. X-ray Detectors in Astronomy
    by G. W. Fraser
16. Pulsar Astronomy
    by A. Lyne and F. Graham-Smith
17. Molecular Collisions in the Interstellar Medium
    by D. Flower
18. Plasma Loops in the Solar Corona
    by R. J. Bray, L. E. Cram, C. J. Durrant and R. E. Loughhead
19. Beams and Jets in Astrophysics
    edited by P. A. Hughes
20. The Observation and Analysis of Stellar Photospheres
    by David F. Gray
21. Accretion Power in Astrophysics 2nd Edition
    by J. Frank, A. R. King and D. J. Raine
22. Gamma-ray Astronomy 2nd Edition
    by P. V. Ramana Murthy and A. W. Wolfendale
23. The Solar Transition Region
    by J. T. Mariska
24. Solar and Stellar Activity Cycles
    by Peter R. Wilson
25. 3K: The Cosmic Microwave Background Radiation
    by R. B. Partridge
26. X-ray Binaries
    by Walter H. G. Lewin, Jan van Paradijs and Edward P. J. van den Heuvel
27. RR Lyrae Stars
    by Horace A. Smith
28. Cataclysmic Variable Stars
    by Brian Warner
29. The Magellanic Clouds
    by Bengt E. Westerlund

THE MAGELLANIC CLOUDS

BENGT E. WESTERLUND
Professor emeritus, Uppsala University

PUBLISHED BY THE PRESS SYNDICATE OF THE UNIVERSITY OF CAMBRIDGE
The Pitt Building, Trumpington Street, Cambridge CB2 1RP, United Kingdom

CAMBRIDGE UNIVERSITY PRESS
The Edinburgh Building, Cambridge CB2 2RU, United Kingdom
40 West 20th Street, New York, NY 10011-4211, USA
10 Stamford Road, Oakleigh, Melbourne 3166, Australia

© Cambridge University Press 1997

This book is in copyright. Subject to statutory exception
and to the provisions of relevant collective licensing agreements,
no reproduction of any part may take place without the written
permission of Cambridge University Press

First published 1997

Printed in the United Kingdom at the University Press, Cambridge

Typeset in Times 10/12.5pt

A catalogue record for this book is available from the British Library

Library of Congress Cataloguing in Publication Data
Westerlund, Bengt E.
The Magellanic Clouds / Bengt E. Westerlund.
  p. cm. – (Cambridge astrophysics series ; 29)
Includes bibliographical references and index.
ISBN 0 521 48070 1
1. Magellanic Clouds. I. Title. II. Series.
QB858.5M33W47 1997
523.1'12–dc20   96-21791   CIP

ISBN 0 521 48070 1 hardback

To
Hillevi
with thanks for 50 years of untiring support

and to the memory of
Bart J. Bok
Daniel Chalonge
J. Jason Nassau
and
Carl Schalén

MY PURPOSE IS TO TELL OF BODIES WHICH HAVE BEEN TRANSFORMED INTO SHAPES OF A DIFFERENT kind. You heavenly powers, since you were responsible for those changes, as for all else, look favourably on my attempts, and spin an unbroken thread of verse, from the earliest beginnings of the world, down to my own times.

(From Ovid's *Metamorphoses*, translated by Mary M. Innes; Penguin Books Ltd)

Contents

Preface	page xiii
Acknowledgements	xv

1 Introduction	1
2 The distances of the Clouds	6
2.1 The reddening, foreground and internal	6
2.2 Distance determinations with the aid of Cepheids	9
2.3 Distance determinations with the aid of RR Lyrae variables	12
2.4 Distance determinations with the aid of Mira variables	14
2.5 Distance determinations with the aid of clusters	15
2.6 Distance determinations with the aid of field stars	15
2.7 Summary of distance determinations	19
3 The Clouds as galaxies	21
3.1 Models of the Magellanic System	21
3.2 Proper motions of the LMC and the SMC	26
3.3 Transverse motions of the LMC and the SMC	26
3.4 Integrated properties of the LMC and the SMC	27
3.5 The geometry of the Clouds	27
3.6 The structure of the LMC	30
3.7 The structure of the SMC	32
3.8 Gaseous coronae?	33
3.9 The Magellanic Stream	35
3.10 The InterCloud region	38
3.11 The generations in the Magellanic Clouds	42
4 The cluster population	43
4.1 Surveys	46
4.2 The classification of Cloud clusters	48
4.3 The surface distribution of clusters in the Clouds	49
4.4 The age distribution of Cloud clusters	51
4.5 The structure of Cloud clusters	70
4.6 Binary clusters	72
4.7 Individual clusters	75

5 The youngest field population — 79
5.1 O stars and B–G supergiants — 79
5.2 The most luminous stars — 81
5.3 Wolf–Rayet stars — 85
5.4 The red supergiants — 85
5.5 Supergiant stars with circumstellar shells — 87
5.6 Supergiant stars in small HII regions — 88
5.7 Multiple stars in the LMC — 90
5.8 Multiple stars in the SMC — 91
5.9 Infrared protostars — 92
5.10 The stellar associations — 93

6 The superassociations and supergiant shells — 99
6.1 Superassociations in the LMC — 100
6.2 Superassociations in the SMC — 109

7 The intermediate-age and oldest field populations — 111
7.1 Main-sequence stars — 111
7.2 Red-giant stars — 113
7.3 The carbon stars — 124
7.4 The CH stars — 125
7.5 Long-period variables — 127
7.6 OH/IR stars and *IRAS* sources — 130
7.7 Planetary nebulae — 132
7.8 Novae — 141

8 The interstellar medium — 143
8.1 The neutral hydrogen — 143
8.2 The ionized hydrogen — 146
8.3 Some well-studied emission nebulae in the LMC — 150
8.4 The molecular content — 155
8.5 The thermal and non-thermal components — 165
8.6 The dust content — 170
8.7 Interstellar absorption lines — 175
8.8 Diffuse interstellar bands — 177

9 X-ray emission and supernova remnants — 179
9.1 X-ray observations of the LMC — 180
9.2 X-ray observations of the SMC — 188
9.3 Supernova remnants in the Clouds — 190
9.4 SN 1987A — 199

10 The 30 Doradus complex — 202
10.1 The extinction in the 30 Doradus region — 203
10.2 The morphology of the 30 Doradus nebula — 206
10.3 The kinematics of the 30 Doradus nebula — 210
10.4 The stellar content of the 30 Doradus region — 213

Contents xi

10.5 Comparison of the R136 cluster with galactic clusters 219

11 Chemical abundances 221
11.1 Element abundances in the interstellar medium 221
11.2 Element abundances in clusters 222
11.3 Element abundances in hot field stars and Cepheids 227
11.4 Element abundances in cool field stars 229
11.5 Element abundances in planetary nebulae 231
11.6 Summary of abundance determinations 232

12 The structure and kinematics of the Magellanic System 235
12.1 The structure and kinematics of the LMC 235
12.2 The structure and kinematics of the SMC 244
12.3 The motions in the Magellanic Stream and in the Bridge 248

Appendix 1 **Acronyms and abbreviations used frequently in the text** 250

Appendix 2 **Reviews and proceedings** 252

Bibliography 253
Object index 271
Subject index 275

Preface

Interest in the Magellanic Clouds has grown tremendously over the past four decades. During this period they have been exposed to investigations, interpretations, and speculations with regard to their origin, evolution, structure and content. At times, they have been viewed as more spectacular than they perhaps really are, e.g. suggested to have supermassive stars and peculiar structures; at other times they have been wished far away. Shapley once said (in Galaxies, Harvard University Press, most recent edition 1973, ed. P.W. Hodge) that 'The Astronomy of galaxies would probably have been ahead by a generation, perhaps by 50 years, if Chance, or Fate, or whatever it is that fixes things as they are had put a typical spiral and a typical elliptical galaxy in the positions now occupied by the Large and Small Magellanic Clouds.... But we must make the best of what we have, and it will soon appear that the best is indeed good. It's marvelous.' This has indeed been shown to be true. The two irregulars, which differ in so many aspects from our Galaxy, have in particular shown their value as two excellent astrophysical laboratories near at hand.

The study of the Magellanic Clouds has in many ways become more 'galactic' than 'extragalactic'. It is therefore equally impossible to cover all Magellanic Cloud research in detail in one monograph as it would be for our Galaxy. I have attempted to include an equal amount of 'extragalactic' and 'astrophysical' topics, i.e. topics dealing with the structure, motion and evolution of the galaxies on one side and those dealing with stellar evolution and the interstellar medium on the other. Unavoidably, some important questions have either been treated too summarily or have been left out. This is, for instance, the case for the supernova SN 1987A. It has been, and frequently is, considered in reviews, and observations, analyses and modelling of it continues. Also, some interesting new approaches, such as the monitoring of the light curves of millions of stars in the Clouds, primarily for the detection of MACHOs (massive compact halo objects) in our Galaxy, have regrettably had to be left out.

The dominating theme in all Magellanic Cloud research is their evolution and present structure. Ten to twenty years ago the leading question may have concerned their spiral or non-spiral structure. Today, with a, not negligible, line-of-sight depth acknowledged in both Clouds, other questions appear more fundamental. Stellar evolution studies are stimulated by the detection of hypergiant stars and supernovae, and by infrared protostars at one end of the stellar lifeline and by the red-giant–planetary nebula link at the other.

It has been exceedingly difficult to order the available material in a, to me, satisfactory way. The more time I have spent on the various topics, the more have

I found them all tied together. Evolution, chemical composition, and kinematics are good single topics for reviews, but one has to consider clusters, HII regions, planetary nebulae, etc., in all of them. At the same time the latter deserve, in their own right, to be treated extensively.

In order to avoid repeating vital information frequently, I have been forced to leave out chapters devoted to variables, luminosity functions, initial mass functions, etc. Cepheids, RR Lyrae stars, Miras, as well as luminosity and mass functions, are treated in their 'natural surroundings', e.g. in distance determinations, in clusters and associations, etc. However, as a consequence, some types of variable, e.g. eclipsing binaries, have been left out.

Apart from the problems of selecting and arranging the material, most of which appeared before the middle of 1995, I have enjoyed seeing the Magellanic Clouds from a number of different angles. I stand impressed by what I have seen of research achieved by astronomers using ground-based and space telescopes and by those who are continuously improving models and theory.

The division into chapters that follows, is the best solution that I could find.

In Chap. 1, the Introduction, some glimpses of the history of Magellanic Cloud research are given. Recent determinations of the distances of the Clouds are considered in Chap. 2. The Clouds as galaxies, i.e. their history as members of the Local Group and satellites to our Galaxy, are discussed in Chap. 3, together with the overall structure of the Magellanic System. Chap. 4 is devoted to the stellar clusters which have played, and continue play, a dominating role in the analyses of the evolution of the Clouds. The most important characteristics of each generation identified among the field stars are described in Chaps. 5 and 7. The superassociations and supergiants shells are treated separately in Chap. 6 as particularly important features in the youngest generation. The interstellar medium in the Clouds is considered in Chap. 8 and, as X-ray emission, in Chap. 9. The 30 Doradus complex, the most investigated part of the LMC, is described in Chap. 10. I have chosen to present the exciting amount of new *HST* data on the 30 Doradus core region without detailed intercomparisons. It appears too early to do this now and it is sufficient to enjoy the promising research underway. Chemical-abundance determinations are treated in Chap. 11, and the structure and kinematics of the Magellanic System are discussed in Chap. 12. Acronyms and abbreviations frequently used in the text are gathered in Appendix 1, and some reviews and symposia are listed in Appendix 2.

Acknowledgements

I am grateful to the following for permission to reproduce copyright material: *American Scientist*, an extract from Harlow Shapley, 'The Clouds of Magellan, A Gateway to the Sidereal Universe'; a number of astronomers with copyright to figures and tables published in the *Astronomical Journal* and the *Astrophysical Journal*, all duly identified in the text; *Astronomy & Astrophysics*, a number of figures and tables, identified in the text; *Blackwell Science Ltd*, Figures 3.1, 8.2, 8.3, and 10.5 from *Monthly Notices of the Royal Astronomical Society*; *Harvard University Press*, quotations from Harlow Shapley's 'Galaxies' (edition revised by P.W. Hodge 1973); *Kluwer Academic Publishers*, Figure 4.2; *Penguin Books Ltd*, an extract from Ovid's 'Metamorphoses' (translated by Mary M. Innes, 1955); *Springer Verlag*, Figures 8.7 and 11.1; *Yale University Press*, an extract from Antonio Pigafetta's 'Magellan's Voyage' (translated and edited by R.A. Skelton, copyright Yale University 1969).

1
Introduction

The Magellanic Clouds have been known for thousands of years to the inhabitants of the southern hemisphere. The natives on the South Sea Islands called them the Upper and Lower Clouds of Mist. The Australian aborigines, who referred to the Milky Way as a river or track along which the spirits travelled to the sky-world, considered the Magellanic Clouds as two great black men who sometimes came down to the earth and choked people while they were asleep (McCarthy 1956, p.130). Al Sufi, in his description of the stellar constellations from the 10th century, told about a strange object, Al Bakr, the White Ox, which is now identified as the Large Magellanic Cloud. Many of the mariners of the Middle Ages noticed the two Clouds. The Italian Corsali described them: 'We saw two clouds of significant size which move regularly around the pole in a circular course, sometimes going up and sometimes down, with a star midway between them at a distance of 11 degrees from the pole and participating in their movements.'

The two objects were called the Cape Clouds for hundreds of years; they were the most striking objects appearing in the sky when ships approached the Cape of Good Hope. They were of importance for the navigators of that time for localizing the South Pole, where there is no star corresponding to Polaris in the North (see Allen 1980).

The Clouds became connected with the name of the Portuguese seafarer Magalhaes through Antonio Pigafetta's valuable narrative of the first circumnavigation of the globe in 1519–1522 (Pigafetta 1524).: 'The Antarctic Pole is not so marked by stars as the Arctic. For you see there are several small stars clustered together, in the manner of two clouds a little separated from one another, and somewhat dim. Now in the middle of them are two stars not very large, not very bright, and they move slightly. And these two stars are the Antarctic Pole.' Even if Pigafetta's definition of the Pole shows some of his limitations, his interpretation of the Magellanic Clouds as consisting of stars is worthy of praise.

The unique character of the Magellanic Clouds has been recognized since the days of John Herschel (Herschel 1847), who gave coordinates and brief descriptions of 244 objects in the Small and 919 in the Large Magellanic Cloud. Their nature as the two nearest external galaxies was first surmised by Abbe (1867): 'The study of the foregoing Tables may lead to the following conclusions or suggestions:

1. The Clusters are members of the *Via Lactea*, and are nearer to us than the average of the faint stars.

2. The Nebulae resolved and unresolved lie in general without the *Via Lactea*, which is therefore essentially stellar.

3. The visible universe is composed of systems, of which the *Via Lactea*, the two Nubeculae, and the Nebulae, are the individuals, and which are themselves composed of stars (either simple, multiple or in clusters) and of gaseous bodies of both regular and irregular outlines.'

Detailed investigations of the two Clouds began when the Harvard College Observatory established its southern station, first at Arequipa in Peru (1889–1927) and then at Bloemfontain, South Africa. Beginning in 1914, the Lick Observatory measured radial velocities of bright-line nebulae in the Small (SMC) and the Large Magellanic Cloud (LMC).

The most significant result of these early studies was Miss H. Leavitt's discovery of the period–luminosity (PL) relation for Cepheid variables, described in *Harvard Circular No. 173, 1912*. The subsequent calibration of this relation made it a fundamental mean for measuring extragalactic distances. The PL relation, and other important results from studies of the Clouds, earned them the title 'Gateway to the sidereal universe' (Shapley 1956).

The first more extensive review on the Magellanic Clouds is the one by Buscombe *et al.* (1954), where references to earlier articles may be found. In the early 1950s the two Clouds were found to be, within observational errors, at the same distance, of 45 kpc. The population of the LMC was considered to be predominantly, though not entirely, type I, whereas the SMC was found to be intermediate in its evolutionary characteristics between type I and type II.

Shapley, in 1956, selected the following 'as the four most important astronomical contributions associated with the Clouds of Magellan:

1. Discovery of hundreds of giant cepheid variables in both Clouds (Harvard 1904+).

2. Measurement of high positive radial velocities for emission-line objects associated with the Clouds, suggesting their independence of the Milky Way (Lick).

3. Discovery and development of the period–luminosity relation of classical cepheids (Harvard).

4. Detection with radio telescopes of neutral hydrogen in and around the Clouds, and the measurement of its distribution (Sydney).'

Among nine 'less important but still significant contributions' that Shapley listed, the following may be worth recalling as the problems that they define continue to occupy the interest of most Cloud researchers:

'Presentation of preliminary evidences of rotation and internal turbulence in the Large Cloud (Lick and Sydney).

Conclusion that at least two types of stellar population occur in both Clouds: the primitive stars, with their presence indicated by the associated globular clusters, and the recently evolved stars, represented by the blue supergiants and suggested by the abundant bright and dark nebulosity (Pretoria, Harvard, *et al.*).

The deduction that the 30 Doradus bright nebulosity in the Large Cloud is more radiant by a hundred times than the brightest globular cluster known anywhere, and in fact more luminous intrinsically than many of the nearby dwarf galaxies with their millions of stars (Harvard).

Discovery of a large 'wing' of the Small Cloud (1940), and recently acquired (1955)

Introduction

evidence from the occurrence therein of only long-period supergiant cepheids that the wing may be, in a sense, a structure distinct from the Small Cloud; but the distances of the two must now be essentially the same since the wing's variables fit the Small Cloud's period–magnitude relation (Harvard).'

During the 1950s and 1960s Magellanic-Cloud research was intense but restrained by the small number of astronomers active in the field and the relatively small optical telescopes available in the south. Important radioastronomical studies of the structure of the Clouds were made early in this period in Australia with the 11 m parabolic aerial and the 460 m Mills Cross near Sydney. In spite of the exciting results derived, the HI observations of the Magellanic Clouds appear only in a brief comment by Pawsey at the 1958 *IAU/URSI Symposium on Radio Astronomy*: 'A most interesting feature of the HI observations is that certain galaxies (the Clouds of Magellan were the first observed) appear to lie in vast clouds of neutral hydrogen', and in a table by Volters and van der Hulst. over 'Extragalactic system observed for 21-cm emission at Dwingeloo'. At the same symposium Shain reported on the 19.7 MHz observations of the Clouds and the absorption of extragalactic radiation by 30 Doradus. Three years later, at the *IAU Symposium No.15, Problems of Extra-Galactic Research*, only two papers dealt with the Clouds.

Two very powerful radio telescopes became available in Australia in the 1960s, the 64 m steerable parabolic telescope at Parkes and the new 460 m Mills Cross at Hoskinstown.

The first symposium to deal with the Magellanic Clouds more extensively was the IAU/URSI Symposium in Canberra and Sydney in 1963, *The Galaxy and the Magellanic Clouds*. Most of the speakers were southern hemisphere astronomers, but the effect of the European sitetesting at the Zeekogat Station, South Africa, was noticeable.

The opportunities offered by the Magellanic Clouds for fundamental astronomical research became generally recognized about 30 years ago when the new southern observatories were built with the particular aim of opening up the southern skies to northern astronomers. The importance of, and interest in, the Clouds increased in the mid-1970s when the large southern telescopes were brought into operation, and still more with the introduction of the, nowadays continuously improved, fast detectors. Conferences and symposia on Magellanic Cloud topics have been arranged rather frequently and reviews on a variety of Cloud topics have appeared (see Appendix 2). There is hardly any conference on any topic today without one or more contributions based on observations of Cloud objects.

The *IUE, IRAS, EINSTEIN, ROSAT* and a number of other satellites have contributed to our knowledge of these galaxies by making available the far ultraviolet, the infrared, and the X-ray spectral regions.

Detailed studies of the Magellanic Clouds in the (sub-)millimetre range are now possible with the aid of the 15 m *SEST* (Swedish/ ESO/ Submillimetre Telescope) on La Silla, Chile.

In the same way as the arrival of the large telescopes and the satellites opened new eras in Magellanic Cloud research, so has the entry of the *Hubble Space Telescope (HST)*. Results from observations with the *HST* have started to appear and have already provided much valuable information.

During the past 25 years the dominating questions in Magellanic Cloud research have concerned their distances, structure, kinematics and composition, with the aim of eventually understanding their evolution. Recent advances in observational techniques as well as in theory have increased the possibilities of tackling these problems. This has, perhaps not so surprisingly, also led to discrepancies, most obviously in the interpretations of the effects of stellar metallicities and ages on the determination of distances. Two distance moduli, $(m - M)_0$, have for some time been in use for the Clouds, a 'short' and a 'long' modulus, differing by about 0.4 mag. Recently, the distance determinations appear to be more in agreement. Further observations and analyses will, hopefully, soon lead to complete unanimity.

A reversal of the 'distance problem' is the application of theoretical isochrones to colour–magnitude sequences for clusters in the Clouds for the determination of their ages and metallicities once their distances are known. Similar work on the field stars has revealed the existence of a number of generations and contributed to our knowledge about the chemical evolution of the Clouds.

All the objects in each Cloud, the LMC or the SMC, have until recently been considered equidistant, a fact of importance for calibration purposes. The higher accuracies achieved today make it possible to consider the tilts and the extensions of the Clouds in the line of sight in determining their mean distances. The positions of the objects in the Clouds thus become significant. Omission of these effects may introduce significant errors.

It is tempting today, 40 years after Shapley's listing of the most significant early contributions associated with the Magellanic Clouds, to present a similar selection of contributions rendered during the last four decades. That new vast fields of research have been opened up, thanks to satellites and a new family of ground-based optical and radio telescopes, has already been pointed out. The aim of Cloud research is also widened. The Clouds are no longer treated as galaxies and 'gateways' only, but serve as astrophysical laboratories in the sense that theories for stellar evolution are tested on their stellar and non-stellar components.

In the following selection of important discoveries made during the past four decades an attempt is made to give equal weight to extragalactic and astrophysical research:

1. The discovery of the Magellanic Stream. This has led to intense theoretical work on the evolution of the Magellanic System through interaction between the SMC, the LMC and the Galaxy, and possibly also M 31.

2. The observations of the distribution of stars and gas in the SMC indicating an extension in depth, possibly a fragmentation of the Cloud.

3. The detection from radio and optical observations of multiple layers also in the LMC.

4. The observations showing that bursts of star formation have occurred in the Clouds and that star formation has propagated through their interstellar medium, leading to superassociations and supergiant shells.

5. The identification of X-ray sources and diffuse X-ray emission in both Clouds.

6. The observations showing that the interstellar reddening laws are different in the far-UV in the SMC, the LMC and the Galaxy.

7. The stellar evolution studies of the most luminous stars, including the Wolf–Rayet stars, and of the red-giant stars: the red-giant branch, the asymptotic-giant branch, and the horizontal-branch stars.

8. The investigation of the dust in the Clouds by *IRAS* observations.

Introduction

9. The detection of CO and other molecules in the Clouds and detailed studies of their distribution and physics.

10. The supernova SN 1987A, offered by the LMC as its own particular tribute to science in this century.

A comparison of observational data obtained in the 'pre-large-telescope period', shows that although higher resolution is achieved, fainter limiting magnitudes are reached, and improved theories are available for the interpretation of the data, most of the early results are confirmed – and most of the problems remain, though they appear closer to acceptable solutions. This is in particular due to the new wavelength ranges that have become available to the astronomers thanks to the astronimical satellites: the millimetre, the submillimetre, the far-infrared, the far-ultraviolet, and the X-ray regions of the spectrum.

2

The distances of the Clouds

Cepheids, RR Lyrae stars, Mira variables, clusters and OB stars have dominated in the attempts to determine accurate distances to the Magellanic Clouds in recent years. The subject is of vital importance not only for Magellanic Cloud research but also for the determination of extragalactic distances (see Feast 1988, van den Bergh 1989, de Vaucouleurs 1993). With the accurate photometry possible nowadays the uncertainties in the distance determinations arise mainly through their dependence on the metallicities and ages of the objects involved and on the interstellar absorption in the line of sight, be it in the Galaxy or in the Clouds themselves.

2.1   The reddening, foreground and internal

The interstellar reddening affects virtually all studies of the Clouds. It is frequently difficult to determine its magnitude from the material available and mean values found in the literature are then, by necessity, applied. This may occasionally lead to erroneous results, in particular as the objects used to determine the mean may not be representative of the whole galaxy. In the following, some investigations will be referred to either as giving general ideas about the mean reddening or as examples of typical deviations.

The galactic foreground colour excess towards the Magellanic Clouds has been investigated by Schwering and Israel (1991) on a scale of 48 arcmin from galactic HI maps, using $E_{B-V} = 0.17 \; 10^{-21}$ N(HI) H cm^{-2}. Their results are given in Table 2.1 and indicate rather small fluctuations over the Clouds. They note, however, that on smaller scales the amplitude of the foreground reddening may be much greater.

Bessell (1991) has summarized reddening determinations from photometry, stellar polarization, and HI column densities, and has also used CaII K lines and UV extinction to derive foreground and intrinsic mean reddenings in the Clouds (Table 2.1). He finds lower values than Schwering and Israel. The average reddening within the SMC is probably about 0.06 mag but there are regions with reddenings up to 0.3 mag. The average reddening in the LMC is similar but the variations are larger, with many regions with higher as well as lower than average reddening. This is indeed confirmed by Greve et al. (1990) in their investigation of dust in emission nebulae in the LMC; the maximum value reported is $E_{B-V} = 1.1$. A typical region with increased and varying reddening is the NW part of the LMC Bar, rich in clusters, OB stars and emission nebulosity. For the NGC 1850 cluster, embedded in the nebula N 103, the mean extinction is $E_{B-V} = 0.18 \pm 0.02$, with a variation over the cluster of ± 0.04 mag

2.1 The reddening, foreground and internal

Table 2.1. *The average reddening of the Magellanic Clouds*

LMC foreground	LMC internal	SMC foreground	SMC internal	Bridge	Ref.
0.07–0.17		0.07–0.09			1
0.04–0.09	0.06	0.04–0.06	0.06		2
				0.00–0.10	3

E_{B-V} (mag)

References: 1. Schwering and Israel 1991; 2. Bessell 1991; 3. Grondin et al. 1991, 1992.

according to Gilmozzi et al. (1994). A similar result was derived by Vallenari et al. (1994b), $E_{B-V} = 0.18 \pm 0.06$ and Fischer et al. (1993), $E_{B-V} = 0.17$ mag.

In the region of 30 Doradus the extinction is highly variable and will be discussed further in Chap. 10. Also to the south of 30 Doradus, in the supergiant shell LMC 2 (See Sect. 6.1.2) extreme variations in reddening are seen. Thus, the small cluster SL 639 near 30 Doradus (marked by a dashed circle in Fig. 4 in Westerlund 1961a) has $A_V \approx 1.60$ mag, measured from four main-sequence stars brighter than $V = 15$.

The reddening in the region of NGC 2257, 9° NE of the LMC Bar, has been estimated as $E_{B-V} = 0.08$ (Stryker 1983), a value frequently used, and not far from the mean reddening found for LMC Cepheids, $E_{B-V} = 0.074 \pm 0.007$ (Caldwell and Coulson 1985). However, using 13 RR Lyrae variables of type *ab* and applying the relation

$$E_{B-V} = -0.24 P - 0.347 - 0.056 [Fe/H],$$

Walker (1989) derived

$$E_{B-V} = -0.087 (\pm 0.007) - 0.056 [Fe/H].$$

E_{B-V} is the colour excess at minimum light and P is the period in days. Assuming $[Fe/H] = -1.8 \pm 0.3$, $E_{B-V} = 0.014 \pm 0.02$. The reddening in the region of NGC 2257 is evidently very low. (Walker prefers to use $E_{B-V} = 0.04 \pm 0.01$ as a likely mean of available estimates.)

The galactic foreground reddening can be surprisingly large in the outskirts of the Clouds. Thus, it amounts to $E_{B-V} = 0.18 \pm 0.02$ at NGC 1841, 15° from the centre of the LMC (Walker 1990). In the outlying cluster GLC 0435-59 (Reticulum), 11° from the centre of the LMC (Walker 1992a), the reddening is only $E_{B-V} = 0.03$, also fully ascribed here to galactic foreground reddening.

Massey et al. (1995) have determined the reddening for 414 OB stars in the LMC and 179 in the SMC using the reddening-free parameter Q ($Q = (U-B) - 0.72 (B-V)$). The derived Qs are compared with the values expected as a function of spectral type from known intrinsic colours. The resulting distribution is shown in Fig. 2.1 (a and b), where the mean reddenings, $E_{B-V} = 0.13$ and 0.09 mag for the LMC and the SMC, respectively, are given. Over 50% of the stars have E_{B-V} between 0.09 and 0.17 mag in the LMC and between 0.05 and 0.12 mag in the SMC. It is, however, obvious that the luminous supergiants in the field do not fall in the most dusty regions. There are

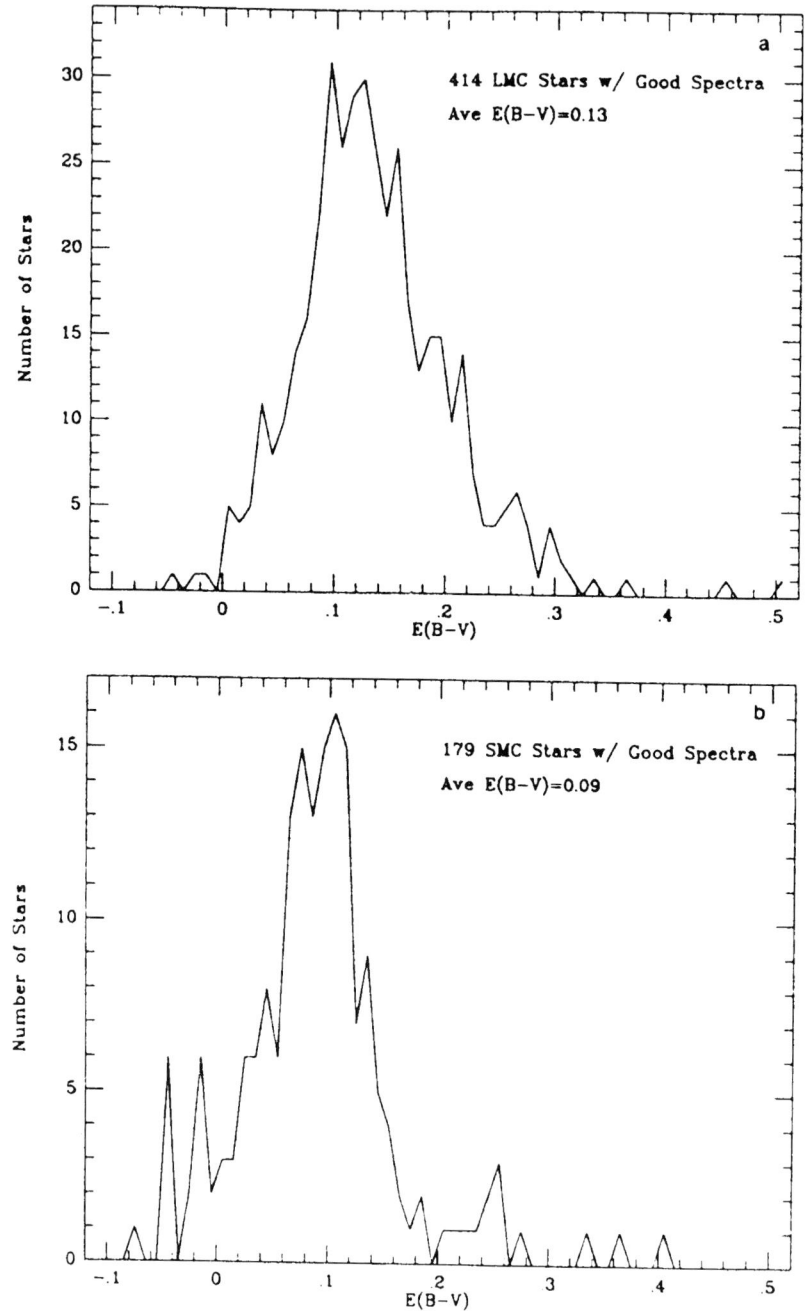

Fig. 2.1. The number of massive stars with various amount of reddenings in the LMC (a) and the SMC (b). Only stars whose reddening-free Q index agreed with that expected from its spectral type are used (Massey *et al.* 1995).

few massive stars with colour excesses ≥ 0.3 mag. The earliest-type stars do not show significantly higher reddening than the later-type supergiants. This suggests that the massive stars emerge from their cocoons more rapidly than is sometimes assumed.

Similar results for the LMC have been obtained by Isserstedt (1975) from an investigation of 702 supergiants using the two-colour diagram for early-type stars and intrinsic colours and MK spectral classes for the intermediate and late-type supergiants. He found a mean colour excess of 0.07 mag and ascribed this value to the foreground reddening. The distributions of the most reddened ($E_{B-V} \geq 0.20$ mag) and the most luminous ($M_V \leq 11.0$ mag) supergiants are similar. This is not surprising as they all belong to the superassociations. In the 30 Doradus region all supergiants are reddened; they appear to be embedded in a giant dust cloud. Almost unreddened stars may be found everywhere in the LMC, showing that the foreground dust is also unevenly distributed. Unreddened and strongly reddened stars appear very close to each other, indicating that a thin dust layer may exist with a fair number of stars on either side of it.

Relatively high colour excesses for the SMC were observed by Hughes and Smith (1994) for a Be/X-ray star connected with the SNR 0101−72.4, $E_{B-V} = 0.23\,^{+0.01}_{-0.04}$.

In the outer parts of the SMC the reddening is small. Walker (1988), using the same method as in the case of NGC 2257 above, found for NGC 121, with [Fe/H] = -1.4 ± 0.1, $E_{B-V} = 0.035 \pm 0.024$.

The reddening is generally small in the Bridge region, and Grondin et al. (1991, 1992) give values on E_{B-V} between 0.0 and 0.10 mag for their stellar associations; they have, however, in most cases kept colour excess and distance as free parameters in fitting their CMDs to theoretical isochrones (Table 2.1).

2.2 Distance determinations with the aid of Cepheids

Much weight is given to distance moduli, $(m - M)_0$, determined with the aid of Cepheids though several sources of uncertainty exist in the use of their period–luminosity (PL) and period–luminosity–colour (PLC) relations. The uncertainty in the zero-point determination is about 0.15 mag (Feast 1988). The PL relation is rather insensitive to metallicity but strongly dependent on reddening, whereas the opposite applies to the PLC relation. Feast suggests that the PL relation in V may be combined with the PLC relation in V and $B - V$ to give $(m - M)_0$ and a metallicity, [Fe/H].

Detailed discussions of the Cepheid distance scale have recently been presented by Madore and Freedman (1991) and Mateo (1992). Here, only a few comments will be made.

PL and PLC relations have been derived with the aid of Cepheids in the LMC clusters NGC 2031 (14 Cepheids), NGC 1866 (9) and NGC 2157 (3) (Mateo 1992):

$$V_0 = const - 3.37(\pm 0.13)\log P + 2.10(\pm 0.14)(B - V)_0;$$

$$V_0 = const - 2.74(\pm 0.32)\log P.$$

The coefficients in these relations agree reasonably well with those summarized by Feast and Walker (1987).

The dispersion of the PLC relation is remarkably smaller than that of the PL relation. The formal scatter of the former, $\sim \pm 0.05$ mag, is about what may be expected from the formal photometric errors. However, a clear offset of -0.07 mag exists between the individual PLC relations of NGC 1866 and NGC 2031. This is hardly due to photometric uncertainties. It may be due to a large error in the relative reddening of the two clusters, an abundance difference of 0.1–0.2 dex (Stothers 1988),

Table 2.2. *The distance to the LMC from observations of Cepheids*

Author	Method	$(m - M)_0$
Cester, Marsi (1984)	PL	18.62 ± 0.22
Visvanathan (1985)	PL	18.82 ± 0.07
Visvanathan (1986)	$(m - M)_{0,Hyad.} = 3.09$	18.5
Mathewson et al. (1986)		18.45 ± 0.05
Caldwell, Coulson (1986)	$(m - M)_{0,Pleiad.} = 5.57$	18.65 ± 0.07
Caldwell, Coulson (1987)	$(m - M)_{0,Pleiad.} = 5.57$	18.45
Revised (Feast 1988)		18.52
Laney, Stobie (1986)	PL, PLC, $(m - M)_{0,Pleiad.} = 5.57$	18.73 ± 0.05
Welch et al. (1987)	PL, PLC in JHK	18.57 ± 0.05
Walker (1987)		18.44 ± 0.15
Stothers (1988)	PL, PLC, abundances	18.51 ± 0.06
Visvanathan (1989)	$(m - M)_{0,Pleiad.} = 5.57$	18.42 ± 0.04
Chiosi et al. (1992)	Ceph. ME method, NGC2157	18.4–18.5
Bertelli et al. (1993)	Ceph. ME method, NGC1866	18.51 ± 0.21
Bertelli et al. (1993)	Ceph. ME method, NGC2031	18.32 ± 0.20
Gieren et al. (1994)	VSB method, NGC1866	18.47 ± 0.20

The Table is arranged in a chronological order, meaning that revisions of values may have occurred.

or differences in distance. The tilted disk model of Caldwell and Coulson (1986) predicts that NGC 1866 is 0.06 mag further away than NGC 2031.

Madore and Freedman (1991) give PL relations for all normal passbands from B to K, based on 25 LMC stars, of which we quote the following here:

$$M_{\langle V \rangle} = -1.24 (\pm 0.22) - 2.88 (\pm 0.20) \log P;$$

$$M_{\langle K \rangle} = -2.28 (\pm 0.10) - 3.42 (\pm 0.09) \log P;$$

The slope of the PL relation in V agrees reasonably well with that given by Mateo, and that for K is in good agreement with that derived by Welch et al. (1987):

$$M_{\langle K \rangle} = -2.38 (\pm 0.07) - 3.374 (\pm 0.006) \log P.$$

The latter also give

$$M_{\langle J \rangle} = -2.15 (\pm 0.07) - 3.294 (\pm 0.006) \log P; \text{ and}$$

$$M_{\langle H \rangle} = -2.36 (\pm 0.07) - 3.319 (\pm 0.005) \log P.$$

The moduli derived with the Welch et al. expressions are not reddening corrected, nor are they corrected for metallicity effects.

The adoption of a mean reddening and mean metallicity for the Cepheids in each Cloud from studies of other objects may introduce errors. The reddening is certainly not uniform. The best reddening determinations presently available for Cloud Cepheids with periods $> 10^d$ come from BVI photometry. This is because the reddening line intersects the intrinsic line, which is narrow and well calibrated, at a favourable angle. Shorter-period Cepheids do not have the same advantages. The

2.2 Distance determinations with the aid of Cepheids

BVI reddenings of Cloud Cepheids with periods $> 10^d$ are accurate to better than 0.04 mag, for well-observed stars probably of the order of 0.02 mag (Caldwell and Laney 1991).

Metallicity differences also exist; Harris (1981) found that the short-period Cepheids in the SMC are more metal-poor than the others.

There are indications that small differences in $(m-M)_0$ exist for LMC Cepheids of different periods so that a flat plane may not be defined for all Cepheids. The long-period ($\log P = 0.9$–1.7) Cepheids have $(m-M)_0$ 0.2 larger than the short-period ($\log P \leq 1$) Cepheids (Feast 1988), corresponding to over 4 kpc at the distance of the LMC if it is a pure distance effect. Laney and Stobie (1986) remark that the tilt of the LMC increases with increasing period of the Cepheids.

We conclude from these results that much of the scatter in the PL and PLC relations for the Cepheids in the LMC is due to the geometry. In other words, the Cepheids may be found over a rather great depth in the LMC (not only caused by the tilt). Visvanathan (1989) reached the same conclusion from his observations of Cepheids at V and IV (1.05 μm). When the geometry of the LMC is considered the Cepheids have a dispersion in the PL, PLC relations of only 0.09 mag.

An independent method for the determination of stellar distances is the *visual surface brightness technique*, VSB. The VSB parameter F_V is defined as

$$F_V = 4.2207 - 0.1\, V_0 - 0.5 \log \Phi,$$

where V_0 is the apparent visual magnitude corrected for extinction and Φ is the stellar angular diameter in milliseconds of arc (Barnes et al. 1993). F_V can be determined from its colour $(V-R)_0$,

$$F_V = b + m(V-R)_0.$$

As a Cepheid satisfies this relation throughout its pulsation cycle, the two equations permit the determination of the instantaneous stellar angular diameter from VR photometry. It is related to the instantaneous linear diameter through

$$\Delta D + D_m = 10^{-3}\, r\, \Phi,$$

where ΔD is the instantaneous linear displacement from the mean linear diameter D_m (both in AU) and r is the distance in parsecs. Integration of the radial velocity yields ΔD. Other steps in the application of the method are the conversion of radial velocity to pulsational velocity and correction of the surface brightness distance modulus for metallicity effects. The technique has been successfully tested by Barnes et al. (1993) on the Cepheid HV 829 in the SMC (Table 2.5). For galactic Cepheids Gieren et al. (1993) have derived what they consider is the currently most precise PL(V) relation from the VSB technique:

$$M_V = -1.371\,(\pm 0.095) - 2.986\,(\pm 0.094) \log P.$$

It agrees well with the PL(V) relations given above for LMC Cepheids.

A new approach to distance determinations with the aid of Cepheids is *the mass-equivalency (ME) method* introduced by Chiosi et al. (1992). It requires the Cepheid evolutionary and pulsational masses to be equal. The moduli derived in this manner depend, naturally, on the evolutionary models employed and on the adopted effective temperature–colour transformations. Chiosi et al. exemplify this by showing that the derived distance modulus to NGC 2157 and its three Cepheids is 19.02 ± 0.10, if a model based on classical semiconvective mixing and transformed to the $V/(B-V)$ CM diagram with the 'Yale' temperature–colour transformation is used, but 18.47 ± 0.08 if the Padova overshoot evolutionary models (Alongi et al. 1993) and the Padova temperature-colour transformations are used. As the latter value is closer to most of the present determinations it may be concluded that the ME method is promising for distance determinations and that the overshoot models have resolved the long-standing Cepheid mass discrepancy between pulsational and evolutionary masses. It is then important to understand how variations in the physical properties affect the distance moduli determined with the ME method. The chemical composition is likely to play an important role. The range in abundances in the LMC may be large (Chap. 11). The importance of this uncertainty has been investigated by Bertelli et al. (1993), using recent observations of the Cepheids in the LMC clusters NGC 1866 and NGC 2031. Applying the Padova evolutionary models and the pulsational models of Chiosi et al. (1993) they derive the moduli 18.51 ± 0.21 and 18.32 ± 0.20 for NGC 1866 and NGC 2031, respectively. The errors reflect (i) the range in pulsational and evolutionary masses of the Cepheids in each cluster, and (ii) the large uncertainties (0.3 dex) of the abundances in each cluster. For Cepheids in the mass range corresponding to those found in these two clusters (4–5 $M_\odot$) the metallicity sensitivity of the distance moduli derived with the ME method is $\Delta(m - M)/\Delta Z_1 = 0.69$, where $Z_1 \equiv \log(Z/0.016)$. Cluster abundances determined to an accuracy of 0.15 dex will then give cluster distances to ± 5 %. As there are numerous LMC clusters containing Cepheids (Mateo 1993, Welch 1992) the ME method may eventually present an accurate distance to the LMC together with its line-of-sight geometry.

Distance moduli derived with the aid of Cepheids are given in Tables 2.2 and 2.5 for the LMC and the SMC, respectively.

2.3 Distance determinations with the aid of RR Lyrae variables

In distance determinations with the aid of RR Lyrae stars an absolute magnitude of $M_V(RR) = 0.6$, independent of the metallicity, is frequently used as typical for the horizontal-branch stars. However, a metallicity dependence may exist. Feast (1988) gives several reasons for using the relation

$$M_V(RR) = 0.92 + 0.2\,[Fe/H].$$

Distance moduli derived with the aid of RR Lyrae variables are presented in Table 2.3 and Table 2.5. As seen there, cluster RR Lyrae stars and RR Lyraes in the surrounding field do not always give the same moduli.

Nemec et al. (1985) have studied the RR Lyrae variables in and around the LMC globular cluster NGC 2257, 9° NE of the Bar. Its RR Lyraes indicate that it is more or less in the plane of the LMC if $i = 45°$. Some of the RR Lyrae stars in this remote 'halo' field have a higher metallicity than the cluster stars. The scatter in mean B magnitude is so large that a mean distance does not appear meaningful. The same scatter around about the same mean magnitude is found for the RR Lyraes in the field of NGC 1783, 4° NNW of the Bar. This does not support the assumption that the field RR Lyrae stars all lie in a thin tilted disk. If the tilt is 45° the difference

2.3 Distance determinations with the aid of RR Lyrae variables

Table 2.3. *The distance to the LMC from observations of RR Lyrae variables*

Author	Method	$(m-M)_0$
Stothers (1983)	$\langle M_{V,Ceph}\rangle = -1.21 - 3.01 \log P$; $\langle M_{V,RR}\rangle = 0.61 \pm 0.15$	18.50 ± 0.10
Walker (1985)	NGC 2210; $\langle M_{V,RR}\rangle = +0.6$ $E_{B-V} = 0.06$ at LMC disk if $i = 45°$	18.42 ± 0.10 18.35
Nemec et al. (1985)	NGC 2257; $\langle M_{V,RR}\rangle = 0.60$; $E_{B-V} = 0.060$; $(m-M)_{0,2257} =$ at LMC disk if $i = 45°$ field at NGC 2257	18.13 ± 0.25 18.15 18.0–18.7
Reid, Strugnell (1986)	$\langle M_{V,RR}\rangle = 0.75$ $E_{B-V} = 0.06$	18.28 ± 0.15
Walker, Mack (1988b)	NGC 1786; $\langle M_{V,RR}\rangle = 0.6$ $E_{B-V} = 0.073 \pm 0.009$ at LMC disk if $i = 45°$	18.44 ± 0.05 18.52
Clementini, Cacciari (1989)	$\langle M_{V,RR}\rangle = 1.0 + 0.17 [Fe/H]$	18.26 ± 0.20
Carney (1990)	$\langle M_{K,RR}\rangle = -0.84 - 2.24 \log P$ $\langle M_{V,RR}\rangle = 1.01 + 0.16 [Fe/H]$	18.3 ± 0.2
Walker (1990)	NGC 1841; $\langle M_V\rangle = 0.5$ $E_{B-V} = 0.18 \pm 0.02$; $[Fe/H] = -2.2 \pm 0.2$; $(m-M)_{0,1841} =$	18.19
Walker (1992a)	GLC 0435-5; $\langle M_V\rangle = 0.6$ $E_{B-V} = 0.03 \pm 0.02$; $[Fe/H] = -1.7 \pm 0.1$; $(m-M)_{0,0435-59}$ if $\langle M_V\rangle = 0.43$	18.38 18.55
Reid, Freedman (1994)	In and near NGC 2210. 2 kpc in front of LMC; $A_V = 0.19$; $i = 27.5°$; $[Fe/H] = -1.9$; $M_{V,RR} = 0.20 [Fe/H] + 1.04$	18.3–18.5

should be 8 kpc, or $\Delta(m-M)_0 = 0.35$ mag. Metallicity effects may explain part of the wide spread in magnitudes of the RR Lyrae stars in both fields, but there is probably also a depth effect, a possibility mentioned by Graham (1977) and Nemec et al. (1985).

The field RR Lyraes seen near the true globular cluster NGC 1841 and possibly belonging to a LMC–SMC halo (Kinman et al. 1976) are 0.3 mag fainter than those just discussed, and thus are at a 7.5 kpc larger distance. This fits well with the position of the NGC 1841 field, about equidistant to the LMC and the SMC. The RR Lyrae stars in NGC 1841 do not differ much from those in the field, so that the cluster may well be in the plane of the LMC disk. This has been suggested by Andersen et al. (1987) but their proposed distance modulus, $(m-M)_0 = 18.1 \pm 0.3$, places the cluster in front of a tilted disk, be it 45° or 27°.

The RR Lyrae variables in the SMC cluster NGC 121 give a true distance modulus

Table 2.4. *The distance to the LMC from observations of Mira variables*

Author	Method	$(m-M)_0$
Feast (1988)	If $\langle M_{V,RR}\rangle = 0.8$	18.28
	$\langle M_{V,RR}\rangle = 0.92 + 0.2\ [Fe/H]$	18.48
Feast (1989a)	$\langle M_{V,RR}\rangle = 0.6$	18.47
Wood et al. (1985)	Pop. II Miras/ 47 Tuc	18.6 ± 0.25

Table 2.5. *The distance to the SMC from observations of Cepheids and RR Lyraes*

Author	Method	$(m-M)_0$
	Cepheids	
Caldwell, Coulson (1986)		18.97 ± 0.07
Welch et al. (1987)	PL, PLC in JHK	18.93 ± 0.05
Laney, Stobie (1986)	PL, PLC	19.05 ± 0.05
Stothers (1988)	PL, PLC, abundances	18.80 ± 0.06
Barnes et al. (1993)	distance to HV 829	18.9 ± 0.2
Sebo, Wood (1994)	P, mass, L	18.92
	RR Lyrae stars	
Reid, Strugnell (1986)	$\langle M_{V,RR}\rangle = 0.75$	18.78 ± 0.15
Walker, Mack (1988a)	NGC 121; $\langle M_{V,RR}\rangle = 0.6$	18.86 ± 0.07
	SMC field	19.2
Clementini, Cacciari (1989)	$\langle M_{V,RR}\rangle = 1.0 + 0.17\ [Fe/H]$	18.85 ± 0.20

for the cluster of $(m-M)_0 = 18.86 \pm 0.07$, if $M_V(RR) = 0.6$ (Walker and Mack 1988a). Using the same technique Walker (1985, 1986) had derived the modulus $(m-M)_0 = 18.43 \pm 0.06$ for the LMC from the RR Lyraes in the clusters NGC 1786 and 2210, indicating a difference of 0.43 ± 0.09 mag in the moduli of the two galaxies. This is greater than the results for the Cepheids in the two Clouds; Feast and Walker (1987) found the difference to be 0.31 ± 0.05 mag and Welch et al. (1987) found 0.36 ± 0.07 mag. The disagreement may be caused by NGC 121 not being at the mean distance of the SMC Cepheids. It may also result from the metallicity dependence of the luminosity of the RR Lyraes.

2.4 Distance determinations with the aid of Mira variables

Feast (1989a) has used the Mira variables in relatively metal-rich galactic globular clusters to calibrate their luminosities. If $M_V(RR) = 0.6$ holds for all RR Lyrae stars, the revised zero point for the Miras gives $(m-M)_0 = 18.48$ for the LMC (Table 2.4). $M_V(RR)$ may then depend rather little on $[Fe/H]$.

A slightly larger distance modulus for the LMC has been derived by Wood et al. (1985). They found $(m - M)_0 = 18.6 \pm 0.25$ for a group of short-period Population II Miras by comparing them with similar stars in the galactic globular cluster 47 Tucanae, with an adopted distance modulus of 13.34.

Using the slope of the infrared (K) PL-relation for the Miras in the LMC on Miras in the SMC Feast (1989a) finds a zero-point difference of 0.30 ± 0.06 mag, which may be seen as the difference in the moduli of the two Clouds. He found about the same difference in the case of the Cepheids. Differences in age–metallicity relations between the two Clouds thus have no significant effect on the zero point of the Mira PL-relation.

2.5 Distance determinations with the aid of clusters

The intermediate-age clusters in the Magellanic Clouds have become very attractive for distance determinations. Main-sequence fittings to Zero-Age Main-Sequence (ZAMS) lines, to galactic clusters, or to theoretical isochrones are now possible as well as fits to other features in the colour–magnitude diagrams (CMD). The parameters which influence the positions of the stars in a CMD are reddening, metallicity, and helium abundance. Preferably, no field stars should be included in the cluster sequence as this tends to lower the distance modulus. This has been hard to avoid as the photometry has usually had to be carried out in the outskirts of the clusters in order to diminish the effects of crowding. Modern techniques, such as the use of CCDs, have reduced this problem.

Table 2.6 contains examples of determinations of cluster distance moduli, ordered chronologically and with the method used indicated: isochrones, fits to galactic clusters, to a ZAMS, to the AGB or the HB, etc.. The large differences seen reflect the still ongoing discussion of a 'short' or a 'long' distance scale. The two values suggested for the LMC are usually $(m - M)_0 = 18.2$ or 18.7 mag, respectively. The uncertainties leading to this situation have been discussed a number of times. We refer to the discussion by Chiosi and Pigatto (1986) of the distance moduli of NGC 2162 and NGC 2190. Incorporating convective overshooting in the stellar models of intermediate-age stars, they found that plausible choices of reddening, hydrogen and metal content, and the use of the luminosity-function method, gave $(m - M)_0 = 18.4$ for the LMC. They also showed that the 'classic' value of 18.6 may be obtained with a solar metallicity. For these clusters Schommer et al. (1984) derived $(m - M)_0 = 18.2 \pm 0.2$.

Considering the existing uncertainties in isochrone fittings, distances determined by that method must be given a low weight. The choice between the 'long' and the 'short' distance scale cannot be made on the basis of the LMC clusters. We will return to the question of cluster distances in Chap. 4.

2.6 Distance determinations with the aid of field stars

2.6.1 Distances from early-type stars

It was once considered rather straightforward to determine distances with the aid of a two-dimensional classification of early-type stars or with their luminosities determined with $H\beta$ or $H\gamma$ photometry. With the progress made in theoretical stellar physics many peculiarities ignored earlier became important clues to the nature of the stars.

Table 2.6. *The distance of the LMC determined with the aid of CM diagrams of clusters*

Ref.	Method	E_{B-V}	$(m-M)_0$ mag
Andersen et al. (1984)	AGB and HB fits in H11; AGB to M 92, $(m-M)_{0,92} = 14.36$ HB, $V_{HB} = 19.0$, $M_{V,HB} = +0.6$	0.02 0.06	18.1 18.4
Schommer et al. (1984)	MS fits of NGC 2162 and 2190 to Hyad.; $(m-M)_{0,Hyad.} = 3.29$, corr. for abundances, see Chiosi, Pigatto	0.06	18.2 ± 0.2
Walker (1985)	MS fits of NGC1866 to Pleiad. Z = 0.016, Y = 0.273 Z = 0.012, Y = 0.25	0.06 0.06	18.52 18.42
Andersen et al. (1985)	AGB fits to M 3 and M 92: for NGC1841 for NGC2210		18.7–18.8 18.1 ± 0.2
Chiosi, Pigatto (1986)	isochrone fittings of NGC2162 and NGC2190 with new models and choices of reddening and metal content		18.4 - 18.6
Andersen et al. (1987)	CMD fit of NGC1841 to M 92	0.15	18.1 ± 0.3
Geisler (1987)	MS fits of 42 giants in NGC2162 and NGC2213		18.2 ± 0.2
Gratton, Ortolani (1987)	Theoretical isochrones fitted to E 2 with Z = 0.01 with Z = 0.001	0.06	18.37 18.52
Jones (1987)	CM diagram fits of NGC 2241 to NGC 2420 and NGC 2506(Gal.)	0.08	18.4
Mateo (1988a)	Isochrone fits to 5 clusters: NGC1711 LW 79 NGC1831 NGC2010 H 4	0.09 0.08 0.04 0.09 0.05	18.25 18.31 18.17 18.20 18.12
Mateo, Hodge (1986)	Fit of MS and AGB of H 4 to NGC2420	0.07	18.1 ± 0.3
Mateo, Hodge (1987)	Fit of MS and AGB of LW 79 to NGC752 and NGC7789	0.08	18.4 ± 0.2

Feast (1993) gives for the Hyades $m - M = 3.33 \pm 0.05$ and for the Pleiades $m - M = 5.62 \pm 0.05$ from observations by various sources between 1988 and 1991.

Thus, the lower metallicity of the SMC is clearly seen in the spectra of its B stars. Their peculiar $U - B$ colours (Garmany *et al.* 1987) may at least partly depend upon their composition. The use of spectroscopic parallaxes, where the calibration of the absolute visual magnitude, M_V, is based on stars in the Galaxy, for the determination of the distances to the Magellanic Clouds then becomes less certain.

2.6 Distance determinations with the aid of field stars

Fittings to a theoretical ZAMS are occasionally also attempted. The effects of metallicity differences are then usually considered. For this method to work, the relation between spectral type, temperature and bolometric correction has to be known.

Conti et al. (1986) have determined $(m - M)_0 = 18.3 \pm 0.3$ for the LMC, using spectroscopic parallaxes as well as a ZAMS fitting. Garmany et al. (1987) derived a spectroscopic parallax distance modulus of 19.1 ± 0.1 for the SMC, and a ZAMS related one of 18.9 mag. They recommended as a compromise that the mean of the two is used.

The use of the OB stars leads to a larger difference in mean $(m - M)_0$ between the LMC and the SMC, of the order of 0.6–0.8 mag, than the use of the Cepheids for which 0.3 mag was found. To what extent this reflects the distribution of the objects in depth in the two Clouds, or simply uncertainties, is not clear. It is not unlikely that objects of different age occupy different volumes of space in galaxies which have suffered as severe disturbances as the two Clouds.

Recently, Massey et al. (1995) have updated existing catalogues of the massive star population in the LMC and the SMC. Using these data they have, via spectroscopic parallaxes for stars earlier than B0.5, determined the distance moduli of the two Clouds, finding 18.4 ± 0.1 and 19.1 ± 0.3 mag for the LMC and the SMC, respectively. These moduli are about the same as those derived by Conti et al. (1986) and by Garmany et al. (1987) and referred to above. As previously, the results depend upon the assumption that the calibration of M_V as a function of spectral type is the same in the Clouds as in the Galaxy.

Arellano Ferro et al. (1991) have used luminous F supergiants to derive the distance moduli 18.19 ± 0.22 and 19.33 ± 0.31 mag for the LMC and the SMC, respectively. Supergiants are difficult to use for distance determinations and these values may be given less weight than the others.

2.6.2 Distances from red-giant stars

The luminosity function for the intermediate-age population red giants in Shapley's Constellation III has been studied by Reid et al. (1987). The reddening was estimated to be $E_{B-V} = 0.07$. With the aid of the $V-I$ colours they derived $[Fe/H] = -1.1 \pm 0.2$. They located the red-giant tip to $I = 14.60 \pm 0.05$, with a mean colour of $V - I = 1.8$. This led to an absolute bolometric magnitude $M_{bol} \sim -3.5 \pm 0.1$, and, eventually, to $(m - M)_0 = 18.42 \pm 0.15$.

2.6.3 Distance determinations with the aid of novae

In 1991, 24 novae were known in the LMC and 4 in the SMC.

The maximum magnitude versus rate of decline (MMRD) relationship between the magnitude at maximum of a nova and its rate of decline offers the possibility to use these objects for distance determinations. The MMRD of galactic novae is well represented by the equation

$$M_V(max) = -7.89 - 0.81 \times \arctan\{[1.32 - \log(t_2)]/0.19\},$$

where t_2 is the time interval required by the nova to decline 2 magnitudes from maximum, and 'arctan' is in radians. The rate of decline is $v_d = 2/t_2$.

The method has been used by Capaccioli et al. (1990) on 14 LMC novae with good photometric data. The novae magnitudes were corrected for reddening, E_{B-V}

= 0.09 ± 0.03, and V(max) plotted against v_d. Fitting the MMRD to the data points gave for the LMC $(m - M)_0 = 18.71 \, ^{+0.27}_{-0.32}$.

The MMRD may, at the level of accuracy of ~ 0.25 mag, be a universal relation. If so, no obvious differences in the nova populations of galaxies of different morphological types and bulge-to-disk ratios may exist, with the possible exception of the frequency distribution along the MMRD sequence and the relative percentage of the bright and fast novae. About 90% of the novae fall within a ±0.5-mag-wide strip along the MMRD; the scatter is due to an intrinsic scatter in the white dwarf luminosities before outburst. However, a number of novae have been observed to deviate from the classical MMRD by deviating towards brighter magnitudes. They may define a sequence of super-bright novae by being ~ 1 mag brighter than suggested by their rate of decline.

One object of the super-bright class is Nova LMC 1991. It was discovered by Liller (1991) on April 18 at $V \sim 12.3$. The maximum, inferred from the light curve, was reached on April 25 at $V \simeq 9.0$, $(B - V)_0^{max} \simeq -0.05$ mag (if $E_{B-V} = 0.09 \pm 0.03$). This is bluer than the $(B - V)_0^{max} \simeq 0.23$ mag, typical for novae at maximum. The early photometric evolution of the nova yielded a rate of decline of $v_d = 0.5$ mag day^{-1}, and the application of the MMRD, given above, gave M_V(max) $= -8.95$. With V(max) $= 9.0$, $A_V = 3.3 E_{B-V}$, $(m - M)_0 = 17.65$ for the LMC. It is, however, evident that in the V(max), v_d diagram nova LMC 1991 is shifted by ~ 1 mag above the mean relation. This is the first super-bright nova detected in the Clouds. It may be compared with the five super-bright novae known in M 31, two in the Galaxy and one in NGC 4472 in the Virgo cluster (Della Valle 1991). It may be assumed tentatively that $\leq 10\%$ of the novae are super-bright. Nevertheless, they can have important effects on distance determinations.

2.6.4 Distance determinations with the aid of SN 1987A

The distance to SN 1987A was determined by Panagia et al. (1991) to 18.55 ± 0.12 from its circumstellar ring; its angular size in an *HST* image in the [O III] 5007 Å line was compared with its absolute size derived from an analysis of the light curves of narrow UV lines (NV $\lambda 1240$, NIV] $\lambda 1486$, NIII] $\lambda 1750$, and CIII $\lambda 1909$) measured with the *IUE*. The relative position of SN 1987A in the LMC was determined on the basis of radial velocity data. Velocity components are seen in the interstellar UV and optical spectra of the SN at ~ 282 and 292 km s^{-1}; strong interstellar absorption lines have been observed in R136, R139, and R145 in the 30 Doradus region at 250, 280 and 300 km s^{-1} (Blades 1980). SN 1987A is then behind two of these gaseous layers, but on the near side of the 30 Doradus complex. Considering also the HI distribution along the line of sight, the major velocity components are between 240 and 300 km s^{-1}, Panagia et al. conclude that SN 1987A must be behind about two-thirds of the main body of the LMC. This leads to a distance modulus of the centre of the LMC of $(m - M)_0 = 18.50 \pm 0.13$.

Observations of absorption line velocities of the expanding shell of SN 1987A have, using Baade's method, given $(m - M)_0 = 18.47 \pm 0.18$ (Hanuschik and Schmidt-Kaler 1991).

Schmidt-Kaler (1992) has reconsidered the assumptions made in applying Baade's method, (i) that the envelope is spherically symmetric, and (ii) that the emitted flux of the expanding envelope can be approximated by a black body. He arrives at

2.7 Summary of distance determinations

Table 2.7. *The distance to the LMC from observations of SN 1987A*

Author	Method	$(m-M)_0$
Panagia et al. (1991)		18.50 ± 0.13
Hanuschik, Schmidt-Kaler (1991)	shell expansion	18.47 ± 0.18
Schmidt-Kaler (1992)	Baade's method, ring expansion and ring light propagation parallax	18.38 ± 0.07
McCall (1993)	based on Panagia et al.	18.52 ± 0.13
Crotts et al. (1995)	light echoes, circular ring	18.61 ± 0.11

corrections for deviations from these assumptions in the actual case of 0.94 and 0.95, respectively, leading to $(m-M)_0 = 18.26 \pm 0.10$. He further derives the distance of the SN from the light parallax of the circumstellar ring, $(m-M)_0 = 18.46 \pm 0.18$, and from the expansion parallax of the SN, 18.48 ± 0.22. Using a tilt of $i = 38°$ Schmidt-Kaler finds that the disk plane at the SN is 0.9 ± 0.15 kpc (at 30 Doradus 1.1 kpc) nearer to us than the geometrical centre, leading to a correction of the moduli of 0.04 mag. By giving higher weight to the distance derived by the Baade method than to the others, he arrives at a most likely value of $(m-M)_0 = 18.38 \pm 0.07$.

McCall (1993) has used the Panagia et al. (1991) results for SN 1987A to redetermine the distance modulus of the centre of the LMC. Assuming a tilt of the LMC disk of $i = 28°.5 \pm 4°$, the position angle of the line of nodes $\theta_0 = 168° \pm 4°$, and determining the position angle of SN 1987A relative to the centre of rotation to $\phi = 90°.956 \pm 0°.006$, he finds that the centre of mass of the LMC is 0.03 ± 0.03 mag nearer than the SN. The uncertainty in the value reflects the possibility that the SN is at any depth within the layer confining 99% of the light of the young disk. Consequently, $(m-M)_0 = 18.52 \pm 0.13$.

Crotts et al. (1995) have used light echoes from the inner few arcseconds of the region around SN 1987A to define the three-dimensional geometry of its circumstellar shell. Following the Panagia et al. (1991) technique they derive a distance modulus of the SN of 18.57 ± 0.12, increased to 18.61 ± 0.11, if the ring is simply circular. Assuming that the SN is 0.43 ± 0.11 kpc behind 30 Doradus, representing the plane of the LMC, and taking the tilt of the LMC to be $i = 32°.6$, they find the distance modulus of the centre of the LMC to be 18.58 ± 0.12 and 18.62 ± 0.10 for the two cases, respectively. The fact that recent observations of 21 cm absorption line velocities tend to place 30 Doradus behind the LMC main plane (Chap. 12) does not change these results; it simply deletes '30 Dor, representing', in the sentence above.

2.7 Summary of distance determinations

Table 2.8 summarizes the distance determinations.

The most likely values on the distance moduli of the LMC and the SMC appear to be $(m-M)_0 = 18.45 \pm 0.10$ and 18.90 ± 0.10, respectively. De Vaucouleurs (1993) has suggested a slightly lower value for the LMC, $(m-M)_0 = 18.38$ and a standard

Table 2.8. *Distance moduli of the Magellanic Clouds*

Object	LMC $(m-M)_0$	n	SMC $(m-M)_0$	n
Cepheids	18.53 ± 0.03	17	18.93 ± 0.04	6
RRLyraes	18.37 ± 0.04	10	18.83 ± 0.03	3
Mira stars	18.46 ± 0.08	4		
OB stars	18.35 ± 0.07	2	19.05 ± 0.07	2
Red stars	18.42 ± 0.15	1		
SN 1987A	18.50 ± 0.04	5		
Novae	18.70 ± 0.23			
Mean values	18.48	7	18.94	3

The mean values for each type of object are unweighted means of data discussed above; the errors indicate the scatter between the individual values. Clusters are omitted because of the uncertainties still prevailing in the fitting procedures.

deviation of 0.19 mag, from 62 sources, many of which are based on cluster MS fittings.

The SMC is $\sim$ 0.4 mag more distant than the LMC when Cepheids and RR Lyrae variables are considered, but 0.7 mag more distant when the early-type stars are used. This discrepancy is not very surprising, considering that at least the SMC is markedly extended in depth. We accept $(m-M)_0 = 18.9$ mag as the most likely value for the SMC. The linear distances of the two Clouds are then 50 kpc for the LMC and 60 kpc for the SMC.

In view of all the problems involved in the distance determinations it is necessary to admit that the distances of the two Clouds are still not sufficiently well known.

3

The Clouds as galaxies

It is essential for our understanding of the evolution of the Magellanic System, comprising the LMC and the SMC, the InterCloud (IC) or Bridge region and the Magellanic Stream, to know the motions of its members in the past. The Clouds have a common envelope of neutral hydrogen. This indicates that they have been bound to each other for a long time. It is generally assumed, but not definitely proven, that the Clouds have also been bound to our Galaxy for at least the last 7 Gyr. Most models assume that the Clouds lead the Magellanic Stream. The Magellanic System moves in the gravitational potential of our Galaxy and in the plane defined by the Local Group It is also exposed to ram pressure through its movement in the galactic halo. The influence of our Galaxy ought to be noticeable in the present structure and kinematics of the System.

The interaction between the Clouds has influenced their structure and kinematics severely. It should be possible to trace the effects as pronounced disturbances in the motions of their stellar and gaseous components. Recent astrometric contributions in this field show great promise for the future if still higher accuracy can be achieved. It should be kept in mind in all analyses that results of interactions may be expected everywhere.

3.1 Models of the Magellanic System
It is desirable that models of the Magellanic System explain the following features:
1. the Magellanic Stream;
2. the formation of the Bar of the LMC and of the bar-like structure of the SMC;
3. the origin of the Wing of the SMC;
4. the concentration of gas and dust in the region of 30 Doradus in the LMC;
5. the formation and evolution of the superassociations/supergiant shells;
6. the East–West asymmetry in the LMC (seen e.g. in dust/plasma ratios).

So far the main interest has been concentrated on the Magellanic Stream and the connected question about the orbits of the LMC and the SMC.

The past history of the Magellanic Clouds is still veiled in obscurity. As have all other members of the Local Group, they may have emerged from near its barycentre. Whereas many of the dwarf galaxies in the Local Group may have had close encounters with either the Galaxy or M 31 but escaped being caught (see Mishra 1985), several, including the LMC and the SMC, may have been less fortunate and are now satellites of either of the two giant galaxies in the Group. Shuter (1992) has suggested that the Clouds set out in a Hubble flow, together with M 31 and the

Galaxy, were subsequently attracted to M 31, and suffered a close collision with it ~ 6 Gyr ago. The tidal acceleration in this collision projected them back toward the Galaxy and produced the velocity differentials required to establish the Magellanic Stream. Byrd *et al.* (1994) consider that the Magellanic Clouds may have left the M 31 neighbourhood ~ 10 Gyr ago and may have been captured by our Galaxy ~ 6 Gyr ago. Whether the Clouds had to visit M 31 before being caught by our Galaxy may be an open question. Murai and Fujimoto (1980) examined the possibility that a single Magellanic Cloud was split into two independent objects, the LMC and the SMC, by tidal disruption at close passage to our Galaxy, but they found no orbits coalescing into a single one towards the past. The two Clouds may then be considered as having always been separate objects, and satellites to our Galaxy for up to 10 Gyr.

Several models for the orbital behaviour of the Magellanic Clouds as satellites of our Galaxy have been presented, generally with the aim of explaining the characteristics of the Magellanic Stream. Einasto *et al.* (1976) discussed our 'Hypergalaxy', consisting of the Milky Way galaxy and its satellites and having a total mass of $(1.2 \pm 0.5) \times 10^{12}$ $M_\odot$. Included with the satellites as permanent members are the Magellanic Stream and the high-velocity hydrogen clouds; they all move around the Galaxy in elliptical orbits. Elliptical dwarf galaxies populate the inner regions in all hypergalaxies except our own. In ours the irregular galaxies, SMC and the LMC, are closest to the Galaxy.[*] It is difficult to see how the Magellanic System could have survived in its present configuration if all parts of it had existed in about the present position from the time of galaxy formation.

Early models in which the Galaxy and the Clouds were treated as a two-body problem, were presented by Lin and Lynden-Bell (1977), Davies and Wright (1977), and Kunkel (1979). They have been further developed by Tanaka (1981). He considered a triple system problem in which the Clouds were revolving clockwise around the Galaxy as seen from the Sun, and producing the Stream by particles pulled out from the LMC by the tidal interaction of the Galaxy. In order to find a reasonable model for the Magellanic Stream, Lin and Lynden-Bell (1982) placed the LMC in an eccentric orbit of $a = 125$ kpc moving in a galactic halo extending to a radius of > 70 kpc. Their model galaxy had a mass of $\approx 3 \times 10^{12}$ $M_\odot$.[†] The LMC would have completed four to five orbits in a Hubble time and be near perigalacticon at present. The Stream would have been detached at the perigalacticon passage ~ 2 Gyr ago, and the SMC probably freed from the LMC at the current passage.

It is essential that a model for the orbital behaviour of the Magellanic Clouds as satellites of our Galaxy succeeds in reproducing all the the main characteristics of the Magellanic System, including the internal structures of the Clouds. The model presented by Murai and Fujimoto (1980) and Fujimoto and Murai (1984) has been developed in that direction by Fujimoto and Noguchi (1990), Gardiner *et al.* (1994) and Gardiner and Noguchi (1995). A basic postulate in the Murai, Fujimoto model is that the LMC and the SMC are gravitationally bound, a reasonable assumption based on the fact that the two Clouds have a common envelope of neutral hydrogen.

The model has the following main characteristics:

---

[*] Recently, a spheroidal dwarf galaxy was detected in Sagittarius, 16 kpc from the centre of our Galaxy, and under tidal disruption (Ibata *et al.* 1994).
[†] Most authors need a galactic mass of ~ 3×10^{12} $M_\odot$ and a galactic halo extending ≥ 200 kpc for their models.

3.1 Models of the Magellanic System

—The LMC and SMC have been bound gravitationally throughout the past 10 Gyr.
—The orbital plane of the LMC is perpendicular to the galactic plane.
—The sense of revolution around the Galaxy is counterclockwise as seen from the present position of the Sun.
—The distance between the Clouds has varied appreciably, being as small as 2–7 kpc 7.5, 2.6, 1.5 and 0.2 Gyr ago, at other times reaching 50 kpc, most recently 0.8 Gyr ago. The present perigalactic distance of the LMC is 50 kpc; the most recent passage through it occurred about 50 Myr ago. It was its closest location to the centre of our Galaxy through its entire history.
—The Magellanic Stream is the tail of the gas pulled out from the SMC by the tidal force of the LMC, and its motion was controlled by the Galaxy after it left the Magellanic Clouds System.
—The distance between the Galaxy and the Clouds decreases in an oscillatory way due to the dynamical friction. The Clouds will merge into our Galaxy in a not too distant future.

In this model the Magellanic Stream has a distance from the Sun of 30–60 kpc and a total hydrogen mass of about 10^8 $M_\odot$. The total mass of the Galaxy, inside 50 kpc from the centre, has to be 7×10^{11} $M_\odot$ and its rotation velocity is 250 km s^{-1}.

In the close encounters between the two Clouds they have temporarily been partially merged several times in the past. These merging events form, together with the close approaches to our Galaxy and the initial collapse of the protoclouds, the epochs in the history of the Magellanic System when bursts of star formation may be expected to have occurred. Rich stellar populations should then exist in the Clouds with upper ages corresponding to these epochs. Thus, for example, most of the intermediate-age clusters have ages between 1 and 2.5 Gyr, a less numerous field population is a few times 10^8 yr old, and the maximum age of the youngest population is $\leq$ 100 Myr. The collision about 0.2 Gyr ago may be responsible for the severe fragmentation of the SMC.

The influence of our Galaxy on the Magellanic Clouds depends to a high degree on the length of time during which the interaction has occurred. If this period is $\sim$ 10 Gyr, the LMC may have exchanged up to 30% of its mass with the galactic corona (Burkert et al. 1990). Significant interactions between the LMC and the SMC as well as between the Clouds and our Galaxy have led to a number of structural and kinematical peculiarities. Valtonen et al. (1984) suggested, from numerical simulations, that globular clusters may have been detached from the Clouds. Using a mass ratio LMC/Galaxy of 0.05, they found a tidal radius of the LMC of $\leq$ 5 kpc, meaning that the three globular clusters, NGC 1466, 1841, and 2257, should be extra-tidal, and no longer gravitationally bound to the LMC. The tidal radius of the LMC may, however, be appreciably larger. Values around 8 kpc are often used.

The close approaches between the Magellanic Clouds and the Galaxy may also have caused disturbances in the latter. The distortions of the outer regions of the galactic HI plane have been ascribed to gravitational effects of the former (Elwert and Hablick 1965, Avner and King 1967, Tanaka 1981). Indications have also been found of at least two star bursts in the Galaxy with ages that are remarkably synchronous with bursts of star formation in the LMC, 5 and 0.3 Gyr ago (Scalo 1987). At these epochs the LMC was at perigalactic distances according to the model just described. There are in the model other close

24 The Clouds as galaxies

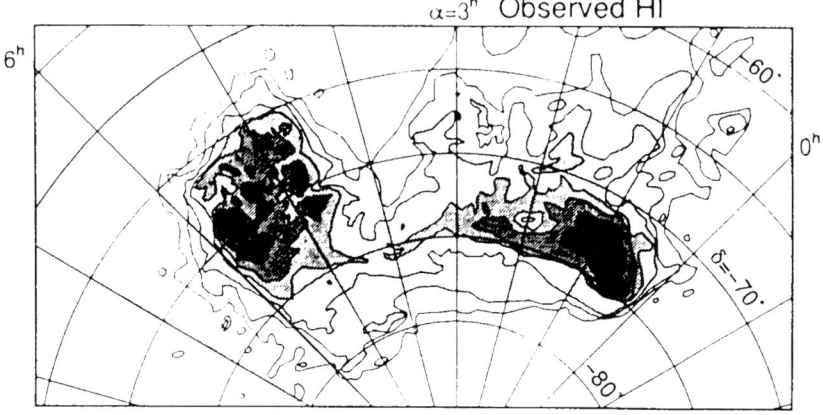

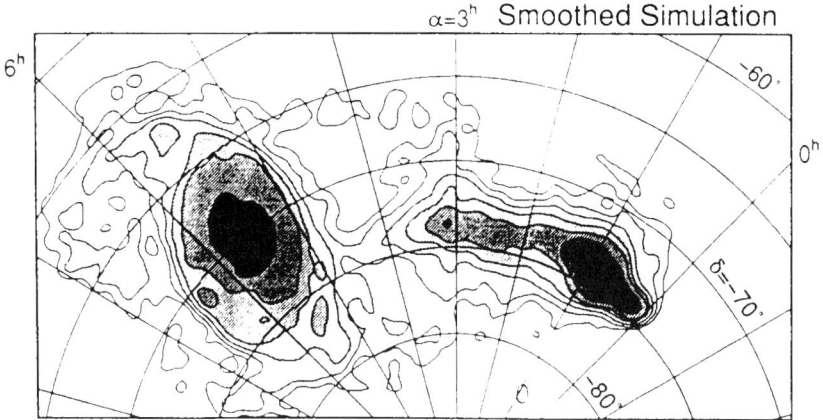

Fig. 3.1. Comparison of the test-particle distribution (bottom panel) derived by Gardiner et al. (1994) and the HI surface density distribution (Mathewson and Ford 1984). (Reprinted from Gardiner et al. (1994) by permission of Blackwell Science Ltd.)

approaches by the Clouds; more synchronous events may be detected once the resolution in time is sufficiently improved. It is also possible that some galactic outer globular clusters have been ejected from the Galaxy by encounters with the Clouds.

The global gas distribution in the Magellanic System has been reproduced by Gardiner et al.(1994) in a test particle simulation.The LMC disk remains largely undisturbed despite a close encounter with the SMC $\sim$ 0.2 Gyr ago. Most test particles are confined to a 1-kpc-thick disk. An important result is that there is no pronounced concentration of test particles to be seen in the 30 Doradus region (Fig. 3.1). Distinct gas masses with different radial velocities are seen in the InterCloud region, in broad agreement with HI studies. The SMC extends $\sim$ 15 kpc in depth.

3.1 Models of the Magellanic System

A further set of N-body simulations on the gravitational interaction of the SMC with the Galaxy and the LMC has been carried out by Gardiner and Noguchi (1995). In the calculations the Galaxy and the LMC are represented by rigid spherical potentials and the SMC by a self-gravitating particle system, composed of a rotationally supported disc and a spherical halo with the mass ratio 1 : 1. The total mass of the SMC is 3×10^9 $M_\odot$. The spatial orientation of the SMC at the beginning of the interaction, 2 Gyr ago, is chosen so that the orientations of the Bar and the major axis today agree with the observations. The features produced by the interactions are the Magellanic Stream, a leading arm, a tidal bridge and a tidal tail. The Magellanic Stream and the leading arm, on the opposite side of the Clouds to the Stream, originated ~ 1.5 Gyr ago, at the time of the previous perigalactic passage of the SMC which coincided with a close encounter between the Clouds. The leading arm mimics the overall distribution of several hydrogen clouds observed in that part of the sky (see e.g. Mathewson *et al.* 1987). The bridge and the tail were generated at the close encounter between the Clouds 0.2 Gyr ago.

Gardiner and Noguchi succeed in reproducing the morphology of the Magellanic Stream, including the general variation of its width, as well as the velocity profile of the HI gas. Features corresponding to the scattered HI clumps at $260° < l < 310°$, $-30° < b < +30°$ (Mathewson *et al.* 1974) are reproduced as a 'leading arm'. The morphology of the SMC Bar on the sky, as well as its spatial orientation, are likewise in reasonable agreement with observations, though full agreement with the Caldwell and Coulson (1986) Cepheid-defined structure is not achieved. The model also explains successfully some major trends in the kinematics of the young populations in the SMC Bar and the older populations in the 'halo' of the SMC as well as the overall pattern in the InterCloud region.

The halo component in their model shows an increasing depth from SW to NE. This is due to the appearance of the tidal tail and the bridge in the NE, and only the tail in the SW region. Observations of the HB/RGC stars by Hatzidimitriou and Hawkins (1989) point in the same direction (Chap. 12). The HB/RGC stars are assumed to be mainly connected with the tidal bridge (0.2 Gyr old), the Cepheids with the tidal tail.

The simulated velocity pattern for the disk and halo components shows in the InterCloud region some correspondence with the velocity pattern observed for neutral hydrogen, young stars and carbon stars. The tidal tail and bridge developed by the disk component provides an explanation for the wide velocity range observed among the young objects. The tidal bridge developed by the halo, where the tail is weak, may explain the concentration of low galactocentric radial velocities of the carbon stars (Kunkel *et al.* 1995). In the model the particle distribution over the whole region shows a rather sharp cut-off on the SW side of the SMC, in agreement with the observed HI distribution.

Kunkel *et al.* suggested that the SMC is not a satellite bound to the LMC. Several hundred recently discovered carbon stars on the periphery of both the LMC and the SMC lead to this dynamical interpretation of the triple system Galaxy, LMC, SMC, unless the LMC has an exceptionally high mass. However, most of the structural problems encountered in the InterCloud region, and accounted for by Kunkel *et al.* as debris left over from a passage of the independently moving SMC close to the LMC, seems to be equally well accounted for by the Gardiner and Noguchi model.

Table 3.1. *Model motions of the LMC*

Model	μ mas yr^{-1}	$v_{t,g}$ km s^{-1}	$V_{c,\odot}$ km s^{-1}	$R_\odot$ kpc
Murai and Fujimoto (1980)	1.7	288	250	10.0
Lin and Lynden-Bell (1982)	2.0	373	244	9.0
Shuter (1992)	1.9	355	220	8.5
Liu (1992)	1.7	310	220	8.5
Gardiner et al. (1994)	1.8	287	220	8.5

$v_{t,g}$ is galactocentric tangential velocity, $V_{c,\odot}$ circular velocity of the Sun and $R_\odot$ Sun–galactic-centre distance.

Table 3.2. *Proper motions of the Magellanic Clouds*

$\mu_\alpha \cos \delta$ mas yr^{-1}	μ_δ mas yr^{-1}	Ref.
0.91 ± 2.34	-0.23 ± 2.77	Tucholke, Hiesgen (1991), LMC
1.3 ± 0.6	$+1.1 \pm 0.7$	Kroupa et al. (1994), LMC, SMC
1.37 ± 0.25	-0.18 ± 0.25	Jones et al. (1994), LMC

3.2 Proper motions of the LMC and the SMC

All models proposed so far assume that the LMC and the SMC lead the Magellanic Stream and predict a proper motion for the LMC of between 1.5 and 2.0 mas yr^{-1}, the smallest value follows from a galaxy with no halo (Table 3.1).

Tucholke and Hiesgen (1991) have measured absolute proper motions relative to 498 reference galaxies, epoch difference 15 yr. Kroupa et al. (1994) used Sanduleak lists to identify 35 PPM stars as members of the LMC and 8 as members of the SMC. Proper motions in the PPM catalogue (Röser and Bastian 1993) are derived from data spanning almost a century. The SMC has suffered significant tidal perturbation so that the interpretation of its measured proper motion is difficult. The present values are consistent with bound as well as unbound orbits of the Magellanic Clouds. Jones et al. (1994) determined proper motions for 251 LMC members (92 reference galaxies, epoch span 14 yr) near NGC 2257, 8.5° from the LMC centre, in $PA = 61°$. The mean absolute proper motion of the LMC stars in this region is $\mu_\alpha = 0.120 \pm 0.028$ arcsec century^{-1}, $\mu_\delta = 0.026 \pm 0.027$ arcsec century^{-1}. Corrections for the rotation of the LMC and the effects of solar motion were applied and the total galactocentric transverse velocity given in Table 3.2 was derived.

3.3 Transverse motions of the LMC and the SMC

The motions of the Magellanic Clouds have received extensive attention over the years. Hertzsprung (1920) examined the scanty radial velocity data available at that time from observations of emission nebulae in the LMC (17) and the SMC (1). He found no clear rotation of the LMC but signs of a common translational motion of the two Clouds. There have been other attempts in this direction, and with limited material.

3.5 The geometry of the Clouds

Table 3.3. *Transverse motion of the LMC*

Heliocentric velocity km s^{-1}	Galactocentric velocity km s^{-1}	Ref.
	100	Feast et al. 1961; stars and nebulae
275	143	Feitzinger et al. 1977
275 ± 65		Meatheringham et al. 1988; gradients in RV
	150	Prévot et al. 1989; late-type supergiants
	200	Hughes et al. 1991; HI, CO, PN, CH stars, clusters
	236	Lin 1993; proper motions
	215 ± 48	Jones et al. 1994; proper motions

Feast et al. (1961) concluded from their analysis of radial velocities of stars and nebulae in the LMC and the SMC that the translational velocity relative to the centre of the Galaxy could hardly exceed 100 km s^{-1}. Feitzinger et al. (1977) used the difference in position angle between the kinematical (188.°0 ± 2.6°) and photometric lines (168° ± 4°) of nodes ($i = 33.0° \pm 3°$) to determine the transverse velocity. The same method was used by Meatheringham et al. (1988). Prévot et al. (1989) analysed the radial velocities of late-type supergiants in the LMC under the assumption that the direction of the transverse motion is defined by that of the Magellanic Stream. The kinematical major axis is then aligned in the direction of the line of nodes, at a position angle of 168°. Hughes et al. (1991) found that the dynamics of the LMC is dominated by a single rotating disk, and that all major populations of the Bar have solid-body rotation. Most authors find the galactocentric radial velocity of the LMC small, < 50 km s^{-1}. Meatheringham et al., for instance, derived 42 ± 10 km s^{-1}, and concluded that the LMC is now close to its perigalacticon, with a space velocity relative to the galactic centre of 278 km s^{-1}.

The individual radial-velocity measurements are accurate to $\sim$ 10-15 km s^{-1}, whereas the individual proper motions are accurate to $\sim$ 3 mas yr^{-1}. The latter corresponds to $\sim$ 700 km s^{-1} at the distance of the LMC. If a 20 micro-arcsec accuracy can be achieved (see the *GAIA* concept and other contributions at *IAU Symposium. No.166*) annual proper motions could be determined for LMC stars to better than $\sim$ 5 km s^{-1}. This would mean an extreme improvement in our knowledge about the internal motions in the Magellanic System.

3.4 Integrated properties of the LMC and the SMC

Fundamental data for the LMC and SMC are presented in Table 3.4. As up-to-date values as possible have been used.

3.5 The geometry of the Clouds

The position of an object in the Clouds was previously ignored in determinations of their distances but has now, as a consequence of improved observational techniques, become important enough to be considered in deriving the distances to their centres.

Table 3.4. *Fundamental data for the SMC and the LMC*

	LMC	SMC
Type	SB(s)m	SB(s)mp
Galactic coord. l,b	$280°.5, -32°.9$	$302°.8, -44°.3$
Distance moduli	18.5 ± 0.2 mag	18.9 ± 0.2 mag
RV_{hel}	275 km s^{-1}	148 km s^{-1}
Tilt	$33°$–$45°$	$90°$
Integrated magnitudes and colours		
B	0.9 mag (100 deg^2)	2.9 mag (12 deg^2)
$B - V$	0.5 mag	0.4 mag
Central surface brightness B	21.2 mag arcsec^{-2}	21.4 mag arcsec^{-2}
$B - V$	+0.46	+0.25
Masses		
Total masses	2×10^{10} (6×10^9?)	2×10^9 M$_\odot$
HI mass	$\leq 8\%$ of total mass	$\sim 30\%$
HI (21 cm)	7.0×10^8	6.5×10^8 M$_\odot$
H$_2$ (CO line)	1.4×10^8	3.0×10^7 M$_\odot$
Warm dust (IRAS)	4–50×10^3	8–10×10^4 M$_\odot$
Cold dust (mm)	8–90×10^5	9–100×10^5 M$_\odot$

The galaxy types are from de Vaucouleurs and Freeman (1973).
The mass of the old, metal poor ([Fe/H] ≤ -1.4) population in the LMC is $\sim 6\%$ of its total mass (Frogel 1984); the mass of the old, slightly more metal rich ([Fe/H] $\sim -1.0 \pm 0.5$) population, represented by the old long-period variables, is $\sim 2\%$ of the total mass (Hughes et al. 1991).
The average Magellanic normal dust is hotter than the galactic due to the high interstellar radiation field (Lequeux 1989). Small grains may have been destroyed by the abundant UV radiation.

The LMC has long been regarded as a thin flat disk seen nearly face-on (see de Vaucouleurs and Freeman 1973). The tilt of the LMC plane to the plane of the sky was determined to be $i = 27° \pm 2°$ on the basis of optical and 21 cm HI isophotes and the surface distribution of clusters, assuming that the untilted distributions are circular, and of individual distances of a limited number of Cepheids (see de Vaucouleurs 1980 for references). In the former determination, the outer HI isophotes as well as the outmost clusters were not considered. A solution for the outer clusters gave $i = 45°$ (Lyngå and Westerlund 1963).

Several new determinations of i are now available (Table 3.5). The highest value so far, $i \sim 48°$, is derived from surface photometry (Bothun and Thomson 1988) and applies to the main body of the LMC, i.e. the star forming regions Constellation III and 30 Doradus are excluded in the ellipse fitting routine. In this way strictly local effects do not influence the determination of the tilt of (the main body of) the LMC. For similar reasons, low-frequency solutions, referring mainly to non-thermal emission, may be the most reliable among the radio observations.

3.5 The geometry of the Clouds

Table 3.4. *Fundamental data for the SMC and the LMC, cont.*

	LMC	SMC
Structure:		
Disk:	Has a major disk made up of intermediate-age and young population (stars and gas)	Layers of HI are seen in depth; if there is a disk it will have to be in the line of sight.
HII regions	Large number	Few large
Diffuse Hα emission	Yes, extended	Yes, extended
Molecular clouds	Several; strongest south of 30 Doradus and SW parts of the Bar	Few, mainly in NE
Bar:	Red stars ; superposed HII and starforming regions	Predominantly blue stars with a scatter of red stars
Halo: Stellar	Probably	Probably
: Plasma	Diffuse X-ray emission, gets softer away from the centre. Optically thin thermal plasma with $\sim 2 \times 10^6$ K in western part; $\sim 10^7$ K near the active star forming region 30 Doradus	Diffuse X-ray emission, optically thin thermal plasma of $\sim 10^6$ K; halo plasma ?

The central regions of the SMC are considerably bluer than the outer parts. This effect is less obvious in the LMC.
The high percentages of gas still unused in the Clouds show that they are far behind the Galaxy in evolutionary status. This is also evident in their lower metal content.
The SMC is much less evolved than the LMC.
No cold molecular clouds exist. SMC lacks HI and H_2 clouds heated by the far UV radiation of hot stars.

The thermal radio continuum emission receives strong contributions from discrete sources, mainly HII regions (see Haynes *et al.* 1986), and as such star forming regions. The use of the neutral hydrogen (HI) distribution for determinations of i is also questionable as it is now evident that HI has a complicated structure (Meaburn *et al.* 1987, Luks and Rohlfs 1992). From the data in Table 3.5 we conclude, therefore, that $i \sim 45°$ is the most likely value for the tilt of the main body of the LMC.

It is now well established that the east side of the LMC is closer than the west. Caldwell and Coulson (1986) found the closest part at the position angle $PA = 52° \pm 8°$, Laney and Stobie (1986) arrived at $PA = 55° \pm 17°$, and Welch *et al.* (1987) arrived at $77° \pm 42°$, all from the distribution of Cepheids.

In the SMC the geometry is more complicated. Its tilt and the *PA* of its line of nodes have been determined (Table 3.6) but may have little relevance with regard

Table 3.5. *Tilt (i) and position angle (Θ) of line of nodes in the LMC*

Method	$i°$	$\Theta°$	Ref.
A. Geometrical methods			
Optical isophotes	27 ±2	170	de Vaucouleurs, Freeman 1973
Surface photometry, 30 Dor, SGS 4 excluded	48		Bothun, Thomson 1988
Inner clusters	25 ±9	171	McGee, Milton 1966
Outer clusters	45	187	Lyngå, Westerlund 1963
HI, HII, Supergiants, PN		168	Feitzinger et al. 1977
Cepheids, PLC relation	31 ±8		de Vaucouleurs 1980
	29 ±6		Caldwell, Coulson 1986
	45 ±7		Laney, Stobie 1986
	37 ±16		Welch et al. 1987
HI brightness contours	35	185	Hindman et al. 1963
HI distribution	27		McGee, Milton 1966
408 MHz isophotes	45		Mathewson, Healey 1964a
45 MHz isophotes, non-thermal	45	180	Alvarez et al. 1987
Optical isophotes	38 ±2	168 ±7	Schmidt-Kaler & Gochermann 1992
B. Kinematical methods			
HI,HII, Supergiants, PN	33	188	Feitzinger et al. 1977
Globular clusters, young		181	Freeman et al. 1983
Globular clusters, old		221	Freeman et al. 1983
HI iso-velocity contour		208	Rohlfs et al. 1984

to its structure as the SMC is markedly extended in depth with the Wing and the northeastern section closer to us than the southern part. The nature and amount of its depth is not yet fully agreed on. A large number of investigations have been carried out; they will be discussed in Sect. 12.2.

3.6 The structure of the LMC

The structure of the LMC has been described in many ways. A basic feature is the Bar, covering an area in the sky of $\sim 3° \times 1°$, with its major axis in PA 120°. Its centre is at $5^h24^m.0$, $-69°47'$ (1950). The pattern of rich OB associations and large HII regions has been interpreted as a spiral structure; support for this has been searched for in the neutral hydrogen distribution. A faint outer loop has been looked upon as an extension of the main spiral arm. It was detected by its unresolved luminosity on small-scale long exposure photographs (de Vaucouleurs and Freeman 1973) and is devoid of bright supergiants and emission regions and rather weak in HI. The feature can be traced by carbon stars to the east of the main body of the LMC, from $-68°$ to $-73°$ at $06^h 15^m$, and by carbon stars and old clusters to the south, near $-75°$ from $4^h 50^m$ to $6^h 10^m$. The 30 Doradus complex does fit poorly into most structural

3.6 The structure of the LMC

Table 3.6. *Tilt (i) and position angle (Θ) of line of nodes in the SMC*

Method	i°	Θ°	Reference
Isophotes	60	45	de Vaucouleurs, Freeman 1973
Surface photometry	56		Bothun, Thomson 1988
Small open clusters		55	Brück 1975
HII regions		52	Torres, Carranza 1987
Optical infrared		55	Johnson 1961
HI		55	Hindman 1967
408 MHz		59	Loiseau *et al.* 1987
45 MHz		38	Alvarez *et al.* 1989

Torres and Carranza used HII radial velocity gradients.

interpretations except in those where it is considered as the centre, or nucleus, of the spiral structure of the LMC.

Several authors, the first one probably being Johnson (1959), have described the LMC as a late spiral, with its nucleus at 30 Doradus in front of an elliptical structure (galaxy) with its major axis along the Bar. Undoubtedly, the LMC must have looked like a dwarf elliptical before the most recent burst of star formation, though more gas-rich than the dwarf elliptical satellites of our Galaxy. The question of a possible spiral structure in the LMC will be further considered in Chap. 12. Here, the overall structure of the LMC will be described by a set of subsystems, containing most observed features (Table 3.7). As there is no galactic nucleus in the LMC, the Bar will have to take that position, figuratively speaking The Central System with the youngest population replaces the spiral system in our Galaxy.

3.6.1 The centroids of the various populations in the LMC

In an irregular galaxy like the LMC, exposed to multiple severe interactions, the positions of the centroids of the various populations may give some information about its history of evolution. The centre of gravity of the galaxy or its centre of rotation may serve as a reference point for the centroids.

The most frequently used centre of the LMC has been the 'radio centre of rotation'. It was originally defined by the condition that the velocity curve resulting from the 21 cm observations should be symmetrical (Kerr and de Vaucouleurs 1955a) so that the mass of the LMC could be derived. The position most frequently used today is 5^h21^m, $-69°18'$ but other values are also in use (Table 3.8).

The radio centre of rotation is displaced from the optical centre of the Bar, at 5^h24^m, $-69°.8$, from the centroid of the projected neutral hydrogen distribution, at $5^h34^m.5$, $-68°30'$, and from all other centroids of the populations of the LMC (Table 3.8).

The two cluster systems, defined by Kontizas *et al.* (1990b), represent a young (inner) and an intermediate-age (outer) population. The planetary nebulae also belong to the latter and some similarity in their centroids may be expected.

The distribution of the Population I objects over the LMC surface is probably too uneven to define any meaningful centre of symmetry. Most of the young clusters, OB associations and supergiants are concentrated in superassociations (see Chap. 6)

The Clouds as galaxies

Table 3.7. *The subsystems of the LMC*

	Bar 3° × 1°	Central System 6° flat	Disk 14° flat?	Halo 24° spheroidal?
Observed content:	*Young Population:* OB associations, clusters, HII regions (from Central System) superposed on intermediate-age red giants.	*Young Population:* OB associations, young clusters, supergiants, HII regions, molecules, dust, SNRs; mostly in superassociations. HI, HII froth.	*Old disk population:* open and globular clusters of intermediate age, RGB and AGB stars, PN, RRLyraes LPVs	*Very old population:* true globular clusters, RR Lyraes, old red giants? old LPVs

It is not known if the Bar contains a very old red-giant population.
The Central System is flat, its content is, however, not restricted to a narrow plane.
The 30 Doradus complex is most likely behind the major HI plane; interstellar lines as well as HI reveal a stratified structure of the ISM.
The Halo is probably disklike with a maximum height of ~ 3 kpc.

of quite different richness. Also the neutral hydrogen has a very uneven surface distribution.

Traditionally, all determinations of the centre of rotation of a population have begun by attempts to fit it to the centre of the HI rotation curve, and success has frequently been claimed. It should thus cause some worries that the centre of rotation of the HI main disk (Luks and Rohlfs 1992) is rather much displaced from the 'classical' centre of rotation and also from the centre of gravity of the Bar.

3.7 The structure of the SMC

The structure of the SMC has proved to be difficult to define, mostly because of its large inclination angle. It has, like the LMC, a bar, elongated in PA 45°, and with its optical centre at $0^h 51^m$, $-73°.1$ (1950). The brightest part of the Bar is in the immediate neighbourhood of this centre. There is also the highest concentration of neutral hydrogen, in itself a remarkable fact when compared with the LMC. A spiral structure has been recorded (See de Vaucouleurs and Freeman 1973). Johnson (1961) divided the SMC into an elliptical and a spiral component. As will be discussed in Chap. 12, the structure of the SMC is more complex than that of the LMC. Here, we will use, as for the LMC, a set of subsystems to describe the *observed* main features, disregarding e.g. the fact that the 'Bar' is far from a uniform feature (Table 3.9).

3.7.1 The centroids of the various populations in the SMC

As the SMC is markedly extended in the line of sight, the importance of centroids based on the apparent surface distribution of a class of objects is difficult to estimate.

3.8 Gaseous coronae?

Table 3.8. *Population centroids in the LMC (1950.0)*

Population	R.A.	Decl.	Ref.
Optical centre of Bar	5^h24^m	$-69°48'$	De Vaucouleurs and Freeman 1973
Yellow light isophotes	5^h20^m	$-69°30'$	De Vaucouleurs and Freeman 1973
408 MHz peak of non-thermal emission	5^h25^m	$-69°36'$	Mathewson and Healey 1964a
45 MHz	5^h28^m	$-69°24'$	Alvarez et al. 1987
Planetary nebulae:			
entire system	$5^h23^m.7$	$-69°01'$	Morgan 1994
outer envelope	$5^h31^m.4$	$-68°47'$	Morgan 1994
rotation centre	$5^h20^m.8$	$-69°06'$	Meatheringham et al. 1988
Novae	$5^h24^m.6$	$-69°06'$	Van den Bergh 1988
outer cluster system	5^h32^m	$-69°18'$	Kontizas et al. 1990b
inner cluster system	5^h21^m	$-68°48'$	Kontizas et al. 1990b
H I mass	$5^h34^m.5$	$-68°30'$	De Vaucouleurs and Freeman 1973
H I rotation curve: centre of symmetry	5^h21^m	$-69°17'$	Freeman et al. 1983
of outer parts	5^h24^m	$-69°48'$	Rohlfs et al. 1984
centre of disk	$5^h12^m.6$	$-69°03'.6$	Luks&Rohlfs 1992
Supergiants	$5^h22^m.0$	$-68°24'$	Isserstedt 1975
Population I rotation centre	5^h20^m	$-68°48'$	Feast 1964

The HI rotation curves are fictitious; the apparent velocity gradient along the Bar results from different contributions of the various HI clouds in the north-east and the southwest (see Chap. 12).

3.8 Gaseous coronae?

Evidence for hot gaseous coronae around the Magellanic Clouds was presented more than ten years ago by de Boer and Savage (1980). Their analysis of high-dispersion spectra of five LMC and two SMC stars, obtained with the *IUE* covering the spectral range 1160–2100 Å, led to the detection of strong interstellar lines of Al III, Si IV and C IV at Magellanic Cloud velocities in addition to the usual lower ionization-stage lines associated with HI region absorption. The high-ionized lines were interpreted as probably coming from regions of hot gas in front of the Clouds. The lines may, however, have formed in hot bubbles in the neutral medium.

Their conclusions concerning the LMC were criticized by Feitzinger and Schmidt-Kaler (1982) who suggested that a warped, two-layer structure of the LMC also

Table 3.9. *The subsystems of the SMC*

	Bar 2.5° × 1°	Wing 6° × 1°	Central System 7° (flat?)	Halo 7° (sphere?)
Observed content:	*Young Population:* OB associations, young clusters, supergiants, HII regions, cepheids, supergiant shell(s) ? SNRs	*Young Population:* OB associations, young clusters, supergiants, HII regions, cepheids, supergiant shell(s)	*Old disk population:* old blue clusters, red-giant stars, carbon stars, planetary nebulae	*Very old population:* old red clusters, RR Lyraes, very old red stars ?

One degree is ∼ 1.1 kpc. No consideration has been taken to the possible extent in depth of the SMC.
The centre of the SMC Central System is assumed to lie near NGC 419; its extent is based on the identified clusters and field stars.

Table 3.10. *Population centroids in the SMC (1950.0)*

Population	R.A.	Decl.	Reference
Optical centre	$00^h 48^m$	$-73°10'$	Westerlund 1982
Small open clusters	$00^h 54^m$	$-72°58'$	Brück 1975
H II regions	$00^h 58^m$	$-72°53'$	Torres, Carranza 1987
Planetary nebulae	$00^h 49^m.65$	$-73°30$	Dopita et al. 1985a
Novae	$00^h 38^m$	$-72°53'$	Graham 1984
Near IR (optical)	$00^h 52^m$	$-73°18'$	Johnson 1961
408 MHz	$01^h 08^m$	$-73°02'$	Loiseau et al. 1987
45 MHz	$00^h 47^m$	$-72°42'$	Alvarez et al. 1989
H I rotation curve	$01^h 03^m$	$-72°45'$	Hindman 1967
H I rotation curve	$01^h 03^m$	$-72°30'$	Loiseau, Bajaja 1981

The 408 MHz centre is the centre of elliptical rings fitting the distribution.

explained the results with regard to interstellar features in the UV, optical and radio domains. The 'warps' agree mainly with the material above the plane in the 30 Doradus region. Undoubtedly, a multi-layer structure exists at least in that region (see. Chaps. 8 and 10) but there are no clear signs of a corona.

In the SMC there have been suggestions of substantial X-ray emission from a halo (see Sect. 9.2). However, this has also been doubted; much of the diffuse emission may be explained by unresolved low-luminosity sources.

3.9 The Magellanic Stream

The Magellanic Stream is a narrow band of neutral hydrogen gas along a small circle 7° from, and parallel to, a great circle that passes through the south galactic pole and cuts the galactic plane at $l = 280°$, ending with the HI clouds detected by Wannier et al. (1972). If the reasonable assumption is made that the Stream forms part of a great circle when seen from the galactic centre, then this parallax places the Stream at about the distance of the Magellanic Clouds (Mathewson et al. 1977). It extends about 100° from the InterCloud region as an essentially continuous filament. It is bifurcated for most of its length into horseshoe-shaped structures. There are six main concentrations, MS I–MS VI. The surface density is greatest at MS I and weakest at MS VI, with each successive cloud having a lower density than the previous one (Mathewson et al. 1987).

The radial velocity of the Stream with respect to the galactic centre becomes increasingly more negative from 0 km s^{-1} at MS I to -200 km s^{-1} at MS VI. The variation within each cloud is independent of this systematic variation, with the radial velocity being essentially constant along the ridge lines of the individual clouds. There is a sharp discontinuity in velocity of ≈ 100 km s^{-1} between the top of the IC region and the beginning of the Stream at MS I (Mathewson et al. 1979). The mean column density of gas decreases from 2×10^{20} atoms cm^{-2} close to the Clouds to 1×10^{19} atoms cm^{-2} at the tip of the Stream.

The Stream is following behind the Magellanic Clouds, which have a transverse velocity of 275 ± 65 km s^{-1} (Table 3.3) and, as seen on the sky, are moving up towards the galactic plane. This follows both from the velocity gradient across the Magellanic Clouds and the ram pressure compression of the density contours on the leading edge of the Clouds.

No evidence of stars in the Stream has so far been found (Brück and Hawkins 1983, Mathewson et al. 1979), nor any evidence of dust (Fong et al. 1987); there is no *IRAS* 100 μm flux associated with it.

Other HI components are seen as overlapping the main Stream. Several of them appear to be typical high-velocity clouds (Morras 1985).

There are several theories for the origin of the Magellanic Stream: Primordial, Diffuse Ram Pressure, Discrete Ram Pressure, Turbulent Wake and Tidal. Many models have been presented and some will be discussed here along with their pros and cons.

In the *primordial theory* the sequence of clouds building up the Magellanic Stream was formed from the material left over from the original cloud that produced the Magellanic Clouds. The Stream is in the same orbit as the Clouds around the Galaxy. Interstellar lines of CaII caused by gas in MS I at a velocity of $V_{LSR} \sim 193$ km s^{-1} were detected by Songaila (1981) in the spectrum of the Seyfert galaxy Fairall 9. The Ca abundance was found to be between 0.01 and 1.8 times solar and showed that the Stream cannot be composed of primordial (zero-metallicity) gas. It does, however, not permit a decision between solar and MC composition gas. *IUE* observations by Penston (1982) of interstellar lines of HI, CII, CIV, OI, MgII, SiII and FeII in the Stream with abundances slightly below solar, as are those in the Clouds, suggest that the Stream originated in the Clouds. Further support for this assumption comes from the detection by Lu et al. (1994) of absorption at $+170$ and $+210$ km s^{-1} in the SiII lines, at 1260 Å, which they identified as absorption in the Magellanic Stream.

Morras' (1983) 21 cm map reveals two components from the MS in the direction of Fairall 9 with $V_{LSR} \sim 160$ km s^{-1}, N(HI) $\sim 2 \times 10^{19}$ cm^{-2}, and $V_{LSR} \sim 200$ km s^{-1}, N(HI) $\sim 6 \times 10^{19}$ cm^{-2}. The small offset between the UV and HI velocities is easily understood as uncertainty in the zero point of the former and/or smearing of the HI profiles. The abundances found are Si/H ≥ 0.2 solar and S/H ≤ 0.9 solar for the lower velocity cloud and Si/H ≥ 0.7 solar and S/H ≤ 0.3 solar for the higher velocity cloud. These limits are consistent with values for the Magellanic Clouds (see Chap. 11).

It is of particular interest to note that a high-velocity cloud (HVC) complex, HVC 287.5 +22.5 +240 (*l, b, V_{LSR}*), on the opposite side to the MS with regard to the Clouds has Si/H ≥ 0.006 solar and S/H $= 0.15 \pm 0.05$ solar (Lu *et al.*(1994). Considering the uncertainties in the values, in particular due to the sampling differences between the absorption line data (infinitesimal solid angle) and the HI 21 cm data (large beam), it may be suggested that this HVC complex is a part of (a leading counterpart to) the Stream. Mathewson *et al.*(1974) expressed some 'nagging doubts' about the HI clouds on that side being members of the Magellanic System because there was no systematic variation in radial velocity in the HI clouds *between* the Clouds and the galactic plane. However, in the tidal model (presented below) the Gardiner and Noguchi (1995) calculations make it quite possible that the HI clouds up to $b = +30°$ may belong to a leading stream.

The *diffuse ram-pressure theory* states that there is a diffuse halo around the Galaxy which produces a drag on the gas between the Magellanic Clouds and causes weakly bound material to escape and form a tail. However, the energy required to move MS I from the IC region to its present position is greater than the energy which can be given by the drag force of the ram pressure of the galactic halo (Mathewson *et al.*1987).

The *turbulent wake theory* states that the passage of the Magellanic Clouds through an extensive hot galactic halo produces vortices along its path. This would explain the separation of the Stream into discrete clouds and their horseshoe-shaped structures. It also explains the decrease in surface density and the increase in velocity half-width of the HI profile with distance from the Clouds, i.e. with age. The high velocity at the tip of the Stream is also easily achieved with this theory. It is, nevertheless, unsatisfactory in that the density enhancements of the form required are not found (Bregman 1979).

In the *tidal model* the Magellanic Stream was torn off from the SMC in an encounter with the LMC ~ 1.5 Gyr ago which also coincided with a perigalactic passage. This model has been discussed extensively above (Sect. 3.1). Mathewson *et al.*(1987) considered the shortcomings of the tidal model to be that (i) no stars have been detected in the Stream – other interacting systems show gas and stars in their tidal arms (as does the Wing of the SMC and also the IC region); and (ii) there is no evidence of the high transverse velocity (> 200 km s^{-1}) which the model requires. Other possible shortcomings of the tidal model are discussed below, as well as further model calculations in support of it.

According to the *discrete ram-pressure theory* density enhancements in the galactic halo, possibly high-velocity clouds (HVCs) collided with the gas in the InterCloud region;* the mixture of gas then forms the Stream. This event coincided with a

---

* The HVCs are then suggested to be the result of a collision of a galaxy with our Galaxy about 6 Gyr ago.

3.9 The Magellanic Stream

collision between the LMC and the SMC about 0.4 Gyr ago. In the same event the SMC should have been split into two fragments, the SMCRemnant and the MiniMC (Mathewson et al. 1986, Wayte 1989). The IC region covers about 500 kpc^2 so that a large number of HVCs may be expected to collide with it. As the density of the IC region is about the same as that of the HVCs, the momentum of an HVC will be about equal and opposite to the interacting gas. The conglomerate will lose its transverse motion and begin to fall towards the galactic centre.

Recently, the ram-pressure model has been brought back into the picture. Sofue (1994) suggests that stripping and accretion of HI and molecular clouds occurred recurrently during past tidal and gas-dynamical interactions of the Clouds and the Galaxy. Many clouds survived the stripping by being shielded from the halo and intergalactic gas, and many, in particular molecular clouds, obtained an eccentric distribution. A large number of the HI clouds, however, were stripped. They formed the Magellanic Stream, trailing from the LMC. Many of them have fallen into the Galaxy in the past; they are observed as high-velocity clouds. Eventually, the Clouds will be rather completely stripped of gas, and the binary state of the LMC and the SMC will be destroyed by the stronger tidal force as they approach the Galaxy due to the dynamical friction.

Moore and Davis (1994) have carried out numerical investigations to test models of the origin of the Magellanic Stream. They have found that the tidal model fails to reproduce several characteristic properties of the Stream. (i) Material stripped from the LMC would retain the internal velocities of the LMC and would define a thick plane surrounding the Milky Way, not the confined wake observed. (ii) If the Stream is tidally produced there should also be a leading stream. (iii) The observed radial velocities along the Stream are inconsistent with the observed orbital parameters of the Magellanic Clouds. (iv) The uniform variation in column density along the Stream cannot be reproduced by tidal forces. (v) No stars have been observed in the Stream, but stars should have been tidally stripped from the LMC as easily as the gas. The recent refinements of the tidal model by Gardiner et al. (1994) and Lin et al. (1995) address some but not all of these problems.

Moore and Davis suggested instead that the Stream consists of material which was ram-pressure-stripped from the Magellanic System during its last passage through an extended ionized disk of the Galaxy. This 'collision' took place some 500 Myr ago at a galactocentric distance of about 65 kpc, and swept $\sim 20\%$ of the least-bound HI into the Stream. The gas with the lowest column density lost the most orbital angular momentum. At the present time this material is at the tip of the Stream and has fallen to a distance of ~ 20 kpc from the Milky Way, attaining a velocity of -200 km s^{-1}. To prevent the stripped material from leading the Magellanic Clouds and attaining too large an infall velocity, the authors postulate the existence of an extended dilute halo of diffuse ionized gas surrounding the Milky Way. If the halo gas is at the virial temperature of the potential well of the Milky Way (2.2×10^6 K), its thermal emission would contribute $\sim 40\%$ of the observed diffuse background radiation in the 0.5–1.0 keV (M) band. This is consistent with recent *ROSAT* measurements as well as with pulsar dispersion measures. Ram-pressure stripping from this extended disk and gaseous halo would explain the absence of gas in globular clusters and dwarf spheroidal companions to the Milky Way. Some fraction of the observed high-velocity clouds might be the infalling debris from previous orbits of the LMC through the extended disk.

A slightly new approach is found in the Heller and Rohlfs (1994) model. They have determined orbital parameters for the Magellanic System which support the idea that the LMC and the SMC form a binary system which has been stable for the past 10 Gyr. A close collision occurred between the two Clouds 0.5 Gyr ago at an orbital position close to the tip of the Magellanic Stream on the sphere. The stars in the Clouds remained largely unaffected but strong hydrodynamical interaction affected their gas content. Part of this gas is seen as the common envelope and part of it as gas clouds in either the SMC or the LMC or in the IC region.* Most of the stars in the latter region should then be younger than 0.5 Gyr. Older stars found there may have been pulled out of either Cloud by tidal interaction from the Galaxy, provided that they were originally loosely bound to either. The Stream has formed from gas torn out of the InterCloud region by the strong wind caused by the Clouds moving through a gaseous galactic halo.

As a result of this collision, which, in this model, is the only one in the history of the Clouds, the velocity of the SMC was perturbed in such a way that the SMC is no longer gravitationally bound to the LMC.

3.10 The InterCloud region

Neutral hydrogen was detected in the LMC and the SMC in 1954 (Kerr *et al.* 1954) and nine years later in the region between the two Clouds (Hindman *et al.* 1963). All attempts to detect stellar members of the Clouds between the SMC Wing and the LMC were unsuccessful for a long time. The IC region remained an HI feature (see Mathewson *et al.* 1979). Its structure is extremely complex. Two radial velocity groups, $+214$ and $+238$ km s^{-1}, form the actual HI Bridge between the SMC and the LMC; three other components, at mean radial velocities of $+155$, $+177$, and $+195$ km s^{-1}, are probably integral parts of the SMC, stretching east to $R.A. \sim 4^h$. Another three components at mean velocities (heliocentric) $+253$, $+272$, and $+293$ km s^{-1} appear to be extensions of HI from the LMC (McGee and Newton 1986).

Until 1980, when Kunkel identified a group of young stars at $2^h\ 30^m$, $-74°$, the tip of the (optical) Wing was at $2^h\ 15^m$, $-74°.5$ (Westerlund and Glaspey 1971). An extensive young stellar population has since been discovered; it forms an optical counterpart to the HI Bridge (Irwin *et al.* 1985). The previously known SMC Wing is the brightest part in this link and contains even younger objects. Diffuse emission nebulosity covering most of the IC region has been detected by Johnson *et al.* (1982) on deep Hα image-tube photographs. It is considered almost certain that it originates in the HI clouds in this region. Five nebulous features near the tip of the Wing proper (Meaburn 1986) indicate the presence of hot stars. One of the shells, at $2^h\ 7^m 26^s$, $-74°58'25''$, has a diameter of ~ 150 pc (distance 60 kpc). Meaburn discusses the possibilities for the creation of such a shell in an HI cloud of a mean density of ≈ 0.5 cm^{-3}. He states that 2.4×10^5 M$_\odot$ of HI must have been displaced in the formation. Its age can only be a few million years whether the mechanism is stellar winds from an O star (most likely) or a supernova explosion. Thus, stars with ages of ~ 5 Myr are present in the Wing.

Far-UV observations of the Wing have been performed with the Very-Wide-Field Camera (*VWFC*) on the Spacelab 1 mission (Courtès *et al.* 1984, Pierre *et al.* 1986),

---

* In the Fujimoto, Noguchi (1990) model the gas clouds remain bound to the LMC after the collision.

3.10 The InterCloud region

Table 3.11. *Blue stellar aggregates and clusters in the InterCloud region (Grondin et al. 1992)*

Number	R.A.	Dec. 1950	Distance kpc	Diameter pc	Number $V < 21$
1	$2^h 20^m$	$-73°19'$	58 ± 6	> 150	19
2	$2^h 27^m$	$-73°59'$	65 ± 5	50	57
3	$2^h 31^m$	$-73°57'$	53 ± 6	~ 150	27–30
4	$2^h 55^m$	$-73°30'$			
5	$3^h 32^m$	$-73°09'$	63 ± 8	~ 100	7
6	$4^h 21^m$	$-74°01'$	55 ± 4	~ 80	34

The distances have been determined by fitting a ZAMS interpolated from the Revised Yale Isochrones (Green *et al.* 1987) for $Y = 0.25$, $Z = 0.004$ to the main sequences.

at 1680, 1950 and 2540 Å, and with the *FAUST* telescope on the *Space Shuttle* (Courtès *et al.* 1995), at 1400–1800 Å. During the first experiment global isophotes were obtained over an area of 3.9 deg^2. The slope of the IMF was found to be relatively flat, $x = 0.94 \pm 0.03$ with an upper mass limit of ~ 30 M$_\odot$, assuming a continuous star formation. The *FAUST* observations were more than 4 mag deeper than those with the *VWFC* and had a higher spatial resolution. About 150 luminous blue stars and aggregates were identified, of which only about half were known from ground-based optical work. The slope of the IMF as well as the upper mass limit were confirmed, indicating ongoing star formation since $\approx$ 100 Myr, with a few objects as young as 5 Myr, in agreement with Meaburn (1986). The Wing, as defined by the young stars, appears to end at about R.A. = 3^h.

The LMC halo can be traced to $\sim$ R.A. = $3^h\ 30^{m*}$ (Irwin *et al.* 1990) and adjoins the area occupied by blue stars belonging to the SMC. On the SMC side no halo population was found eastward of the SMC Wing tip. Thus, the halos of the SMC and the LMC do not touch.

Six stellar associations have been identified in the Bridge in an arc between R.A. = $2^h 20^m$ and $4^h 20^m$ at about decl. $-73°.5$ (Table 3.11). In the SMC Wing, between R.A. = $1^h 48^m$ and $2^h 16^m$, there are 18 possible clusters of blue stars. A difference in distance between the associations 2 and 3, only $17'$ apart, of $\sim 12 \pm 8$ kpc indicates a certain depth of the Bridge.

The ages of the Bridge clusters are ≤ 16 Myr (Grondin *et al.* 1990, 1992; Demers *et al.* 1991). The youngest in the chain, No. 6, is < 10 Myr old, and could, as its distance also indicates, be a member of the LMC. It is, however, too young for its position, 6.4° from the optical centre of the LMC and in a part of the LMC where Extreme Population I is normally not seen. Demers *et al.* estimate the minimum age of the LMC field stars to 750 Myr, but it may be appreciably higher. It is therefore more likely that association no. 6 shares the origin of the other OB associations in the IC region. The associations are characterized by an IMF slope of $x = 1.3 \pm 0.1$ for masses in the 1.5–12 M$_\odot$ range ($\xi(M)$

---

* The globular cluster NGC 1466 is at $3^h 45^m$, $-71°41'$ and may well be considered as a halo cluster.

Table 3.12. *The generations in the Magellanic Clouds*

Generation	Content
The oldest generation ≥ 10 Gyr	*True globular clusters* In the SMC: NGC 121; In the LMC: NGC 1466, 1786, 1835, 1841, 2210, 2257, H 11, Reticulum
	RR Lyrae variables They exist in all the true globular clusters except H 11 and in the fields of both Clouds, probably rather evenly distributed. They appear to form thick disks rather than halos.
	Very old field population Red old stars have been observed in both Clouds. In the LMC very old LPVs are known. CH stars are observed; if some of them belong to this age group is not clear.
The intermediate-age generations <10–1 Gyr	*Clusters* Most clusters in the Clouds are of intermediate age. More clusters in the SMC may have ages ≥ 3 Gyr than in the LMC. Many of these clusters have carbon stars at the top of the AGB. *Field stars* Most of the red-giant stars belong to this generation. Among them are most of the AGB stars, including the carbon stars. The number of the latter is estimated to $\sim 12\,000$ in the LMC and ~ 2000 in the SMC.
	It has been noted that the field of the SMC lacks stars older than 1–2 Gyr in contrast to what is observed for the clusters.
	Planetary nebulae Most of the PN belong to this group.

$\propto (M^{-(1+x)})$. With this definition, the Salpeter (1955) mass function has $x = 1.35$. This is flatter than what is normally derived for Magellanic Cloud clusters.

A survey for red stars in the InterCloud region gave 57 stars with magnitudes compatible with LMC carbon stars (Demers *et al.* 1993). Two-thirds of them have

3.10 The InterCloud region

Table 3.12. *The generations in the Magellanic Clouds (cont.)*

The intermediate-age generations 0.9–0.2 Gyr	*Clusters* Many clusters belong to this age group. Their AGBs end with M giants.
	RR Lyrae stars A group of overluminous RR Lyrae stars have been observed in the SMC with [Fe/H] ≈ -0.4.
	Field stars Some of the brightest AGB stars, types MS, J, and Ce belong to this age group as well as a number of M giants.
	Planetary nebulae Type I PN have been found in both Clouds.
The youngest generation ≤ 100 Myr	All the typical Population I objects known in our Galaxy have been found in the Clouds.
	In the LMC practically all the young stars, clusters and stellar associations are found in *superassociations* with diameters up to 2 kpc. Some of them have on their circumferences *supergiant shells*, containing the youngest stars and much of the gas (HI, H_2, H II, CO) and dust.
	In the SMC the youngest stars appear to be relatively evenly distributed in the Bar and in the Wing. There are few stellar associations, mainly in the NE end of the Bar and in the Wing. Only one *superassociation* with a well-defined *supergiant shell* has been identified; it is in the Wing.

been shown by spectroscopy to be C stars. The stars to the east of $R.A. = 3^h40^m$ are considered as likely members of the LMC, those in the 'true' Bridge area could possibly be associated with the halo of the Galaxy. Bothun *et al.* (1991) have identified 32 carbon stars with galactic latitudes $\geq 40°$, with the most remote one at 115 ± 35 kpc.

3.11 The generations in the Magellanic Clouds

The populations referred to in Tables 3.7 and 3.9 in one sense represent a division according to age. As an introduction to the following chapters, where the various components of the Clouds will be discussed in detail with the aim of understanding the evolution of the Clouds, a brief summary of the dominating generations is given in Table 3.12.

4

The cluster population

Stellar clusters occupy a central position in research aimed at the structure and the evolution of our Galaxy and of those of our neighbours in which clusters can be identified. Often the integrated cluster properties, magnitudes, colours, spectra, are the only ones within reach. In the Magellanic Clouds most of the clusters can be sufficiently resolved for the investigation of individual members by photometry and spectroscopy even if the stars in the cores in some cases are too crowded for ground-based observations. As the clusters have a range of age that covers the whole lifetime of the Clouds, this should permit the study of the complete evolution of the Clouds. En route, a number of steps have to be taken. It is necessary to determine their distances, ages, and metallicities, and, before these, their reddening. The latter is difficult to determine for an individual cluster without knowledge of its physical properties, and is, therefore, frequently assumed known. As the reddening is small over most of the Clouds (see Chap. 2), the astronomer may feel entitled to use any low value recommended in one survey or another. However, even a small error in the colour excess, E_{B-V}, may have noticeable effects on the other quantities. Also the distance to the cluster, i.e. to a particular part of the SMC or the LMC, is frequently assumed known or determined by isochrone fittings: isochrones for different compositions and ages are fitted to the main sequences (MSs) and/or the red-giant branches (RGBs) in the colour–magnitude diagrams (CMDs) and the best fitting one is accepted as defining the cluster properties.

Critical masses and evolutionary important points along an isochrone are shown in Fig. 4.1. There, M_{MS} = mass at the point where the isochrone begins to depart from ZAMS; M_{TO} = mass at the turn-off point; M_{TM} = mass at the MS termination stage; M_{HeS} = mass at the start of core He-burning; M_{BL} = mass at maximum extension of the blue loop or clump; M_{HeE} = mass at the end of core He-burning; M_{AGB} mass at maximum AGB luminosity. The latter is mass at TP-AGB if the initial mass is $< M_{up}$, or mass at C-ignition stage if the initial mass is $> M_{up}$.

The models behind the isochrones may be 'classic' or 'overshooting' (see below). The fitting of an isochrone is done in different ways. Rather frequently, it is drawn through the middle of the observed main sequence, probably under the assumption that the observed scatter is due to observational errors. In other cases, where the photometric accuracy may be believed to be higher, the isochrone is fitted on the blue side of the main sequence, thus allowing for evolutionary effects as well as for binaries. In this case the treatment is in line with that applied to galactic clusters (see e.g. Meynet *et al.* 1993). The observed scatter along the MSs in the CMDs for

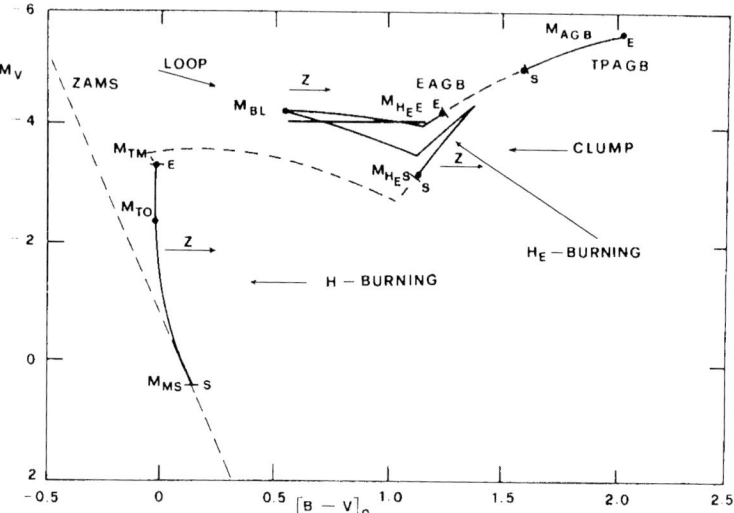

Fig. 4.1. $M_V/(B-V)_0$ diagram showing critical masses and evolutionary stages along an isochrone of given age (Chiosi et al. 1989b). Core H- and core He-burning phases are drawn as heavy lines. The effects of increasing metallicity, Z, are indicated by arrows The definitions of the various points are given in the text.

LMC clusters has also been interpreted as showing that the cosmic scatter of MS stars is larger than previously assumed (Meylan and Maeder 1982), though this is now questioned.

Another problem facing the observer is the elimination of field stars and the proper definition of the periphery of a cluster. In the star-rich regions of the Clouds, the danger of including field stars in the cluster population is great, particularly in those clusters for which the CMD has been based solely on the outer cluster regions. Many young clusters are embedded in rich unbound halos (Elson et al. 1987, Lupton et al. 1989), and up to 50 % of the mass of some clusters may be in such halos. However, as many clusters have formed in stellar associations, they may as well be surrounded by stars belonging to the latter, and, as their ages may almost be the same, it may be exceedingly difficult to determine the true content of the cluster. The catalogued sizes of the clusters depend also to too high a degree on the instrumentation used in the observations and on the technique used in the measurements. It is essential in all analyses to use a homogeneous material as a mixture may lead to erroneous results.

In the Galaxy the separation between globular and open clusters causes no problems. In the Magellanic Clouds, with a different composition of clusters, the definitions are not so clear. There are true globular clusters in a galactic sense, i.e. with ages $\geq$ 10 Gyr and red-giant stars as the most luminous objects, but they are few. Instead, there is a fair number of large globular-shaped clusters dominated by blue stars; in many cases they also contain evolved yellow and red stars. Hodge (1960) has catalogued 35 rich red clusters, true globulars and populous old clusters. He has also listed 23 blue objects and named them young populous clusters (Hodge 1961). They are all larger and richer than any galactic open cluster. Their ages place most of them in the intermediate-age group, i.e a few times 10^8–10^9 yr old, but there are younger ones, e.g. NGC 2100, about 10 Myr old. The populous clusters in the Clouds are nowadays usually referred to as globular clusters without consideration of their stellar content, and divided into true globulars and intermediate-age clusters to indicate the most important difference between the two groups. This designation* is reasonable if a dynamical definition of a globular cluster is applied, meaning that most stars will not be able to escape from it, whereas an open cluster or a stellar association will disintegrate by time.

The intermediate-age globulars in the Clouds represent a type of cluster not known in the Galaxy and deserve all the interest they do receive. A possible explanation for their existence has been given by Fujimoto and Noguchi (1990) and further developed by Kumai et al. (1993). A strong non-circular motion of ~ 50 km s^{-1} or higher existed in the LMC and the SMC following their hydrodynamical collision ~ 0.2 Gyr ago. Massive gas clouds, $\sim 10^4$ M$_\odot$, collided frequently then, causing large-scale and strong shock compressions which led to the formation of globular clusters. When the velocity–amplitude of this unorganized motion is higher and the heavy elements are less abundant, at least in the range $-2.5 <$ [Fe/H] < 0, these conditions lead to the formation of more massive globular clusters in the shock-compressed regions. Galaxies with globular clusters younger than ~ 100 Myr are likely to be in strong interaction either with a companion galaxy or with a cooling flow of gas or a merger.

The conditions necessary for the formation of globular clusters existed in the Galaxy more than 10 Gyr ago but died out; the turbulence decayed as there was no lasting external disturbance. This is in contrast to the situation in M 31, similar to the Galaxy in morphological type, size, mass and ordered rotation, but still generating globular clusters. M 31 is, however, exposed to external disturbances. M 32, with a mass of $\sim 10^9$ M$_\odot$, i.e. similar to the SMC, crosses the M 31 gas disk at ~ 4 kpc from the centre every 10^8 yr (Byrds 1976). This causes a tidal disturbance of the disk gas at that place with an amplitude of non-circular velocity of ~ 50 km s^{-1}, enough for gas compression and globular cluster formation.

Other explanations have been attempted for the existence of young globular clusters in the Magellanic Clouds but the lack of such in the Galaxy. They have, to a large extent, been based upon the fact that the Clouds are metal-deficient. The MC clusters cover, however, such a wide range in [Fe/H] that it appears unlikely that metal abundance can play any major role except in making the less abundant clusters larger.

Kennicutt and Chu (1988) have argued that the populous blue clusters have formed in super-HII regions, such as the 30 Doradus complex. There are few regions in the LMC which may once have looked like the present 30 Doradus complex; the only likely ones are the superassociations/supergiant shells. Some of the young globulars have ages similar to those of the superassociations. The question is whether they could have formed in the stochastic processes occurring in these giant formations. Several blue globulars are, however, outside the superassociations and can hardly have formed in that way. It appears, therefore, more likely that all of them have resulted from other processes, among which that of external disturbances appears most likely for the moment.

---

* We may recall Baade's (1963) comment: 'globular cluster' is a good Herschel term, but it tells nothing about the colour–magnitude diagram.

Table 4.1. *Comparison of CCD photometry of NGC 2004 and NGC 2100*

Cluster NGC	ΔV mag	$\Delta(B-V)$ mag	Author – BJ
2004	-0.182 ± 0.008	-0.070 ± 0.006	Bencivenni et al. 1991
	$+0.099 \pm 0.014$	$+0.037 \pm 0.009$	Elson 1991
	-0.036 ± 0.006	$+0.038 \pm 0.007$	Sagar et al. 1991b
2100	$+0.100 \pm 0.016$	$+0.038 \pm 0.016$	Elson 1991
	-0.011 ± 0.004	-0.013 ± 0.005	Sagar et al. 1991b

11 stars appear to be in common to BJ and Bencivenni et al., 8 to BJ and Sagar et al.; in the other cases no information is given.

Much trust is nowadays placed in CCD photometry as being more accurate and reliable than the classic combination of photoelectric (*pe*) photometry of standard stars and photographic (*pg*) photometry of the majority of the cluster stars. This is certainly true as far as the intrinsic accuracy is concerned, in particular because of the possibility of avoiding the inclusion of faint stars in the 'aperture', a *pe* problem, and to reach faint stars which were previously outside the reach of *pg* (and *pe*) photometry. Nevertheless, there are large zero-point differences between CCD observers. Thus, Balona and Jerzykiewicz (1993, BJ) report the differences between their data and others for stars brighter than V = 17.00 in two clusters (Table 4.1).

The reason for the zero-point differences may in some cases be the transfer techniques: nightly zero-points are determined from standard stars and applied to the programme fields by taking into account the differences in exposure time and atmospheric extinction coefficients, as well as between aperture and PSF magnitudes.

4.1 Surveys

Among the 919 objects recorded by Herschel (1847) in the LMC and the 214 in the SMC, there are a fair number of clusters referred to by their NGC numbers. A large number of surveys on the Clouds have since been carried out and have resulted in catalogues. Examples are: for SMC clusters, Kron (1956, K), Lindsay (1958, L), Brück (1975,1976, Br) and, for LMC clusters, Shapley and Lindsay (1963, SL), Lyngå and Westerlund (1963, LW), Hodge and Sexton (1966, HS), and Hodge (1988c, H). The SL, LW and HS clusters are identified in the Hodge–Wright (1967) Atlas to the extent that they are inside the 125 deg² covered by the Atlas. A search for clusters outside the Hodge–Wright Atlas by Olszewski et al. (1988), added 37 clusters, mainly to the north of $-64°$ and south of $-73°$. It confirmed the ellipticity of the cluster system, noted by Lyngå and Westerlund and covering about $15° \times 11°$.

In deep surveys of selected regions in the Clouds, using large-scale plates, Hodge reached limiting magnitudes between $B = 22$ and 23 mag . This corresponds to $3 \leq M_B \leq 4$ mag, i.e. the samples should contain all surviving clusters in the regions searched. The total number of catalogued clusters amounted to 2053 in the LMC (Hodge 1988c) and 601 in the SMC (Hodge 1986), leading to the conclusion that the total number of clusters is about 4200 in the LMC and 2000 in the SMC. A similar deep survey of three selected regions in or near the LMC Bar was carried out by

4.1 Surveys

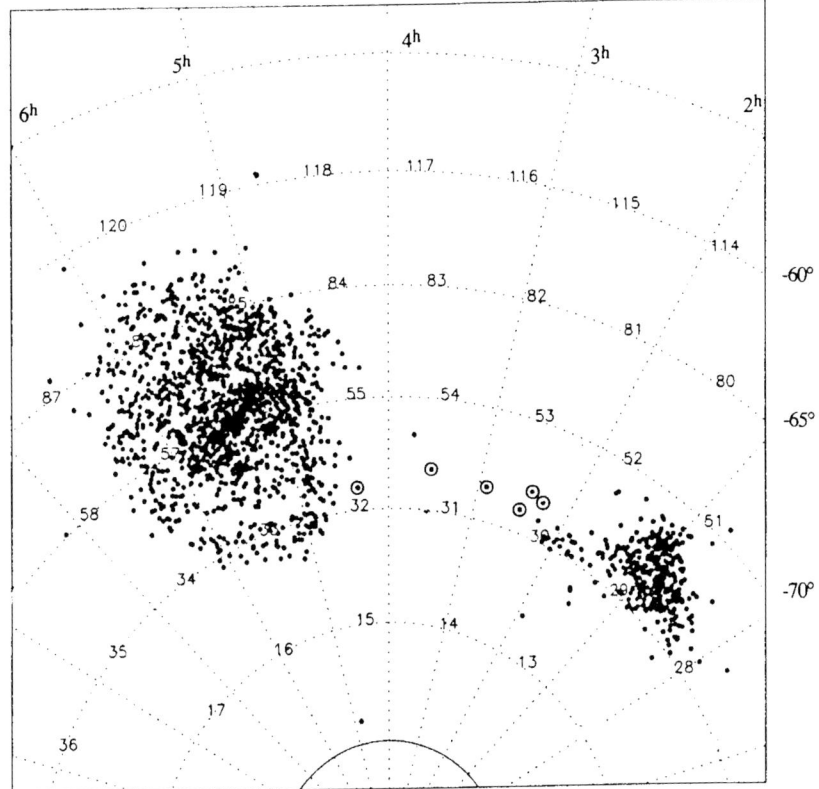

Fig. 4.2. The distribution of Cloud clusters and InterCloud associations (dots in circles) on a grid of UKST field numbers. (Reprinted from Irwin (1991) by permission of Kluwer Academic Publishers.)

Kontizas et al.(1988), adding a large number of clusters to those previously known from low-scale surveys.

The distribution of the clusters in the Magellanic Clouds is shown in Fig. 4.2 from a compilation by Irwin (1991).

Statistical analyses of the catalogued clusters show that there are relatively many more old open clusters in the SMC and the LMC than in the Galaxy. The median ages of the clusters in the three galaxies are 0.9, 1.1 and 0.2 Gyr, respectively. Most of the clusters in the Clouds thus belong to the intermediate-age generations. The cluster-formation histories in the two Clouds may show non-uniform rates depending upon the positions in the Clouds (Hodge 1988c).

The most recent comprehensive catalogue of LMC clusters has been presented by Kontizas et al.(1990a). It is based on the ESO/SERC Southern Sky Atlas and gives positions and measured diameters for 1762 clusters in a $25° \times 25°$ area of the sky centred on the LMC but excluding the crowded 3.5 deg^2 around the Bar. Comparisons with previous catalogues based on low-scale photographs showed a reasonable agreement: $\sim$ 1150 of the 1762 clusters were identified in the Hodge–Wright Atlas. Some previously catalogued objects were not detected, others were rejected as asterisms, galaxies, associations, nebulae or dust clouds. The total cluster population in the LMC is estimated to be $\approx$ 4600; the technique used by Hodge was applied. The number is unavoidably uncertain and may be corrected in the near future when the faint end of the cluster distribution becomes more complete. A step in this direction is the automatic identification of clusters in the Clouds with the aid of objective, machine-produced data, developed by Bhatia and MacGillivray (1989). Preliminary results for a 6 deg^2 field in the NE end of the LMC Bar resulted in the detection of 284 clusters down to a size limit of $\sim$ 4 pc, quadrupling the number of clusters known in this area.

A bibliography of colour–magnitude diagram studies of Magellanic Cloud clusters has been presented by Sagar and Pandey (1989). Sixty-four such studies are available for 44 SMC clusters and 191 for 121 LMC clusters. The age–metallicity relation derived in the analysis of the data shows a large scatter for clusters younger than 10 Gyr; the oldest clusters show appreciably lower [Fe/H] values. The LMC clusters show no correlation between age and size; they resemble the open clusters in our Galaxy of ages 0.05–1 Gyr. In the SMC, on the contrary, the older clusters have larger radii than the middle-aged clusters. The authors suggest that this may indicate a different type of evolution in the two Clouds. The comparison is, however, incomplete, as there are no LMC clusters with ages over 2 Gyr in their comparison. Moreover, the radii quoted are measured on plates of totally different scales: the LMC clusters have radii estimated on CTIO 4 m telescope plates (Hodge 1980), whereas the SMC clusters have their dimensions determined from UK Schmidt telescope plates (Kontizas 1984). There is an appreciable zero-point difference between the two investigations, indicating that the dimensions are overestimated for the SMC clusters. This is supported by looking at individual clusters. Thus, the red SMC cluster K3 has a tidal radius of $4'.6$ in the Kontizas catalogue, but Rich et al.(1984) found from CTIO 4 m telescope CCD photometry that 'there was no contribution from the cluster outside of a $2'.5$ annulus inscribed on the cluster center'. Kontizas et al.(1986) have also noted that 'the tidal radii of the studied (LMC) clusters are in the same range of ...the SMC clusters'. In this case the same telescope was used for the two galaxies. Obviously, great care has to be taken in comparing dimensions derived from vastly different material.

Data have also been compiled for Magellanic Cloud clusters by Seggewiss and Richtler (1989). They briefly discuss cluster masses, mass-to-light ratios, initial mass functions (IMFs) and abundances, and conclude, correctly, that the uncertainties in all areas are great.

4.2 The classification of Cloud clusters

The clusters in the Magellanic Clouds may be divided into old (globular) clusters and young (open) clusters according to the same criteria as that applied for clusters in our Galaxy. However, whereas in the Galaxy the old and the young clusters are easily separated by their appearance, this is not true in the Clouds. As discussed above there are young blue globulars in the Clouds. Thus, there is a broad range of age among the (metal-poor) clusters in the Clouds, contrary to that seen in the Galaxy. The properties of the cluster systems in the Clouds have long been considered of fundamental value for understanding the dynamical and chemical evolution of the two galaxies. The number of clusters with known properties needed for a complete image of the evolution of the Clouds is, however, huge. It may therefore be unrealistic to expect that information about a sufficient number of clusters can be derived by photometric and spectroscopic studies of their stellar members. Instead, the integrated characteristics of the clusters, photometric and/or spectroscopic, must be used, and therefore quantified, so that the clusters can be arranged in one- or two-dimensional arrays.

A very useful classification system of this kind was introduced by Searle et al. (1980, SWB). Its original basis consists of the reddening-free parameters $Q(ugr)$ and $Q(vgr)$, defined

$$Q(ugr) = (u - g) - 1.08(g - r);$$

$$Q(vgr) = (v - g) - 0.68(g - r);$$

where u, v, g, r are magnitudes obtained through filters centred on relatively line free regions, g, r, on a region heavily blanketed by Balmer lines, v, and on a region on the short wavelength side of the Balmer discontinuity, u. The quantities measure the extent to which the flux in the u and the v passbands fall below the flux in a pseudo-continuum, defined by a g,r baseline. The Magellanic Cloud clusters define a physically significant sequence in the $Q(ugr), Q(vgr)$ plane, showing increasing age and decreasing metal abundance. It is divided into segments, SWB types I–VII; the oldest clusters are in SWB type VII, the intermediate-age clusters in SWB types IV–VI if they have ages ≥ 1 Gyr, and in II–III if they are between 0.7 and 0.2 Gyr. The youngest are in SWB type I. The Cloud clusters of type VII occupy the same area as galactic halo clusters and may be expected to have age and metallicity in the same ranges as the latter. There are many Cloud clusters of types V and VI that have no counterparts in the Galaxy. Similarly, there are 10-Gyr-old clusters in the Galaxy that are more metal-rich than the halo clusters, and that have no counterpart in the Clouds.

IUE observations of 17 clusters in the LMC (Cohen et al. 1984) confirmed that the SWB classification is one of increasing age accompanied by decreasing metallicity towards later SWB classes. In the UV the giants do not dominate the integrated light so that the blueward shift of the giant branch with decreasing metallicity is negligible in comparison with the changes in main-sequence location with age.

The SWB classification scheme has been applied to the UBV two-colour diagram by Elson and Fall (1985). They introduced a parameter 's', by dividing the one-dimensional SWB sequence into 51 intervals of equal length. Their classification agrees with that of SWB for 90% of the LMC and 75% of the SMC clusters. By using available age determinations they derived the relation

$$log(\tau/yr) \approx (0.087 \pm 0.004) s + 5.77 \pm 0.12.$$

The uncertainty in determining an age, τ, with this relation is estimated to a factor of 2 in the ages af individual clusters. The corresponding expression for SWB types is

$$log(\tau/yr) \approx 0.5 \, (\text{SWB type}) + 6.6.$$

4.3 The surface distribution of clusters in the Clouds

When the more luminous clusters in the LMC are considered, the following conclusions may be drawn (van den Bergh 1981): the distribution of the youngest clusters, < 4 Myr, in the LMC is very similar to that of the Shapley Constellations; the young

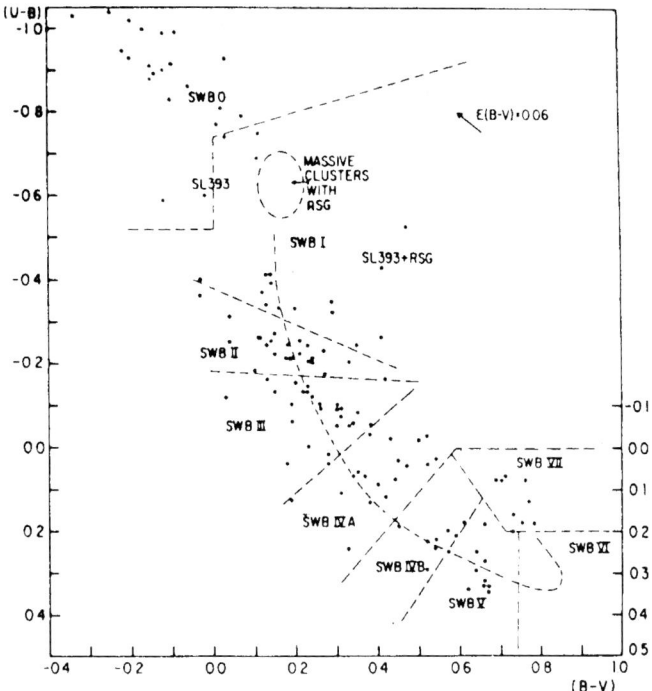

Fig. 4.3. The SWB sequence in a $(U-B),(B-V)$ diagram. In addition to the zones corresponding to the original SWB types I–VII the one of type 0, introduced by Bica et al.(1992), is identified in the sequence. The clusters plotted are from Bica et al.. Note the area occupied by massive clusters with red supergiants (RSG) and the shift in position of SL 393 when the RSG is subtracted.

clusters, 4–200 Myr, appear at the western end of the Bar and beyond it, and to the east of 30 Doradus; clusters older than 1 Gyr are scattered over the LMC with the centroid well to the west of the centre of the Bar.

For the same brightness category, the following apply in the SMC: the youngest clusters, some of which are members of associations, are found in the north-east end of the Bar and in the Wing; the young objects concentrate along the Bar; the old clusters are widely scattered.

In deeper surveys additional features appear. In the SMC the distribution of the clusters shows a gap of about 40 pc diameter centred at 1^h00^m, $-73°20'$ (Hodge 1974c, Brück 1975) in a region where a minimum in the HI distribution is also seen (Hindman 1967). The distribution of all the clusters is elliptical, if the extension in the direction of the Wing is disregarded. The most numerous class in Brück's classification system[*] is type 5: small blue clusters with diameters < 5 pc and the most luminous stars at $M_V \simeq -2.5$ mag. These clusters are similar to the Pleiades, aged about 50 Myr, and may be expected to define a galactic disk in the SMC. The position angle of the major axis was determined as 55°, agreeing with the direction

---

[*] Brück classified the clusters by colour and structure into six types, from red, old globulars (type 1) to blue globulars (3), and from stellar associations (4) to small blue clusters (5) and probable small clusters (6).

of maximum velocity gradient in the 21 cm radio contours (Hindman 1967) and the ratio b/a was found to be 0.6. Confined to the regions of high cluster density the value of b/a was 0.5, in agreement with previous results. The distribution of the clusters in a disk was to some extent corroborated by the impossibility of fitting the distribution to a model of a spheroidal relaxed system.

Kontizas et al. (1990b) interpreted the distribution of their 1762 clusters in the LMC as showing the existence of two cluster systems: one, an outer ellipse with $b/a \sim 0.7$ and a generally low surface density, and, another, an inner ellipse with $b/a \sim 0.9$ and a high surface density. The outer agrees with the Lyngå, Westerlund system, the axes are estimated to be $13°.2$ and $9°.8$, and leads to an inclination of $i = 42°.3 \pm 3°$, assuming that the system is flat and circular. The centre of the system is at $5^h 32^m$, $-69°.3$, and the position angle of its major axis is $\Theta = 9° \pm 4°$ d. The density of the clusters increases at about $-75°$ following a minimum of about $1°$ further north. This had been noted by Lyngå and Westerlund, and is also confirmed by Olszewski et al. (1988). The inner system, centred at $5^h 21^m$, $-68°.8$, with the major axis in $\Theta = 140° \pm 8°$ and $i = 30° \pm 11°$, shows some structure with regions of very high cluster density and one low-density region, centred at $5^h 10^m$, $-67°$. The high-density regions fall within the eight large regions defined by OB2 stars and B–G supergiants (Martin et al. 1976) (see Sect. 6.1). Both systems are dominated by intermediate-age clusters, with the outer system containing the older ones. The inner system consists of younger intermediate-age clusters and very young clusters. The latter are probably mainly in the high-density regions. It is important in understanding the evolution of the LMC to determine the density and structure of the two intermediate-age systems. For this, it is necessary to identify and subtract the very young clusters as well as the older ones from the inner system, a task that may not be impossible.

The results just presented are based on surveys using low-scale telescopes. Recent surveys of limited fields in the Clouds using higher-scale telescopes have more than doubled the numbers of clusters in the fields searched. The added clusters are either faint, small and old, or bright, but small and missed in the previous surveys because of unsufficient scale (Hodge 1986, 1988c). According to Hodge there is a conspicuous correlation between cluster size and distance from the centre of the LMC, with relatively more large clusters in the remote outer parts of the Cloud. This effect had been noted earlier for the 'populous' clusters and was then, as now, ascribed to tidal effects. It is in agreement with the distribution of open clusters in our Galaxy (Burki and Maeder 1976), where those in the inner parts have significantly smaller sizes than those in the outer parts. Kontizas et al. (1990b) argue, however, on the basis of their extensive material, that there is no difference in size distribution between the population inside and outside a radial distance of $4°$ from the LMC centre. This smooth distribution of the small clusters may be compared with that for Brück's (1975) class 5 clusters in the SMC, discussed above. The previous correlations found for the LMC are, according to Kontizas et al., due to too small a number of clusters in the analyses.

4.4 The age distribution of Cloud clusters

Age–abundance relations, age distributions, and initial mass functions of the star clusters in the Magellanic Clouds have attracted great interest over the past decades. Many interpretations of the compiled data have been presented and reviewed. The aim has been to determine the evolutionary history of the Clouds, their star formation activity and metal enrichment in the past and now. The great uncertainties still affecting these studies are connected with the difficulties in determining the ages of the clusters, in particular of the intermediate-age clusters. Recently, new and better stellar evolutionary models have improved the situation appreciably.

Most of the discussions of the evolution of the Cloud clusters concern the populous clusters. If they, as is possible, are the exception in the star formation activities of the LMC and the SMC rather than the rule, then much remains to be done before a reliable overall image of the evolution of the Clouds is at hand.

Cluster ages are determined in different ways: main sequence (MS) termination magnitudes, red-giant star luminosities, and maximum AGB star luminosities are used but also, more and more frequently, isochrones are fitted to the MSs and to red-giant branches (RGBs) in the CMDs or the HR diagrams. Also used for the age determinations are the locations of the horizontal branches (HB) and red-giant clumps (RGCs) in the HR diagram, the SWB type or the s parameter. All methods are affected by uncertainties due to the determination of the distance modulus, the interstellar reddening, the bolometric correction (BC), the chemical abundance and the cluster membership of the observed stars.

Flower's (1984) investigation of the ages of six intermediate-age LMC clusters, with $-0.5 \leq$ [Fe/H] ≤ -2.0 shows typical discrepancies in age determinations. RGC luminosities and MS termination magnitudes gave ages ≤ 1 Gyr; the ages of these clusters derived from terminal AGB luminosities were ≥ 3 Gyr. The cluster SWB types gave no significant correlation with age. By using new evolutionary models, incorporating overshooting, Chiosi et al. (1986) found good agreement between cluster ages determined by the three methods. The convective overshooting changes the ages determined from MS termination and red-giant luminosities, increasing them by a factor of three as compared with the old estimates. The ages obtained from AGB luminosities hardly change, and the ages based on the three methods are thus brought closer together.

Da Costa (1991) has compared a sample of LMC and SMC clusters for which ages have been determined from MS turn-off photometry. The LMC contains a large population of clusters 1–3 Gyr old as well as a number of (in his sample seven) true globular clusters ≥ 12 Gyr old. Only one cluster, ESO 121–SC03 ($06^h01^m.5$, $-60°31'.3$), $10°$ from the centre of the LMC, is in the age range between the two groups. The sample indicates that the rate of cluster production in the LMC increased significantly about 3 Gyr ago. Mateo's (1988b) investigation of the cluster population in a region in the northern outskirts of the LMC also shows this increase.

The SMC clusters are believed to cover a wide range of age, extending back to ~ 10 Gyr, but with no cluster as old as the oldest LMC group. However, the CMDs used to determine the ages of the SMC clusters in the 3–10 Gyr range are not easy to interpret. It appears possible that the clusters most frequently referred to are all close to 10 Gyr old (see Sect. 4.4.2). The major difference between the LMC and SMC cluster production appears to be that it started a few Gyr later in the SMC but that it has otherwise followed the same course.

In the SMC the mean abundance of the young clusters is about -0.6 dex and that of the oldest is about -1.3 dex. In the LMC the metal abundances of the three age groups differ considerably, the mean abundance for the youngest population being

4.4 The age distribution of Cloud clusters

~ -0.3 dex, for the 1–3 Gyr population ~ -0.5 dex, and for the old group ≤ -1.6 dex (see e.g. Da Costa 1991, Olszewski et al. 1991). The lack of clusters in the 12–3 Gyr range makes it impossible to decide if there has been a sudden increase in abundances or a constant rate of enrichment. The sole cluster in this range, ESO 121–SC03, has an abundance of -0.9 ± 0.2 dex (Mateo et al. 1986), and an age of $\sim$ 9 Gyr. This could indicate that an increase in abundances occurred early in the history of the LMC but also that a continuous increase has taken place. ESO 121–SC03 is in several aspects similar to the cluster L1 in the western extremes of the SMC, in the diametrically opposite part of the Magellanic System. The two clusters may have a similar evolutionary history determined by the conditions in the outskirts of the Clouds which may not be representative for the central regions (see Sect. 4.4.2). Disregarding the peripheral clusters we find that a major increase in metallicity in the clusters occurred about 3 Gyr ago in both Clouds, possibly in connection with a burst of star formation.

The significant gap between the metallicities of the old and the young LMC clusters suggests that the stars and clusters younger than 3 Gyr have all been formed from material enriched in heavy elements during the 'halo' phase of the evolution of the LMC (van den Bergh 1993). A similar proposal has been made for the Galaxy (Ostriker and Thuan 1975). The stars in the LMC halo, flat or spherical, accounting for ≈ 2 % of the visible mass of the LMC, have enriched the gas in the LMC to $\sim 0.1\ Z_\odot$, essentially identical to the metallicity of the most metal deficient galactic stars with disk kinematics.

All the clusters discussed so far are rather large. The diameters of the intermediate-age clusters listed in Table 4.7 are, with two exceptions, over 30 pc. Only 1 % of the galactic clusters have diameters larger than 25 pc; the corresponding values for the SMC and the LMC are 2 % and 5 %, respectively (Hodge 1987). The large intermediate-age clusters, which attract most of the interest in the Clouds, mostly as testing grounds for evolutionary hypotheses, are then hardly representative of the majority of that age group. The mean diameter of the clusters is 7.7 pc in the LMC (Hodge 1988c). In the SMC it is 5.8 pc, quite similar to that of galactic open clusters. About one-third of the 1792 LMC clusters catalogued by Kontizas et al. (1990b) have diameters ≤ 10 pc.

It has been suggested that the LMC clusters show a non-uniform age distribution over the field, with certain areas showing a high production at certain limited periods. This is illustrated by Hodge's field 14, at $5^h48^m, -71°.5$, where 60 % of the clusters have ages between 0.8 and 1.2 Gyr (Hodge 1988b). However, considering the uncertainty in his bright-star magnitudes, ± 0.5 mag, it may be concluded (from his Table I) that this is the dominating age interval among the clusters in most of the fields. (Ages are given for 473 clusters in 12 LMC fields, determined from their brightest blue stars with a calibration based on CMDs for 29 clusters.) There is also a maximum, secondary in some cases, at about $B = 16$–17, corresponding to ages around 0.1–0.5 Gyr. The two dominating age groups in the Hodge fields, the old and the young intermediate-age clusters, match the outer and inner cluster systems in the Kontizas et al. definition (Sect. 4.3). The older clusters are responsible for the relatively smooth decrease in cluster density with increasing distance from the LMC centre that is seen in the number density contours (Kontizas et al. 1990b, Fig. 4).

It is somewhat surprising to find three small clusters 0.1–0.5 Gyr old as far from the main body as in Hodge's field 7, $4^h 56^m$, $-74°.3$, 6 kpc SW of the centre of the LMC. The dominating cluster in that region is NGC 1777, about 1 Gyr old, with a diameter $d = 31$ pc, and a metallicity below solar (Mateo and Hodge 1985). Also the stellar field population is old there, about 1–3 Gyr. It appears necessary to await accurate CMDs for the three clusters in question before accepting their young age.

Few small-dimension clusters have been studied so far in the LMC and these have mainly been in the outer regions. In spite of this limitation they may illuminate the cluster age/location question in that galaxy:

Hodge and Flower (1987) have presented preliminary CMDs for five clusters, which, because of their positions in the LMC, are of particular interest. All of them are in the outer regions of the LMC as defined by the cluster system. SL258, LW127 and LW134, all with $d \sim 9$ pc, are situated at $\sim 5^h.1$, $-65°$, in a region where NGC 1831 is the best-known cluster with $d = 57$ pc and an age of 0.3 Gyr (see Table 4.7). The three 'small' clusters have ages between 0.3 and 0.8 Gyr (± 0.2), thus being about the same as NGC 1831. The other two clusters, SL4 and NC1 (uncatalogued), at $\sim 4^h.5$, $-72°.4$, are likely to be around 1.5 Gyr old, though only SL4 has a reasonably well defined CMD. They are the only clusters in the observed ~ 900 pc diameter field.

Olszewski (1988) has published CMDs for five clusters in the outermost regions of the LMC disk, more than 4.5 kpc from its centre (three clusters to the N and two to the S). The clusters have $13 \leq d \leq 26$ pc. Their ages are 2–3 Gyr and their metallicities are $Z \sim 0.01$. They belong to the older of the two groups of intermediate-age LMC clusters. Also H4 and LW79 belong to the 2–3 Gyr age group: H4 ($\equiv$ SL556), near 4 kpc N of the LMC centre and with $d = 36$ pc, is ~ 2 Gyr old and has a metallicity of [Fe/H] $= -0.7 \pm 0.2$. LW79 ($\equiv$ SL61), almost 7 kpc SW of the LMC centre and with $d = 34$ pc, is also ~ 2 Gyr old with a metallicity of [Fe/H] $= -0.3 \pm 0.3$ (Mateo and Hodge 1986, 1987).

Six relatively small clusters, $d \sim 11$ pc, have been observed by Linde et al. (1995). They represent LMC regions of rather different composition: three of the clusters are in or close to a starforming region, 3.5 kpc NW of the LMC centre; the other three are in a field apparently free from such regions, 3.7 kpc SW of the LMC centre. The former have ages between 0.1 and 0.5 Gyr and metallicities close to solar, the latter are about 1 Gyr old and of lower metallicity, $Z = 0.015$ (Westerlund et al. 1995).

The cluster formation rate (CFR) in a small field 5 kpc NNE of the optical centre of the LMC, outside the limit of detected HI, has been investigated by Mateo (1988b). The metallicities of the 31 clusters observed vary from [Fe/H] close to solar for the youngest clusters to -0.5 for the oldest intermediate-age clusters and -0.9 for the 10 Gyr old. The ratio $CFR_{2Gyr}/CFR_{200Myr} \sim 20$ indicates a non-uniform CFR in at least this part of the LMC. His results confirm that the LMC was rather inactive between ~ 10 and ~ 4 Gyr before a burst of star formation activity occurred.

The small-dimension intermediate-age LMC clusters just discussed show that there are clusters between 3 and 1 Gyr old in all the outer fields. Also clusters belonging to the younger age group are found quite far from the LMC centre but only in the northern areas. At present most data indicate the lack of a population younger than ~ 1 Gyr in the SW, outside about 3.5 kpc from the centre of the LMC. Accurate CMDs for many more small clusters in all parts of the LMC are needed before the age/location relation can be definitely established.

4.4 The age distribution of Cloud clusters

Table 4.2. *Lifetimes of Cloud clusters according to Hodge (1988b)*

Galaxy	Median age, Gyr	Median dissolution time, Gyr	Decay time, Gyr
LMC	1.1	1.0	1.5
SMC	0.9	0.6	1.8
Galaxy	0.2	0.2	0.2

Median age: Half of the clusters are younger, half older.
Median dissolution time: The time at which the number of clusters in a small age interval has fallen to 1/2 of the present value.
Decay time: The ratio of the total number of surviving clusters to the present formation rate, extrapolated over the age of each galaxy.
The data for the Galaxy are from Wielen (1971).

The cluster population in the SMC has been analysed by Hodge (1987) to faint limiting magnitudes. In a material of 327 certain clusters d between 1 and > 30 pc, only about 15 clusters have $d > 20$ pc, confirming that the populous globular clusters are few in the SMC. There is a lack of young clusters in the sky-projected outer regions of the SMC. The age distribution indicates that the cluster production has been uneven or that the cluster lifetimes are finite. The latter may be more likely. In the first case most of the clusters should have formed in the last few hundred Myr. The SMC cluster lifetime (Table 4.2) is almost five times as long as in the Galaxy. The life expectancy of the Cloud clusters is much longer than that of their galactic counterparts. This has been ascribed to a less active dynamic environment in the Clouds where the destruction by interstellar clouds is less frequent.

4.4.1 The oldest clusters

The oldest generation in the Magellanic Clouds, ≥ 10 Gyr, may have formed in a similar way to Population II in our Galaxy during the initial collapse of the proto clouds. Compared to our Galaxy this generation at first sight appears poor in the Clouds and with no clear evidence of spherical halos nor of core regions, though extensions in depth are obvious. It is represented by a few true globular clusters, many RR Lyrae variables and a number of novae. There is evidence of a stellar field population of ≥ 10 Gyr age in the LMC (Westerlund *et al.* 1995a) and also in the SMC (Gardiner and Hatzidimitriou 1992).

In the field of the LMC seven clusters have, for some time now, been accepted as true globulars, i.e. similar to the globular clusters in the galactic halo. Three of them, NGC 1466, NGC 1841 and NGC 2257 have been considered as doubtful LMC members because of their positions in the outskirts of the LMC and, possibly, are extra-tidal. They now appear, however, to be accepted as likely members. NGC 1841, 14°.9 from the LMC centre, is accepted because of its radial velocity, 201–288 km s^{-1} (Schommer *et al.* 1992). In the SMC one cluster was early recognized as a true globular, viz. NGC 121, even if it is slightly younger than the LMC true globulars.

To this 'classical' group of true globulars in the Clouds have been added six LMC clusters (Suntzeff *et al.* 1992), all of SWB type VII and with [Fe/H] < -1.3. Possibly, Hodge 7 should also be included in this group. Its integrated colours agree with those of NGC 2257 and Hodge 11 (Bica et al. 1991). The Reticulum cluster, 11°.4 away, is counted as a LMC cluster because of its radial velocity, ~ 240 km s^{-1}. The SMC cluster NGC 339, SWB type VII, has been considered as a possible member of this class but belongs to the intermediate-age clusters (Table 4.6). It contains carbon stars and its age is now estimated to be ~ 3 Gyr.

The number of true globular clusters in the LMC and SMC may appear small. However, if the number of true globulars is proportional to the mass of a galaxy, then, with an estimated total number of 180 in the Galaxy (mass 3×10^{11} M$_\odot$) we would expect about 12 in the LMC (2×10^{10} M$_\odot$) and 2 in the SMC (3×10^9 M$_\odot$), in reasonable agreement with the observed numbers. It is, however, possible that additional true globulars will be found in the Clouds.

The mean number, N_{RR}, of RR Lyrae stars in the seven LMC globulars found to have such stars, per unit cluster luminosity scaled to $M_V = -7.5$ mag, is $N_{RR} = 14$, very similar to the values observed for galactic halo globular clusters, $10 < N_{RR} < 18$, depending on galactocentric distance (Suntzeff et al. 1992). The mean V_T for the eight certain old globular clusters in the LMC is $\langle V_T \rangle = 11.2 \pm 0.4$, $\sigma = 1.0$. (Adding the possible globulars in Table 4.3 has no significant effect on the mean value or the dispersion, $\langle V_T \rangle = 11.1 \pm 0.3$ mag, $\sigma = 0.9$.) With the distance modulus $m - M = 18.5$, the corresponding absolute magnitude is $M_V = -7.3$. The luminosity functions for all globular cluster systems so far studied are Gaussian with $M_V = -7.4$ and σ in the range 1.1–1.3. Thus, the LMC globular cluster system is in accord with those in other galaxies, considering the halo system for our Galaxy.

Mean magnitudes, abundances and reddenings for a total of 182 RR Lyrae variables in seven LMC clusters have been determined by Walker (1992a), using the CMDs for the clusters themselves (Table 4.4). With a distance modulus of 18.5, the RR Lyraes have in the mean $\langle M_V \rangle = 0.44$ at [Fe/H] $= -1.9$. Accepting the slope for the magnitude–metallicity relation from galactic data (Carney et al. 1992), $\langle M_V(RR) \rangle = 0.15$ [Fe/H] + 0.73.

The LMC zero point is 0.28 mag brighter than suggested by statistical parallax analyses of galactic field RR Lyraes and from parallaxes of subdwarfs, and 0.19 mag brighter than in the relation given in Sect. 2.4. Support for the correctness of Walker's relation is provided by the Mira variables which occur both in the LMC and in metal-rich galactic globular clusters. Calibrating the Mira PL-relation (Feast et al. 1989) and using the galactic Miras gives $M_V(HB) = 0.68 \pm 0.09$ at [Fe/H] $= -0.58$. Then, applying the magnitude–metallicity slope used above, $M_V(HB) = 0.48$ at [Fe/H] $= -1.9$, in good agreement with the mean value used for the LMC RR Lyraes.

A further comparison between the LMC true globulars and the galactic halo clusters is achieved by using the horizontal branch (HB) morphology index, $(B - R)/(B + V + R)$, where $B, V,$ and R are the numbers of blue HB, RR Lyrae variables and red HB stars, respectively (Lee 1989, 1992). Metallicity and age appear to be the two parameters with most influence in determining the HB morphology, i.e. the colour distribution, of galactic globular clusters. Da Costa (1993) has compared the LMC true globulars with galactic halo globular clusters in a HB index, [Fe/H] diagram. The LMC clusters fall among the outer galactic halo clusters ($R_{gc} > 20$ kpc) and avoid the region occupied by the inner galactic halo clusters ($R_{gc} < 8$ kpc). He concludes that the LMC globulars are affected by the second parameter, age, to the same extent

4.4 The age distribution of Cloud clusters

Table 4.3. *Clusters of SWB type VII in the LMC and the SMC*

Cluster	V_T mag	Age Gyr	RR Lyrae stars	[Fe/H]
SMC:				
NGC 121	11.24	12	4	−1.4
LMC:				
NGC 1466	11.59	≥12	24ab, 14c	−1.85
NGC 1786	10.88	≥12	4ab, 5c	−1.87
NGC 1835	9.52	≥12	21ab, 14c	−1.79
NGC 1841	10.98	≥12	17ab, 5c	−2.11
NGC 2210	10.36	≥12	11ab, 1c	−1.97
NGC 2257	11.48	≥12	15ab, 16c, 2d	−1.8
Hodge 11	12.10	≥12	0	−2.06
Reticulum	12.80	≥12	22ab, 10c	−1.71
NGC 1754	11.42	0.8 −1.2	...	−1.54
NGC 1898	11.10		...	−1.37
NGC 1916	9.88		...	−2.08
NGC 2005	11.16		...	−1.92
NGC 2019	10.70		0	−1.81

The total visual magnitudes, V_T, for the SMC cluster and the 8 first LMC clusters are from van den Bergh (1981,1993) and for the remaining 5 from Suntzeff *et al.*(1992).
The numbers of RR Lyrae stars and the metallicities, [Fe/H], are from Suntzeff *et al.*(1992). The search for RR Lyraes is insufficient in 4 clusters (identified by '...'), possibly also in the core of H 11. The latter cluster was once classified as of intermediate age, but an improved colour–magnitude diagram (Andersen *et al.* 1984) shows that it is very old and metal-poor. It is seen against a field containing a younger population which, when partly included in the cluster sequence, caused its age to be underestimated.
The ages are given as ≥ 12 Gyr for the 'classical' LMC globulars. Estimates between 12 and 18 Gyr may be found in the literature.
The age of NGC 1754 is from Jensen *et al.* (1988) and contradicts its inclusion as an old cluster. Its old-globular-cluster character has received support from Da Costa (1993), whose integrated CCD photometry placed the cluster within the region occupied by SWB type VII clusters. He suggested that the young age ascribed to NGC 1754 by Jensen *et al.* could be due to confusion of the cluster main sequence with that of the field. Judging from the technique used by the latter to correct for field stars and from the identification of stars well inside the cluster limits, this appears less probable. The scatter in their CMD is nevertheless large, so a final judgement will have to await further studies.
The nature of NGC 2019 as a true globular is supported by its position in the UV two-colour diagram C(18 − 28), C(15 − 31) (Barbero *et al.* 1990). It falls, together with NGC 2210 and NGC 1835, on the sequence defined by galactic globulars (see Fig. 4.5).

as the outer galactic halo clusters. Their formation may then have coincided in time with that of the latter, i.e. 2–3 Gyr after the formation of the inner galactic halo clusters.

Another similarity between the LMC globulars and the outer galactic halo clusters is that neither group shows any abundance gradient with the distance from their respective galaxy centres.

Table 4.4. *Absolute magnitudes for RR Lyraes in LMC clusters (Walker 1992b)*

Cluster	$\langle V \rangle$ mag	E_{B-V} mag	M_V mag	M_V mag
NGC 1466	19.33	0.09	0.55	0.40
NGC 1786	19.27	0.07	0.55	0.51
NGC 1835	19.37	0.13	0.46	0.43
NGC 1841	19.31	0.18	0.25	0.19
NGC 2210	19.12	0.06	0.43	0.52
NGC 2257	19.03	0.04	0.40	0.56
Reticulum	19.07	0.03	0.48	0.40
mean values			0.45	0.43
std. dev.			0.10	0.12

The number of variables used by Walker agrees with those in Table 4.3 except for NGC 1835 and NGC 2210, where Walker has 35 and 12, respectively. Likewise, the [Fe/H] values agree essentially, except for NGC 1786, where Walker uses -2.3.
The M_Vs in col. 4 are derived with all cluster moduli = 18.50, those in col. 5 with the LMC centre modulus 18.50 and geometric corrections for $i = 29°$, $PA = -9°$. Both columns indicate that NGC 1841 is appreciably closer to us than the other clusters.

Differences between the two systems of globular clusters are:
(i) The mean abundance of the 13 LMC true globulars (in Table 4.3) is $\langle$[Fe/H]$\rangle$ $= -1.84 \pm 0.06$, making them more metal-poor than the 14 galactic halo globular clusters with $R_{gc} > 20$ kpc, for which $\langle$[Fe/H]$\rangle = -1.67 \pm 0.06$. This result is in line with the conclusion that the mean abundance of a globular cluster correlates with the luminosity of its parent galaxy (Da Costa and Mould 1988).
(ii) A more important difference is found in the kinematics of the two systems. The galactic halo clusters form a roughly spherical system with a large velocity dispersion and little rotation, whereas the LMC true globulars form a flattened disk system with a rotation velocity (Schommer *et al.* 1992). In addition, this disk appears to be aligned with the disk defined by HI and the young LMC clusters.

In the SMC, the only cluster (almost) qualifying as a 'true globular', viz. NGC 121 (Table 4.3), has the right abundance, [Fe/H] $= -1.4 \pm 0.1$,[*] for a Population II cluster and possesses a few RR Lyrae variables, but otherwise has a red HB morphology. A low-luminosity, variable carbon star appears to be a member, making the decision about its true status difficult. The C star is just at or above the top of the first giant branch; it may be an AGB star with the 'third dredge-up mechanism' just beginning (Stryker *et al.* 1985). Possibly, NGC 121 should be considered as intermediate between the true globulars and the populous clusters with AGB branches; it has evolved far enough to produce RR Lyrae stars but not enough for blue HB stars.

Its red HB morphology places NGC 121 close to outer galactic halo clusters, like Pal 4 and Eridanus, in the HB morphology/abundance diagram. It is thus younger than the galactic halo clusters, consistent with the suggestion by Stryker *et al.* The

[*] [Fe/H] $= -1.63 \pm 0.31$ from photometry (Suntzeff *et al.* 1986).

4.4 The age distribution of Cloud clusters

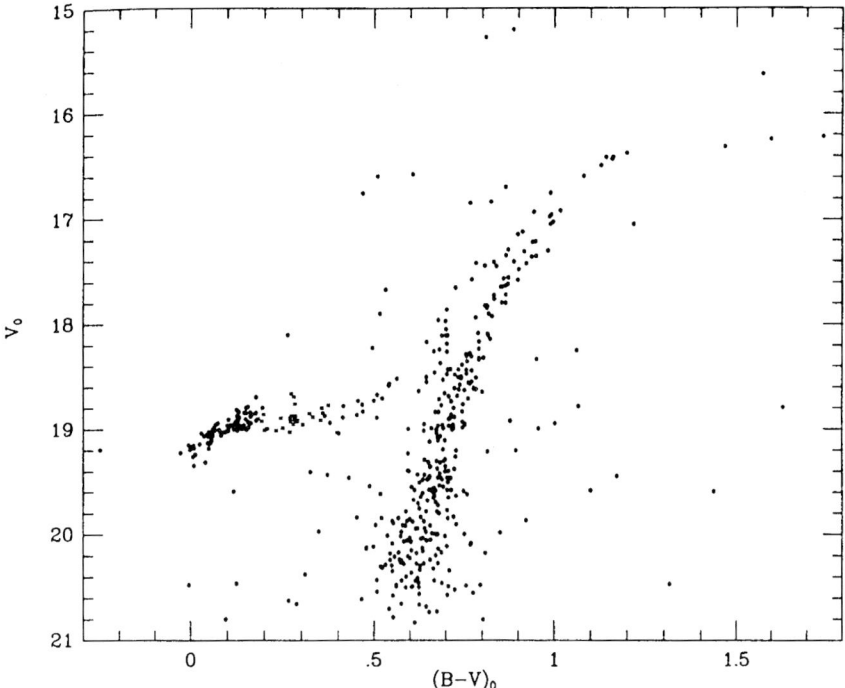

Fig. 4.4. The colour–magnitude diagram of NGC 2257 (Walker 1989). The AGB is well separated from the RGB. The latter has only few stars with $(B-V)_0 \geq 1.2$. RR Lyrae variables are identified by 'x's; they are in the interval $0.2 \leq (B-V)_0 \leq 0.4$. The two metallicity indices, height of RGB over HB at $(B-V)_0 = 1.4$ and colour of RGB at the level of HB suggest [Fe/H] $= -1.7$ to -2.0.

latter also suggested that the age of NGC 121 is comparable to the age of the oldest field stars in its neighbourhood. This conclusion is supported by Walker (1991) who finds the age of the cluster and the surrounding field stars to be 12–14 Gyr. This would make the age comparable to that of the oldest population in the LMC. Most results, however, indicate a delayed initial star formation in the SMC as compared with that in the LMC and in the galactic halo.

For those old clusters for which accurate observations of HB stars and RR Lyrae variables exist, their location relative to the main disk of the LMC may be determined. Presently, the following information is available.

NGC 2257 is located 9° NNE of the Bar. Nemec et al. (1985) recognized 41 cluster variables within 4' of its centre and identified 47 variables in the surrounding field. The completeness of the variable star discoveries was estimated to $\sim 82\%$ in the cluster and $\sim 53\%$ in the field. The mean period for the 16 ab-type cluster variables is $\langle P_{ab} \rangle = 0^d.583$, and for the 22 cd-type cluster variables $\langle P_{cd} \rangle = 0^d.353$. These periods suggest that NGC 2257 is an intermediate Oosterhoff type system (i.e. type I –II). From the period–amplitude diagram it follows that there are no very metal-rich RR Lyraes in the cluster; the variable star pulsation properties give [Fe/H] $= -1.8 \pm 0.3$. The ratios of cd to ab-type stars, and blue to red HB stars suggest that the distribution of stars along the HB is skewed toward the blue.

The mean magnitude of its RR Lyrae stars is 0.17 mag less than that of the field RR Lyraes, meaning that the cluster is 3.7 kpc in front of the field stars at that position (Walker 1989). The (few) field RR Lyraes observed indicate a possible depth of the LMC disk at the location of NGC 2257 of ~ 9 kpc. (The distribution may well be spherical; the material permits no decision.) A recent determination of the distance of the cluster, also using its RR Lyraes, gave 50 ± 10 kpc (Testa et al. 1995), a mean error that may raise some pessimism about distance determinations, generally, as it leaves open the position of the cluster relative to the LMC disk. However, the separation between field and cluster variables seems sufficiently reliable to place the cluster in front.

NGC 1841 is located $\sim 15°$ from the LMC centre and about equally far from the SMC. It has been considered to be intergalactic (Shapley and Paraskevopoulos 1940), extra-tidal to the LMC (Valtonen et al. 1984), and close to or slightly beyond the LMC main plane, but it is considered to be a member of the LMC (Gascoigne 1966). Its kinematics is not consistent with those of the other LMC clusters of similar age (Freeman et al. 1983). Its high age, ≥ 12 Gyr, indicates that, if it belongs to the Magellanic Clouds, it is a LMC member. Observations of its RR Lyrae variables show that it is ~ 7 kpc closer to us than the LMC centre (Walker 1990).

The Reticulum cluster (GLC 0435–59), $11°$ N of the Bar, is probably at the distance of the LMC disk (Walker 1992a). Most of its HB stars are variables. There are eight red and six blue HB stars compared with 32 RR Lyraes. Both NGC 1841 and NGC 2257 have bluer HBs. There are a number of stars brighter and bluer than the MS turn-off. As they follow the cluster distribution more than the field star distribution they are likely to be blue stragglers.

Hodge 11, $\sim 5°$ from the LMC centre, is ~ 2 kpc closer to us than the centre, which places it in the disk when the inclination is taken into consideration (Walker 1993). Star counts show that there are probably no red HB stars and no RR Lyrae stars but three groups of blue HB stars, centred at about $V = 19.35$, $V - I = 0.15$ (14 stars), 19.8, 0.08 (29) and 21.2, -0.15 (4). The 'true' blue HB is 0.15 mag brighter than the first group, leading to $M_V(HB) = 0.44 \pm 0.12$ mag, in agreement with that found for other metal-poor LMC clusters.

NGC 2210 is located $4°.5$ from the LMC centre. Reid and Freedman (1994) have observed 27 RR Lyraes in the vicinity of the cluster and conclude that the majority are cluster members. The mean magnitude of the field RR Lyraes in the neighbourhood is 0.1 mag fainter than the cluster variables. With the Reid and Freedman data of $V(RR) = 19.12$, $A_V = 0.19$, $[Fe/H] = -1.97$, and $M_V(RR) = 0.15 [Fe/H] + 1.01$, and assuming a tilt of the LMC of $i = 45°$, NGC 2210 falls 1.7 kpc in front of the LMC plane at the position of the cluster, or 5.6 kpc closer to us than the LMC centre, at 50 kpc distance.

4.4.2 The intermediate-age clusters

The possible similarity between the cluster systems in the LMC, the SMC and the Galaxy, seen in the Population II clusters, ceases when the populous intermediate-age clusters are considered. There are no such clusters in the galactic halo. In the galactic disk such clusters are found covering a wide range in age, but there are few, if any, as rich and with as high a central concentration as the blue globulars in the Magellanic

4.4 The age distribution of Cloud clusters

Clouds. As mentioned above (Sect. 4.4), the LMC lacks intermediate-age clusters older than ~ 3 Gyr.

In the SMC, the age distribution of the intermediate-age clusters may be open for discussion. It has been claimed that it is rather uniform, with some bright clusters in the 1–10 Gyr range with no counterparts in the LMC. Examples are L1, age ~ 10 Gyr (Olszewski et al. 1987), K3, 5–8 Gyr (Rich et al. 1984) and L113, 4–5 Gyr (Mould et al. 1984). The uncertainty in the ages of K3 and L113 is a consequence of the uncertainty in the distance modulus of the SMC. The ages correspond to the long, $(m - M)_0 = 19.3$, and short, 18.8, scales, respectively. Both K3 and L113 may contain contain C stars at the top of the AGBs; their membership needs to be confirmed. The luminosities of these C stars are around $M_{bol} = -4$. The variable C star, T-V8, in NGC 121 has $-3.6 \leq M_{bol} \sim -2.6$ $(m - M = 18.9)$. All, except T-V8 at its low limit, are in the range observed for the C stars in the LMC cluster NGC 1978, SWB type VI (see e.g. Westerlund et al. 1991).

All the known SMC clusters older than 3 Gyr, i.e. K3, L113, L1 and NGC 121, are well outside the main body of the Cloud but are inside its sky-projected tidal radius, ≥ 5 kpc (Gardiner and Noguchi 1995, Welsh et al. 1987). K3 and NGC 121 are ~ 2° west of the Bar, L113 ~ 4° east of the Bar, and L1, the westernmost known cluster in the SMC, is ~ 3°.5 from the Bar and outside the hydrogen envelope of the Magellanic System. Also NGC 339, ~ 3 Gyr old, is outside the main body, ~1° south of the Bar.

There are no direct determinations of the cluster locations along the line of sight. This is unfortunate as the SMC may have an extension in depth of up to 20 kpc, depending on the objects observed and the techniques used in the analysis. For NGC 121 Stryker et al. (1985) argue that the cluster RR Lyraes ought to have the same $\langle R \rangle$ as the field RR Lyraes and derive the modulus 18.85 in that way. For the other clusters, assumptions about the SMC modulus and geometry have steered the determinations.

Presently available CMDs for K3, L113, and L1 do not show enough main sequences for accurate determinations of their distances or ages. The CMDs of L1 and the LMC cluster ESO 121–SC03 are very similar (Olszewski et al. 1987) and indicate a relative distance modulus of 0.4 mag (L1 being the most remote). ESO 121–SC03, at about 10° from the centre of the LMC and outside its tidal radius of 8 kpc, may not be in the plane of the LMC. Mateo et al. (1986) derived $(m - M)_0 = 18.2 \pm 0.2$ for ESO 121–SC03, which would give $(m - M)_{0,L1} = 18.6 \pm 0.2$.

The RGC/HB magnitude is probably independent of age for these old clusters (Hatzidimitriou and Hawkins 1989) and can serve as a distance indicator. Table 4.5 gives the data for the four clusters and, for comparison, also for the slightly younger cluster NGC 411.

Using $\langle M_R \rangle = 0.22$ (± 0.04) and $A_R = 0.07$ (Hatzidimitriou and Hawkins 1989), the moduli for the five clusters, L1, NGC 121, K3, NGC 339, L113, become 18.51, 18.96, 18.69, 18.71, and 18.41 with the corresponding distances being 50, 62, 55, 55, and 48 kpc. The modulus of L1 thus derived agrees well with that obtained above by comparison with ESO 121–SC03. The main-sequence turn-off points of L113 and K3 are brought into fair agreement if the derived distances are applied. We may ask if these two clusters are not of the same age as L1. NGC 121 is the only one with RR Lyrae variables, and, for that reason, is some Gyr older.

Table 4.5. *Mean RGC magnitudes for SMC clusters*

Cluster	Age Gyr	⟨R⟩ mag	Reference
NGC 411	1.5	18.8	Da Costa, Mould 1986
NGC 339	3.0	19.0	Olszewski et al. 1987
L113	4.5	18.7	Mould et al. 1984
K3	6.0	18.98	Rich et al. 1984
L1	10.0	18.8	Olszewski et al. 1987
NGC 121	12.0	19.25	Stryker et al. 1986
Mean value		18.92 ± 0.04	

NGC 411 has [Fe/H] ~ -0.9, the others have [Fe/H] $= -1.4 \pm 0.2$.

The derived cluster distances are well inside the range of distances of SMC Cepheids, 43–75 kpc (Mathewson et al. 1986). All except NGC 121 are outside the SMC tidal radius and their SMC membership could for that reason be in doubt. Presently, the tidal radius may be a rather meaningless notion if the galaxy is in a state of disruption. The situation is similar for some clusters in the LMC: ESO 121–SC03, E2, $\sim 8°$ north of the LMC centre (Schommer et al. 1986), the Reticulum cluster, NGC 1466, NGC 1841, and possibly NGC 2257, are all outside the sky-projected tidal radius, but still considered to be LMC members. Therefore, the SMC clusters under discussion may be considered to be SMC members. It would, however, be of great interest to know the cluster positions at the time of their formation.

The few field stars observed in the L113 and K3 regions have distributions much like those around NGC 121 (compare Stryker et al. 1985, Fig. 9b, Mould et al. 1984, Fig. 3, and Rich et al. 1984). The dominating population is about 10 Gyr old, and thus of the same age as the one in the L1 field (Olszewski et al. 1987). It may extend over all of the SMC. There is no indication in any of the fields of a significant contribution from an intermediate-age component. Closer to the SMC main body, younger field components appear, as e.g. in the field of the galactic cluster NGC 362, about 2° north of the SMC centre, where a population as young as 0.2 Gyr is seen (Bolte 1987).

Because of the importance ascribed to these SMC clusters in establishing that appreciable differences exist in the evolution of the LMC and the SMC, it is highly desirable that their distances and ages are redetermined and their kinematics studied. A search for clusters of 10–3 Gyr age in the core of the (projected) SMC would contribute to settling this question.

Independent of the possible spread in age of these clusters, the character of peripheral objects in the SMC and the fact that a corresponding field population is very weak or missing indicates that the evolution of the LMC and the SMC may have been more similar than is usually assumed, with an early production followed by a long quiescent period and a burst of star formation a few Gyr ago. The fact that the oldest clusters in the SMC are not as old as the oldest in the LMC only confirms that the SMC 'trails behind' the LMC in the same way as the LMC does relative to our Galaxy. The initial collapse of the SMC was less efficient in star production and

4.4 The age distribution of Cloud clusters

the SMC retained a much larger fraction of its total baryonic mass in the form of gas than did the LMC.

In a review of the cluster systems of the Magellanic Clouds, van den Bergh (1991) suggested that 'the fact that the two largest bursts of star and cluster formation in the LMC appear to have no counterparts in the SMC indicated that the bursts of star formation in the Large Cloud were not triggered by close encounters between the Clouds'. This conclusion will have to be revised if it is confirmed that the peripheral clusters in the SMC are all close to 9 Gyr old, and if it is accepted that the initial star formation in the SMC was slightly delayed relative to that in the LMC. The second burst of star (and cluster) formation, and one due to interaction between the two Clouds, occurred in both galaxies about 3–4 Gyr ago.

The lack of stars and clusters of ages between $\sim$ 10 Gyr and $\sim$ 4 Gyr in the LMC (and probably also in the SMC) may be due to a reversal of the initial collapse of the protogalaxy by supernova-driven stellar winds which ejected a significant fraction of the gas originally present (van den Bergh 1993, following Berman and Suchhov 1991). This will have stopped star formation efficiently, but it may have started again after a few Gyr when the gas retained in the potential well, mostly produced by the dark halo, recollapsed (Mathews 1989) to contribute to the formation of more metal-rich objects.

4.4.2.1 Red-giant and asymptotic-giant branches in intermediate-age clusters

Ferraro *et al.* (1995) have carried out *JHK* photometry for 12 intermediate-age clusters. From the $K, (J-K)$ CMDs it can be seen that the clusters divide into two groups: one with a well-developed giant branch and one without such a branch. To the former belong the well-known clusters NGC 1783, 1806, 1978, 1987, and 2173; to the latter belong NGC 1756, 2107, and 2209. The others are in between the two groups (see Table 4.7).

The clusters without a well-developed giant branch are those in which the masses of the presently evolving stars are larger than the critical mass for production of RGB or AGB stars. The clusters in the other group are older and the evolving stars have masses appropriate for yielding RGB and AGB stars. The intermediate cases represent the transition between the two groups. The AGB stars contribute $\sim$ 60 % of the integrated cluster luminosity at the K band, while the bright RGB stars contribute between 0 % at $s = 0$ and 20 % at $s > 35$.

Many of the SWB type IV–VI clusters in the SMC and the LMC contain carbon (C) stars at the top of the AGB. The C stars produce about half of the bolometric luminosity of these clusters (Persson *et al.* 1983). In the SWB type II–III clusters about half the light at 2.2 μm comes from luminous oxygen-rich red stars, M giants, at the top of the AGB. They have virtually no effect on the integrated visible luminosities. The only cluster of SWB types II and III supposed to contain carbon stars is NGC 1850 in the LMC: $V_T = 8.96$, $(B-V)_0 = -0.03$, $(J-K)_0 = 0.54$, with one C star (Frogel *et al.* 1990). NGC 2136 and NGC 2214 have possible candidates. There are no known SMC clusters of these types with C stars. It appears wise to await the confirmation of the membership of the C star in NGC 1850 before drawing any conclusions.

No LMC cluster of SWB type VII is known to contain C stars, whereas some SMC clusters of this type, NGC 121, NGC 339 and K 3 (VI –VII), may do so. These

Table 4.6. *A sample of intermediate-age clusters in the SMC and the LMC*

Cluster	$(m-M)_0$ mag	SWB type	E_{B-V} mag	Age Gyr	[Fe/H]	Ref.
SMC						
NGC 411	18.8	V–VI	0.04±0.02	1.8±0.3	−0.9	1
	19.3		0.04±0.02	1.2±0.3	−0.9	1
NGC 152	18.8	IV		1.3±0.4		2
NGC 339		VII		3		3
K3	18.8		0.04	8	−1.2	4
L1	19.0		0.03	10	−1.3	5
L113	18.8	VII	0.04	5	−1.4	6
LMC						
NGC 1783	18.2	V	0.03–0.10	1.1±0.4	−0.5	7
	18.7		0.03–0.10	0.7±0.2	−0.5	7
NGC 1831	18.5	V	0.05	0.35–0.55	−0.3	8
NGC 1866	18.6	III	0.06	0.086	−0.1	9
	18.6		0.07	0.2–0.25	0.0	10(a)
	18.6		0.07	0.1–0.2	0.0	10(b)
NGC 1978	18.2	VI	0.19	2.1	−0.7	11
NGC 2134	18.5	IV	0.22	0.19	−0.4	13
NGC 2241	18.4		0.08	3–4	−0.5	12
NGC 2249	18.5		0.25	0.55	−0.4	13
121-SC03	18.5		0.03	9±2	−0.9	14
E2	18.2		0.06	1.5±0.5	−0.3	15

References: 1, Da Costa and Mould (1986); 2, Melcher and Richtler 1989; 3, Olszewski 1988; 4, Rich *et al.* 1984; 5, Olszewski *et al.* 1987; 6, Mould *et al.* (1984); 7, Mould *et al.* (1989); 8, Vallenari *et al.* (1992); 9, Becker and Mathews (1983); 10, Lattanzio *et al.* (1991) (a) overshoot, MS turn-off and red giants; (b) semiconvection, MS turn-off and red giants ; 11, Mould and Da Costa (1988); 12 Jones 1987;13, Vallenari *et al.* (1994b); 14, Mateo *et al.* 1986; 15, Schommer *et al.* 1986.
The distance moduli are those assumed by the authors for the SMC or the LMC except in the case of L1, which ref. (5) takes to be 0.3 mag larger than that of the SMC because of an assumed tilt of the Cloud.
All the SMC clusters in the table are supposed to contain carbon stars, though their membership is doubtful in most cases.

clusters are younger than the true globulars (see above). If the C stars really are cluster members, it is not evident how they have formed. Their luminosities place them in a range where a helium core flash may have occurred, though much evidence indicates other mechanisms in their formation.

There is no difference between the proportion of light due to C stars in clusters in the SMC and in the LMC (Persson *et al.* 1983). The sharp rise in the C/M star ratio for field stars (Blanco *et al.* 1980) in the SMC is due only to the non-detection of the warmer red (K-type) giants in the more metal-poor SMC.

A comparison of the C stars in the LMC and the SMC clusters and fields (Westerlund *et al.* 1989a) shows that in the LMC the faintest known field C stars correspond to those in the SWB type VI clusters. In the SMC numerous field C stars are known that are less luminous than the C stars in the old SMC clusters NGC 411

4.4 The age distribution of Cloud clusters

Table 4.7. *AGB and RGB stars in intermediate-age clusters in the LMC*

Cluster NGC	SWB	Age Gyr	$L/L_\odot$	N_{AGB}	N_{RGB}	d pc	s
1756		0.4	4.74	0.21	0.21	16	32
2107	IV	0.4	10.00	0.10	0.00	31	32
2209	III–IV	0.7–1.2	12.75	1.15	0.00	41	35
1831	V	0.4–2.5	15.00	0.20	0.13	57	31
1868	IV	0.3–1.8	7.59	0.26	0.13	57	33
2108	IV-V	0.8–2.2	6.61	0.15	1.51	29	36
2162	V	0.8–3.8	1.99	1.00	1.50	31	39
1783	V	0.8–3.3	13.80	0.58	1.81	65	37
1806	V	4.3	15.56	0.64	1.41	33	40
1978	VI	1.2–6.6	21.88	0.37	1.69	57	45
1987	IV	0.5–1.5	7.66	0.91	0.91	25	35
2173	V-VI	1.5–6.5	5.70	0.88	1.75	34	42

The ages are the extreme values given in the compilation by Corsi *et al.* (1994). $L/L_\odot$ is the cluster luminosity in units of $10^4 L_\odot$, and N_{AGB} and N_{RGB} are the number of AGB and RGB stars, respectively, normalized to $10^4 L_\odot$ (Ferraro *et al.* 1995). In determining these quantities the separation threshold between AGB and RGB stars has been taken to be $K_0 = 12.3$ ($M_{bol} = -3.6$), and 20% of the stars between $12.3 \leq K_0 \leq 14.3$ are assumed to be E-AGB stars. The diameters, d, are from Kontizas *et al.* (1990b). The *s*-parameter is defined in Sect. 4.2.

and K3. There are field C stars that are even less luminous than the variable C star in NGC 121 (Westerlund *et al.* 1995b, see Chap. 7). The SWB type IV cluster C stars in the LMC and the SMC will coincide in luminosity in the HR diagrams if a difference in $(m-M)_0$ of ≈ 0.5 mag is accepted. The transition M–S–C occurs at a significantly higher temperature in the SMC than in the LMC and also at a lower luminosity. Both differences are understood as effects of the lower metallicity of the SMC.

Several factors may be responsible for the existence of C stars at the tip of the AGB in SWB type IV–VI clusters and of M stars in types II–III. A comparison of the two clusters, NGC 1978, 2 Gyr, and NGC 1866,* 0.2 Gyr, may illustrate the situation. They are typical for their age groups: NGC 1978 has carbon stars at the tip of the AGB, and NGC 1866 has M giants. The latter cluster also contains Cepheids with periods between 2.6 and 3.5 days, i.e. with masses near 5 $M_\odot$. They will eventually evolve into stars which terminate their AGB evolution at the present tip of NGC 1866, at $M_{bol} = -5.9$.

Mould and Aaronson (1986) favour the hypothesis that thermal pulses are not able to establish themselves in stars of 5 $M_\odot$ or more before the stellar envelope is eroded away by rapid mass loss. It is also possible that the carbon is converted to nitrogen in the 'hot-bottom burning' occurring in stars just below 5 $M_\odot$ (Renzini and Voli 1981).

---

* The age of NGC 1866 given here is from Chiosi *et al.* (1986) and determined from MS termination magnitudes, red-giant star luminosities and maximum AGB star luminosities. It is a factor of 2.5 larger than that derived by Becker and Mathews (1983) and used by Mould and Aaronson (1986).

Metallicity differences of any importance do not appear to exist between NGC 1866 and NGC 1978.

4.4.2.2 The helium flash gap in integrated colour–colour diagrams for LMC clusters

The division of the intermediate-age clusters into two groups is identified by the so-called helium flash gap. The gap is seen in the integrated $V,(B-V)$ diagram at $(B-V) \approx 0.5$. It is also seen in the $(U-B),(B-V)$ diagram in the region of SWB type IV. Several explanations have been suggested: age gap, effect of cluster disruption (van den Bergh 1981), RGB or AGB transition, transition of the core-helium-burning phase from loop to clump. Chiosi et al. (1986) considered that the gap 'simply reflects the combined effects of a more or less continuous activity of cluster formation and different chemical compositions'. Other points of view have been presented recently, favouring either the age gap explanation or the RGB or AGB transition model.

Clusters located in the gap-region in the $(U-B),(B-V)$ diagram are listed in Table 4.8. The three clusters on the red side of the gap are the only ones belonging to the 'IR-enhanced' group in the Persson et al. (1983) definition. Their infrared colours and indices are accounted for by their content of luminous carbon stars. These clusters have undoubtedly developed a thermally pulsating AGB (TP-AGB).

In a $(U-B),(B-V)$ diagram for 624 LMC clusters (Bica et al. 1991) the gap is centred at $(U-B) \approx 0.19$ mag, $(B-V) \approx 0.47$, with an amplitude of 0.1 mag in both colours, splitting SWB type IV into a type IV blue and a type IV red. Clusters listed 'in the gap' in Table 4.8 are on the blue side here. Age estimates are available for four clusters near the gap (Sagar and Pandey 1989 and Seggewiss and Richtler 1989): on the blue side, NGC 1868, 0.69 Gyr; NGC 2249, 0.55 Gyr; and on the red side, NGC 1987, 0.89 Gyr; NGC 2209, 1.68 Gyr. These ages are slightly higher than the turn-off ages given by Bica et al. but do not alter their conclusion: the cluster distribution supports the theoretical prediction that the RGB phase transition occurs close to 0.6 Gyr over a time interval of ≈ 0.2 Gyr.

In the UV two-colour diagram, $C(18-28)$, $C(15-31)$, the stellar clusters should arrange themselves along a well-defined sequence (Barbero et al. 1990), with their location along the sequence being a smooth function of age. A plot of the UV colours of 31 LMC clusters, observed by Cassatella et al. (1987), shows rather good agreement with theory. Only the three SWB type VII clusters NGC 1835, NGC 2019 and NGC 2210 deviate significantly from the sequence and fall on the line valid for galactic globulars (Fig. 4.5). Their location is ascribed to the existence of hot HB stars in these clusters,[*] NGC 1835 and NGC 2210 also contain RR Lyrae stars but not NGC 2019.

There is a total lack of clusters in the range $-0.8 < C(18-28) < 0.2$ along the age sequence (Fig. 4.5); it is delimited by the two SWB type IV clusters NGC 1856 and NGC 1987, with NGC 1987 on the 'old' side of the gap. The number of SWB type IV objects may be too small to permit a conclusion about the reality of the gap. However, a comparison of the distribution of the present cluster sample with that observed by van den Bergh (1981) indicates that the sample is representative for the LMC. Therefore, Barbero et al. consider that the gap is real and implies a lack of

---

[*] All true globulars in the LMC have blue HBs (van den Bergh 1991).

Table 4.10. *LMC clusters with Cepheids (Mateo 1992)*

Cluster NGC	N_{Ceph}	Period Range days	V_{Ceph} mag	Age Myr
1755	2	3–6.3	15.1	
1850	1	> 10?	14.2	8
1854	1	> 10?	14.3	
1866	23	1.9–3.5	16.2	80
2010	4	2.1–3.5	16.1	170
2031	14	1.8–4.4	16.2	140
2134	2	3.4–5	16.2	190
2136	8	3.2–10.5	15.7–14.9	
2157	3	2.9–7.7	15.3	30
2164	1	?	15.0	63
2214	1	~ 11	14.3	75

Ten new variables, of which six were inside the annulus studied by Arp and Thackeray, were reported by Storm *et al.* (1988). Welch *et al.* (1991) found that only six of them were Cepheids, one of which is not a cluster member. They determined the periods for 16 of its now identified 21 Cepheids.

Today 11 LMC clusters are known to contain Cepheids (Mateo 1992). They are listed in Table 4.10. The clusters are all young and massive and of a population that does not appear to exist in our Galaxy.

Cepheids in clusters offer a number of advantages over the field Cepheids. The cluster HR diagram can be used to determine the MS turn-off mass. This allows the period–luminosity–mass–colour (PLMC) relation of the pulsation theory to be used to determine the pulsation mass of its Cepheids: The evolutionary turn-off mass (M_e) of the cluster must be equal to the pulsation (M_p) mass of its member Cepheids, and the distance of the cluster can be determined. The method has been successfully applied to NGC 2157 (see Chap. 3).

4.5 The structure of Cloud clusters

Many of the luminous clusters in the Magellanic Clouds are highly flattened. The youngest clusters in the LMC, SWB types I and II, are appreciably flatter than the older ones; the oldest, type VII, are rather similar to galactic globulars (Frenk and Fall 1982). No correlation has been found between the shape of a cluster and its position in the LMC.

The ellipticities of globular clusters vary with respect to the radius. This indicates that their dynamical evolution is different from that of other systems, such as dwarf elliptical galaxies. The ellipticities of globulars in different galaxies should therefore be compared by measuring the axial ratios at a common radius inside the clusters. In this way the problem of defining the outer isophotes is avoided; they may be ill defined as well as disturbed by tidal effects. The half-mass radius, r_h, represents a well-defined region throughout the evolution of a cluster.

A comparison of the globular clusters in the Milky Way, the LMC, the SMC and M 31 shows that the globulars in our Galaxy and in M 31 are much more spherical

4.4 The age distribution of Cloud clusters

Table 4.9. *Data for clusters of SWB type I–III*

Cluster	E_{B-V} mag	SWB type	Age Myr	[Fe/H]	Ref.
SMC					
NGC 330	0.06	I	10–25	−0.4	1(a)
			10–48		1(b)
	(0.03)		30	−1 – −0.7	2
NGC 458	(0.03)	III	100	−1 – −0.7	2
LMC					
NGC 1711	(0.09)	II	32	0.0	3
NGC 1818	0.07	I	20–40	−0.8	4
NGC 1850	0.18	I	8 and 70	−0.4	5
NGC 1858	0.15		8	−0.4	5
NGC 2004	(0.06)	I	8	0.0	6
	(0.09)		16	0.0	3
	0.028 ± 0.005		40	−0.7	7
NGC 2031	0.18 ± 0.05		140	−0.4	8
NGC 2100	(0.19)	I	16	0.0	3
	0.282 ± 0.008		40	−0.7	7
NGC 2164	(0.10)	III	63	0.0	3
NGC 2214	(0.07)	II	63	0.0	3
	(0.07)		90	0.0	9
HDE 269828	0.46 ± 0.06		5		10

(a) semiconvective model; (b) overshoot model.
References: 1, Chiosi *et al.* 1995; 2, Stothers and Chin 1992b; 3, Sagar and Richtler (1991); 4, Will *et al.* (1995); 5, Vallenari *et al.* (1994a); 6, Bencivenni *et al.* (1991); 7, Balona and Jerzykiewicz (1993); 8, Mould *et al.* (1993); 9, Bhatia and Piotto (1994); 10, Heydari-Malayeri *et al.* (1993).
$m - M = 18.85$ for NGC 330.

the Bar and must be considered superposed. A sequence of cluster-forming regions can be identified. Ten coeval clusters are seen in the eastern end of the Bar, centred on NGC 2058 (see also Sect. 4.6) and NGC 2065, with ages of about 100 Myr. Two still younger groups, with ages between 10 and 70 Myr, can be identifed in the Bica *et al.* data, one in the 'Bok region' and one centred on NGC 1913, NGC 1922 just south of S Doradus.

4.4.3.1 Cluster Cepheids

Individual observations of Cepheids in a LMC cluster were first reported by Arp and Thackeray (1967) and Arp (1967) who observed seven Cepheids in NGC 1866 and attempted an interpretation of the CMD based on evolutionary models. Their photometry was limited to the cluster periphery, $1'.4$ from its centre. Arp noted that six of the seven variables fell on the red edge of the period–colour relation for Cepheids. The redness of the mean colours of these Cepheids relative to the complete LMC sample was confirmed by Walker (1987), suggesting that the filling of the instability strip is a function of metallicity and age, as expected from model calculations.

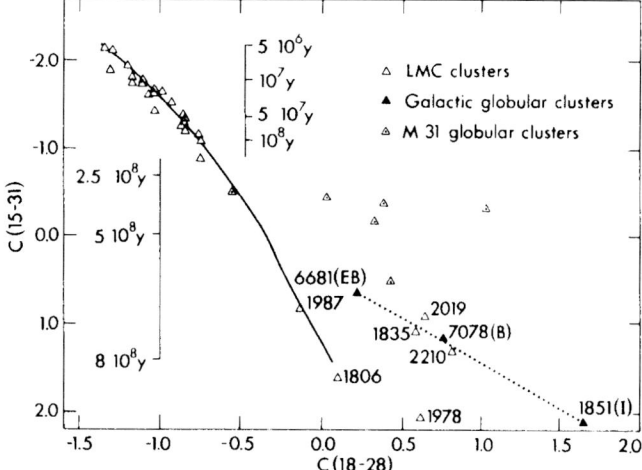

Fig. 4.5. The UV two-colour diagram for clusters in the LMC, the Galaxy and M 31 (Barbero et al. 1990).

CM and HR diagrams have been determined for many clusters in this age group and have been used to determine ages, metallicities, and initial mass functions. A few examples will be treated below (Sect. 4.7) and similar analyses will be presented for a few stellar associations (Sect. 5.9).

Red supergiants supply 80 % of the light at 2.2 μm in SWB type I clusters (Persson et al. 1983). The only difference in integrated infrared light between clusters in the LMC and the SMC is that the youngest supergiant-dominated objects in the SMC have weaker CO absorption, consistent with the lower metal abundance in the SMC.

Integrated spectra in the range 5600–10 000 Å have been used by Bica et al. (1990) to analyse the evolution of young LMC and SMC clusters. They found that the red supergiant phase of a luminous blue cluster is very time-peaked, occurring from 7 to 12 Myr. To this group belong e.g. NGC 2004 and NGC 2100 in the LMC and NGC 299 in the SMC. The latter cluster differs from the LMC clusters by having weak molecular bands due to the SMC metal deficiency.

Only the most massive clusters are able to keep some gas from their formation period. Hα emission seen from clusters aged 10–50 Myr is mainly from Be stars. Typical examples are NGC 1818 and NGC 1850 in the LMC and NGC 330 and NGC 376 in the SMC. Clusters aged $\geq$ 100 Myr again show a red phase, presumably related to AGB stars. Examples are NGC 1866, NGC 2031 and NGC 2134. (See Sect. 4.4.2. The ages used by Bica et al. are appreciably younger than those most commonly used.)

The effect of red supergiants on the integrated colours of the young clusters is shown by Bica et al. (1992) in their Fig. 4 (see Fig. 4.3). The 'clusters with RSG' occupy a central part in the two-colour diagram; the curvature in the van den Bergh (1981) diagram may be explained by those clusters.

The clusters in the 'Bok region', in the NW part of the LMC Bar (Bok and Bok 1969), as well as the majority of clusters observed in the Bar region (Bica et al. 1992, Alcaino and Liller 1987), are significantly younger than the stellar field population in

4.4 The age distribution of Cloud clusters

Table 4.8. *Clusters near the colour gap (from van den Bergh 1981)*

Cluster NGC	SWB type	V_T mag	$(B-V)_0$ mag	$(J-K)_0$ mag	S and/or C stars
Blue side of gap:					
1856	IV	10.06	0.10	0.43	1 C:
2002		10.12	0.25:	0.84	
1872	III–IV	11.04	0.22	0.56	
1831	V	11.18	0.24	0.48	2(C)
2107	IV	11.51	0.19	0.80	no C
1895		11.90	0.22:		
1885		11.97	0.28:		
1849		12.80	0.20:		
In gap:					
1868		11.56	0.31	0.69	1 S
1756		12.24	0.30:		
2249		12.23	0.33:		
Red side of gap:					
1987	IV	12.08	0.40	0.89	2 C, 1 S
2108	(V)	12.32	0.40	1.18	1 C
2209	III–IV	13.15	0.46	1.68	2C

Reddening corrections and $(J-K)_0$ colours are from Persson *et al.*(1983). The corrections used are mean values of E_{B-V} from nearby early-type stars or a mean LMC value; they cannot be considered free from systematic errors in the way of being underestimates (see Cohen *et al.* 1981). For clusters without previously estimated E_{B-V}, 0.10 has been assumed, identified by ':' in col. 4.
The number of C and S stars are from Frogel *et al.*(1990) or Westerlund *et al.*(1991). There is no indication of any star redder than $B-V \sim 1.2$ mag in NGC 2249 (Jones 1987).

clusters in the age range 0.2 to 0.7 Gyr. This supports the interpretation above of two intermediate-age generations in the LMC. Further support is given by the age distribution of red field stars (Chap. 7): two AGB branches can be seen in the LMC field, separated in age by about the same amount as found here. The RGB phase transition, occurring in this interval, is of shorter duration than suggested by the UV diagram. A separation of 0.2 Gyr should, nevertheless, be sufficient to permit the identification of two generations.

The appearance of S and C stars in the clusters on the 'red side' of the gap is a confirmation that the helium flash at degenerate core conditions has occurred.[*]

4.4.3 The youngest clusters

In the youngest cluster population in the Clouds are included those with ages ≤ 100 Myr. The outstanding very young cluster in the SMC is NGC 330. Among the very young clusters in the LMC are NGC 1818, 2004, 2100, all of SWB type I. The age determinations of clusters like NGC 1866, SWB type III, are such that they become borderline cases, but they are, anyhow 'young'.

[*] None of the other SWB IV type NGC clusters listed by Bica *et al.*(1991) appear in either the Frogel *et al.*(1990) or the Westerlund *et al.*(1991) catalogues over LMC clusters with C, M and S stars.

4.5 The structure of Cloud clusters

than those in the Magellanic Clouds (Han and Ryden 1994). The Cloud globulars show a small variation in intrinsic shape whereas the globulars in our Galaxy and in M 31 show comparatively large variations. The rotation and the internal velocity distribution of a globular are responsible for its shape; as they evolve with time, the morphological differences will be due to age differences between the globulars in each galaxy. Globulars in galaxies with similar structure, mass, and age tend to have similar shapes. Han and Ryden find that the globulars in M 31 and the Galaxy are most likely oblate spheroids, whereas those in the Clouds are triaxial ellipsoids. However, the LMC and SMC clusters used (49 and 28, respectively) are dominated by young globulars (age $\leq$ 3 Gyr) whereas, presumably, those in the other galaxies are true globulars. Thus, the conclusion by Frenk and Fall (1982), given above, that the Cloud clusters of SWB type VII are similar to the old galactic clusters, still holds.

The SMC clusters are more elliptical than the LMC clusters when measured at r_h according to Kontizas et al. (1990a). The most massive clusters are the most elliptical, and the inner parts of a cluster are clearly more elliptical than its outer parts.

Surface brightness profiles have been determined by Elson et al. (1987) for ten rich star clusters in the LMC (aged between 8 and 300 Myr) from aperture photometry in the inner parts and star counts in the outer. The profiles extend over 8–10 mag in surface brightness and to radii of $4'$ ($\equiv$ 58 pc). Most of the clusters do not appear to be tidally truncated. At large radii the projected density falls off as $r^{-\gamma}$ with $2.2 \leq \gamma \leq 3.2$, and a median value of $\gamma = 2.6$. The estimated masses* of the clusters are 10^4–10^6 $M_\odot$; much of this range is due to uncertainties in the mass-to-light ratios (derived from stellar population models). The density profiles were used, together with the assumption of hydrostatic equilibrium, to calculate the velocity dispersion as a function of radius within the clusters. Typical central velocity dispersions are 1–8 km s^{-1} along the line of sight. The crossing times in the outer parts of the clusters are of the order of their ages, and two-body relaxation times are generally considerably longer. Thus, the clusters are well mixed but are not relaxed through stellar encounters. The observed profiles may have resulted from expansion of a newly formed cluster either through mass loss or during violent relaxation so that halos of unbound stars have formed. Most of the observed clusters may have up to 50% of their total masses in unbound halos.

The tidal field of the LMC will eventually drip away unbound stars from a cluster. This may be determined from the circular velocity curve of the galaxy or from its mass distribution. The former method is the best as it also reflects the mass distribution; rotation curves are available for the LMC. The tidal acceleration of a star at the distance r from the centre of a cluster is taken as the differential acceleration along a line joining the centre of the cluster with the centre of the galaxy. The result is $(4\Omega^2 - \kappa^2) r$, where Ω and κ are the circular and epicyclic frequencies of the parent galaxy at the pericentre of the orbit of the cluster.

Radial velocities for between 11 and 37 stars in each of the three rich young LMC clusters, NGC 1866, NGC 2164, and NGC 2214, were measured by Lupton

---

* The mass of the stellar population of a cluster is the sum of MS stars, evolved stars and stellar remnants; the mass as a function of time will have to allow for stellar mass loss and, possibly, the escape of stars.

et al. (1989). Due to contamination by field stars and observational errors comparable to the observed velocity, only upper limits to the true velocity dispersions in the clusters could be determined. The limits for all three clusters are $\sigma_v \leq$ 3–4 km s^{-1} at the 95% confidence level. Combining this result with surface brightness profiles, derived by Elson et al. (1987), the upper limits on the total masses of the clusters were found to be 6, 2 and 4 $\times$ 10^5 M$_\odot$, respectively. The corresponding limits on the global mass-to-light ratios were 0.5, 0.7, and 1.3 in solar units.

The tidal radii (r_t) of the clusters were determined with the aid of King's formula (1962) as it was considered to lead to conservative estimates of the masses of any unbound haloes. Also the mass was obtained with the aid of a King model.

NGC 1866 has a small velocity dispersion and a radial extent well beyond the King model and is therefore most likely to be not yet tidally limited by the LMC. The mass bound to the cluster, M_t, is $\leq 2.5 \times 10^5$ M$_\odot$, leaving more than 30% of the total mass in an unbound halo. Also the other two clusters may have unbound halos but the limits on the velocity dispersions were not tight enough to permit definite conclusions.

Further studies of young globulars (0.01–1 Gyr) in the Clouds (Elson 1991, 1992) have revealed that substructures exist; there are bumps, sharp shoulders and central dips. There are also upper limits to the cluster cores, growing from $\approx$ 1 pc for the youngest clusters to $\approx$ 6 pc for the oldest in the group. The core expansion may be driven by mass loss from the evolving stars. The rate of the expansion may depend upon the shape of the IMF, the slope of which is consistent with the Salpeter value ($x = 1.35$).

NGC 1978 is an outstanding intermediate-age cluster, $\sim$ 2 Gyr, located in the NE part of the LMC. It is one of the most elliptical clusters known. The ellipticity of its surface brightness distribution is determined to $\epsilon = 0.30 \pm 0.03$ (Fischer et al. 1992). The major axis position angle is $PA = 152° \pm 7°$. The cluster has a systemic velocity of 293.3 $\pm$ 1.0 km s^{-1}. Radial velocities for 35 red giants did not show any obvious trends, such as decreasing velocity dispersion with increasing projected radius, or, if rotation were present, a sinusoidal functional dependence on position angle. Fischer et al. suspected that the data were too sparse and the uncertainties too large relative to the effects looked for. They determined the central relaxation time to be 0.3–0.8 Gyr and the half-mass relaxation time to be 6–16 Gyr. The age of NGC 1978 was estimated to $\sim$ 2 Gyr; consequently, large portions of the cluster have not had sufficient time to relax. The total cluster luminosity is $L_V = 3.0 - -3.5 \pm 0.2 \times 10^5$ L$_\odot$, where the range results from different model extrapolations of the brightness profile. The total mass of the cluster is of the order of 1–2 $\times$ 10^5 M$_\odot$, the global M/L range over 0.3–1.5 for a sample of mass function slopes. There are no significant differences between the M/Ls derived with oblate spheroid models and those derived using spherical models. Rotating models cannot be completely ruled out. There is no morphological evidence for a merger.

4.6 Binary clusters

The Magellanic Clouds are rich in double clusters. As the probability of the tidal capture of one cluster by another is very small, clusters in a physical pair must

4.6 Binary clusters

have a common origin. If clusters originate in Giant Molecular Clouds (GMCs) in single bursts of star formation, they are parts of large complexes and, whether they are physical pairs or not, they should have similar ages and more or less the same metallicity. The proof for true binary status has to be found elsewhere.

The existence of binary clusters has important consequences for the theory on the formation and evolution of clusters. It has, for example, been suggested that the structure of the blue globular clusters can be explained on the basis of mergers (Bhatia and MacGillivray 1988).

The evolution of a pair or a complex of clusters will depend upon their initial dispersion velocities: if they are small enough a cluster pair may become gravitationally bound. A number of factors can destroy such a binary cluster within < 100 Myr (Bhatia 1990, 1992). Close pairs will merge while wide pairs will be disrupted, mainly by the tidal field of the LMC; passing GMCs may play a minor role. Binary clusters must, consequently, be young, have a common origin and not too large age differences (best seen as colour differences).

The colour differences between the components in nine pairs, with separation ≤ 18 pc, was measured to $\Delta(B - V) = 0.14 \pm 0.07$. Monte Carlo simulations for 147 clusters gave $\Delta(B - V) = 0.35 \pm 0.01$ for nine randomly picked pairs (Bhatia 1992). The observed nine pairs may thus not be a result of random superposition of single clusters. Age estimates were possible for five of them, three were younger than 100 Myr and two younger than 600 Myr. The age differences between the clusters in each pair were between a few Myr to 100 Myr, suggesting that they have a common origin.

Calculated projection effects for 69 pairs of clusters in the LMC with a centre-to-centre separation of less than $\sim 1'.3$ (~ 19 pc) could account for only 31 of the pairs (Bhatia and Hatzidimitriou 1988, Bhatia et al. 1991). A considerable fraction of the remaining objects could be binary clusters. Support for this assumption was found in the fact that the ages available for some of the clusters were less than the computed times of merger or disruption of the binary cluster system.

Kontizas et al. (1989a, b, 1993a) have studied the stellar content of seven young and eight old pairs in the LMC. From the spectral types of individual cluster stars and the radial distribution of the star number density they concluded that both clusters in most pairs have a stellar population of the same age and appear physically bound. Seven of the brightest and most populous double clusters in the LMC have ages between 6 and 80 Myr. For three of these pairs, NGC 2006/SL538, NGC 2011a/b, and NGC 2042a/b, integrated spectra of their cores and the stellar content of the outer regions showed a common origin and a young age, ≤ 20 Myr, for each member (Table 4.11). Dynamically, they were found to be gravitationally interacting. They have not had the time to merge, nor to be destroyed by dynamical friction.

The eight old pairs had a population older than 0.6 Gyr. It could not be decided if these pairs are still gravitationally bound and if binary clusters can survive so long as pairs in a lower density galaxy like the LMC. Bhatia et al. (1987) have suggested that the upper age for binary clusters is ~ 1 Gyr.

Eleven close pairs, two trios, and one compact grouping of clusters have been identified in the LMC Bar region by Bica et al. (1992). Among the pairs, NGC 1938/NGC 1939, SWB types IVA and VII, differ by at least 5 Gyr in age. Nevertheless, distortions indicate that they are interacting. Four other pairs are coeval, having the

Table 4.11. *Sample of binary clusters in the LMC*

clusters	r_t arcmin	Ages Myr	Masses 10^5 M$_\odot$	δ arcmin	Ref.
NGC 1850/a		90±30/6±5	0.6±0.2	0.5	1
NGC 1850/a		50±10/4.3±0.9			2
NGC 1850/a		50–70/8–10			3
NGC 2006/SL 538	2.6/3.3		5.0/2.5	0.89	4
NGC 2011a/b	3.1/2.0		5.0/1.2	0.47	4
NGC 2042a/b	2.9/3.3		17.0/24.0	0.95	4

In col. 4, δ = separation between the clusters.
The mean extinction in NGC 1850 is $E_{B-V} = 0.18 \pm 0.02$ with a variation over the cluster of ± 0.04 mag according to (2), $E_{B-V} = 0.17$ (1), and $E_{B-V} = 0.18 \pm 0.06$ mag (3).
For the three components of NGC 1850, NGC 1850, NGC 1850A, and H 88-159, the integrated magnitudes and colours are (Bica *et al.* 1992):
NGC 1850: $V = 9.57$, $B - V = 0.15$, $U - B = -0.27$, SWB type II;
NGC 1850A: $V = 11.23$, $B - V = -0.14$, $U - B = -0.89$, SWB type 0;
H88-159: $V = 11.84$, $B - V = 0.14$, $U - B = -0.41$, SWB type I.
References: 1, Fischer *et al.* 1993; 2, Gilmozzi *et al.* 1994; 3, Vallenari *et al.* 1994a; 4, Kontizas *et al.* 1993a.

same SWB types in each pair. Ten clusters form a coeval group in the eastern part of the Bar; its centre is the pair NGC 2058/NGC 2059, SBW types III.

The young cluster NGC 1850, in the NW part of the Bar of the LMC, is embedded in the nebula N103 and is responsible for its ionization. It is located in a region rich in star clusters and small HII regions. Robertson (1974) noted that the CMD for the cluster suggested the presence of two clusters. The smaller of the two (NGC 1850A), 30″ to the west of the main body, contains many bright, blue stars. Bhatia and MacGillivray (1987) confirmed the existence of two clusters and noted many signs of interaction: the bigger cluster is very irregular and has a tail in the NE direction. The irregularity indicates that the cluster is not relaxed. The tail is interpreted as a sign of interaction. However, the published photograph does not give an irrefutable proof of its reality. The part closest to NGC 1850 is the cluster H 88-159, superimposed on the former (Hodge 1988c). The remainder of the tail, if any, would need to be established more accurately. The age of NGC 1850 is estimated as 30 Myr, and the merger of NGC1850 and NGC 1850A should occur within $\approx$ 1 Myr from now. The colour difference between NGC 1850 and NGC1850A indicates an age difference of $\sim$ 50 Myr.

Observations, in UV and optical passbands, of NGC 1850 with the WFPC2 camera on the *HST* reveal three distinct populations (Gilmozzi *et al.* 1994):
(i) The bulk of the stars constitute the main cluster NGC 1850* with a globular-like appearance. The turn-off point occurs at $M_V \simeq -1.5$, $(B-V)_0 \simeq -0.16$, indicating that the most massive unevolved star has a mass of 6 M$_\odot$ within $\pm 7\%$.
(ii) The very young cluster NGC 1850A with about 30 stars with $T_{eff} \geq 20\,000$ K, half of which have a luminosity $\geq 3 \times 10^4$ L$_\odot$. The total Lyman-continuum photon

---

* Gilmozzi *et al.* call the main cluster NGC 1850A and the smaller one NGC 1850B. Here the designation used by most authors is followed.

4.7 Individual clusters

flux is $\sim 4 \times 10^{49}$ photons s^{-1}, accounting for the ionization of the nebula N 103A. The B versus $(B-I)$ diagram shows the presence of a pre-main-sequence (PMS) population: stars with masses up to 3 M$_\odot$ are still in the phase of approaching the main sequence. The fairly uniform spatial distribution of the PMS stars as well as the existence of bright OB stars up to 1' from the centre of NGC 1850A is interpreted as showing that this cluster is looser than the main cluster and has an extent at least comparable to that of NGC 1850. However, the clusters are part of the OB2 star-rich region A1 (Martin *et al.* 1976), so that the bright OB stars may be 'field stars'.

(iii) The third population seen in the CMDs is the LMC field population, represented by red-giant stars; the spread in colour and magnitude suggests an age spread from 0.5 to ≥ 5 Gyr.

The two clusters NGC 1850 and 1850A have very different IMF slopes. The main cluster has a flat slope, $f(m) \propto m^{-1.4\pm0.2}$ and the younger one a much steeper slope, $f(m) \propto m^{-2.6\pm0.1}$.

These results agree in all essentials with those derived by Fischer *et al.* (1993) except that the latter do not discuss any PMS stars. Their low cut-off is at $V \sim 19$, the upper limit for the PMS stars that Gilmozzi *et al.* (1994) suggest exist.

Fischer *et al.* used echelle spectra of 36 stars to study the cluster dynamics. A rotational signal in the radial velocities was detected at the 93% confidence level, implying a rotation axis at a position angle of $100° \pm 40°$.

The LMC cluster NGC 2214, at $6^h 13^m.2$, $-68°14'.6$, is considered to be a young (40 Myr) binary star cluster in an advanced stage of merger (Bhatia and MacGillivray 1988). The cluster has an almost spherical halo and a very elliptical core – the ellipticities are 0.15 and 0.5, respectively. The core is dumb-bell shaped, characteristic of a double structure. The western component is compact, the eastern more amorphous as judged from isophotal maps. Star counts confirm the duplicity. The merger of the two coeval components is estimated to be complete within 10 Myr. The CMD of the cluster shows according to Sagar *et al.* (1991a) two well-defined supergiant branches, separated by ~ 2 mag in luminosity, corresponding to ages of 60 and 170 Myr. The older population is more concentrated towards the compact centre, the younger one has a more extended distribution (this is contrary to the situation in NGC 2100, where the younger population is in the centre (Westerlund 1961a)). However, the second supergiant branch seen by Sagar *et al.* may be due to photometric errors (Bhatia and Piotto 1994). If the cluster is a binary cluster, the two components must be of almost equal age. There is, however, a significant variation in stellar density over the cluster. The core shows the two components and, in the outer parts, the west shows an excess as compared with the eastern part. The CMDs rule out a different reddening over the cluster, and variation in the field background density is likewise ruled out. The effects noted are most likely due to gravitational interaction, a merger in progress.

4.7 Individual clusters

Several of the rich clusters in the Clouds have been extensively studied, in particular for testing stellar evolutionary models. Clusters containing a large number of evolved stars have then been looked at in particular. Among those are NGC 1866 in the LMC and NGC 330 in the SMC. Others, like NGC 1850, have attracted attention because of their complicated structure (see Sect. 4.6). Several of the clusters at both ends of the age sequence, the true globulars and the extremely young ones, are especially important for comparisons with our Galaxy and for studies of the evolution of extremely massive stars, respectively. Some of them have been discussed above (Sects. 4.4.1 and 4.4.3).

The young populous cluster NGC 1866 is renowned for its richness, the colour distribution among its brightest stars, and the large number of classical Cepheid variables (see Sect. 4.4.3.1).

The CMD and the MS luminosity function (LF) of NGC 1866 (see Table 4.6) have been studied by Chiosi et al. (1989a,b) 'with the aim of discriminating among possible scenarios for the evolution of intermediate (and low) mass stars'. They use CCD photometry in B and V of 1517 stars in the central region of the cluster and of 640 stars in a nearby field. Care was taken to define the region in NGC 1866 suffering the least from photometric crowding and incompleteness. This region was then corrected for incompleteness as well as for contamination by field stars.

Synthetic CMDs, LFs, integrated magnitudes and colours in the UBVRI bands were constructed as functions of cluster age, chemical composition, initial mass function, and total number of stars. The algorithm used allowed for age dispersion, fluctuations of stochastic nature in the IMF, and simulations of observational errors in magnitudes and colours. All evolutionary phases, from the main sequence to the latest ones, were included. The evolutionary schemes tested were the classical models based on standard assumptions for the extension of convective cores (Becker 1981) and the models incorporating convective overshoot (see Bressan et al. 1986). The latter possess larger convective cores throughout their evolutionary history. The difference between the two schemes is not marginal. It reflects onto different paths in the CMD, different lifetime ratios among various evolutionary phases (core-He-burning, and AGB), hence relative numbers of stars in different areas of the CMD, and different LFs for the MS stars. The comparison between theory and observations allows the determination of the age and composition and the estimation of the distance modulus of NGC 1866 in the two alternatives for the stellar evolution background.

The simultaneous fit of most of the constraints indicates that the appropriate chemical composition is $Y = 0.28$ and $Z = 0.02$ (or slightly less), the true distance modulus is $(m - M)_0 = 18.6$, and the age is 0.2 Gyr for models with overshoot or 0.07 Gyr for classical models. For the latter the fit of the red-giant stars is rather poor; they are fainter than the mean luminosity of the loop and bluer than the theoretical Hayashi limit.

New theoretical models, calculated by Lattanzio et al. (1991) and taking into account the presence of semiconvection in the He-core burning phase, have been compared with a set calculated with convective overshoot where the metallicity, the He content and the other parameters are homogeneous. Both sets have been compared with the CMD and LF of NGC 1866. Neither set of models is capable of simultaneously fitting the observed luminosity of the main-sequence turn-off as well as that of the least luminous evolved stars. Compatibility between the observations and either set of models is achieved if the presence of unresolved binaries is allowed for. Only the convective overshoot hypothesis is in agreement with the observed ratio of main-sequence to evolved stars.

The populous blue globular cluster NGC 2100 contains more M supergiants (~ 20) than any other cluster in the LMC, the most luminous of which are in the inner parts

4.7 Individual clusters

of the cluster (Westerlund 1961a). An ultraviolet study of its most luminous B stars (Böhm-Vitense et al. 1985) shows that the most massive stars follow an isochrone for 6 Myr and have a mass of about 30 $M_\odot$. They are evolved supergiant stars but show no mass loss. At least for these LMC stars mass loss is less than for galactic stars with similar effective temperatures and luminosities.

The blue, populous clusters NGC 330, SWB type I, and NGC 458, SWB type III, (Table 4.9) in the SMC contain a considerable number of evolved stars, more than in comparable clusters in our Galaxy. The brightest stars have $V \sim 12$ in the former and 15 in the latter; their red supergiants have $B - V$ between 1.4 and 1.6 and ~ 0.8, respectively (Arp 1959a,b). The post-MS stars are in the mass range 4–15 $M_\odot$. Arp suggested that the peculiar CMD of NGC 330 indicated abundance anomalies. Spectroscopic work by Feast (1979) made a metal deficiency likely. Feast (1972) had also shown that NGC 330 contained an unusually high frequency of Be stars. Bessell and Wood (1993) suggested that Be stars have rotationally induced mass loss at the equator, leading to the formation of the extended disk in which the Be phenomenon is created. During the H-burning MS evolution of the B star, the star initially spins up, the rotational velocity reaches a maximum and then decreases before the core collapse. It follows that as the stars evolve from the ZAMS the number of Be stars will increase but, beyond maximum rotation, the Be phenomenon will vanish. This is seen in all the clusters studied. The proportion of Be stars reaches a maximum at $M_V \sim -4$, about one magnitude below the MS termination point. The very high proportion of Be stars in NGC 330 may reflect the higher rotation of its pre-cluster material.

Radial velocities for 17 early- and 8 late-type stars in NGC 330 (Feast and Black 1980) gave a velocity dispersion of 1.8 km s^{-1} and an estimated mass of the cluster of $M \leq 4 \times 10^4$ $M_\odot$ and $M/L < 0.1$. There appear to be no binaries among the most luminous stars in the cluster.

An extensive photometric (UBV, DDO, JHK) and spectroscopic investigation of NGC 330 and its surroundings has been carried out by Carney et al. (1985). The field population consists of a mixture of a very old population (galactic globular cluster type) and a rather young one (but older than the cluster). The reddening was found to be small, $E_{B-V} = 0.03$. The measurements indicated that the cluster is metal-poor, possibly as low as $[m/H] = -1.8$. Using synthetic CMDs they estimated its age as 12 Myr; the younger part of the field population is ~ 18 Myr old.

Stothers and Chin (1992a,b) have attempted to reproduce the two most critical observations for NGC 330 and NGC 458: the maximum T_{eff} of the hot stars and the difference between the average M_{bol} of the hot and the cool evolved stars by using theoretical evolutionary sequences of models for stars with low metallicities, using Cox–Stewart as well as Rogers–Iglesias opacities. (For both clusters $(m - M)_0 = 18.8$ and $E_{B-V} = 0.03$ were assumed.) They found that only those sequences with little or no convective core overshooting could reproduce the observations. Metallicity determinations for stars in NGC 330 refer mainly to the red supergiants; derived values vary between $[m/H] = -1.8$ and -0.8 (see Chap. 11).

There are 9 blue and 15 red supergiants in NGC 330. The difference between the average luminosities of the blue and the red supergiants is $\Delta \langle M_{bol} \rangle = -0.60 \pm 0.18$ mag, a purely evolutionary effect. The difference between the brightest stars still on the MS and the brightest evolved supergiants is $\Delta V = -1.7$ mag; their masses ought to be identical. The theoretically predicted gap for stars of ~ 12 $M_\odot$ is -2.5 mag. A better match with the observations could be achieved if the metal abundance were solar.

The observed luminosity spread among the supergiants in NGC 330 was interpreted by Stothers and Chin as due to a dispersion of masses, 9–14 $M_\odot$. For the NGC 458 giants it is 4–5 $M_\odot$. The ages of the red stars in the two clusters are then 15–30 Myr and 80–130 Myr, respectively, with the higher values being close to the likely cluster ages.

For NGC 458 the same low metallicty as for NGC 330 is accepted due to lack of data. In this cluster the hot (10) and cool (22) giants differ in mean luminosity by $\Delta \langle M_{bol} \rangle = -0.48 \pm 0.19$ mag, and the difference between the brightest stars still on the MS and the brightest evolved giants is $\Delta V = -1.7$ mag. The theoretically predicted gap for a star of ~ 5 $M_\odot$ is -1.8 mag for Roger–Iglesias opacities, -2.3 mag for Cox–Stewart opacities.

Whereas the data for NGC 458 may fit the theoretical model relatively well, the positions of the red supergiants in the HR diagram for NGC 330 have so far eluded simple theoretical interpretations. Recently Chiosi et al. (1995) have analysed the CMD and the LF of NGC 330 on the basis of the BV CCD photometry by Vallenari et al. (1995) and the T_{eff} and the M_{bol} for a few supergiant stars from Caloi et al. (1993) and Stothers and Chin (1992a,b). The colour excess is taken as $E_{B-V} = 0.06$ and the distance modulus as $(m-M)_0 = 18.85$. The metallicity is found to be slightly lower than $Z = 0.008$ and the age a few times 10^7 yr. The CMD and the HRD indicate, however, a significant spread in age. Using semiconvective models the spread is 1 to 2.5×10^7 yr; for full and diffuse overshoot models it is 1 to 4.8×10^7 yr. The study of the integrated LF suggests that the slope of the IMF is about $x = 2.35$ for semiconvective and full overshoot models, and $x = 2.00$ for those with diffuse overshoot. No clear indication is obtained about the mixing scheme.

5

The youngest field population

Luminous stars in the Magellanic Clouds are very suitable for the study of evolutionary processes because their distances and, hence, their luminosities can be determined with a reasonably good precision, permitting a direct comparison with theoretical tracks in the HR diagram. The evolutionary paths of these massive stars depend critically upon the mass-loss rates at different stages of the stellar development. The mass lost may determine whether the stars evolve back to the blue or end their lives as red supergiants (e.g. Chiosi and Maeder 1986, Wood and Faulkner 1987). There are at least some stars in the LMC and the SMC showing that a post-red supergiant evolution occurs, resulting in anomalous He-burning A-type supergiants (Humphreys *et al.* 1991). Other stars, more massive, may return to the blue stage as WR stars (Sect. 5.2.3 or, like SN 1987A, as B supergiants (Sect. 9.4).

The youngest population in the Magellanic Clouds (age $\leq$ 100 Myr) contains the most recently formed stars and the remaining or returned gas and dust. Best studied among the stars are the OB stars and a minor number of supergiants and hypergiants of types B–G, suitable for abundance analyses. Massive stars are believed to be born primarily in clusters and in associations. There is, however, a fair number of massive stars so far from any cluster or OB association that they can hardly have moved from those centres of star formation to their present locations during their short lifetimes. They may then have formed in less extensive events than those producing the large associations, but, as they are generally members of superassociations, they are still links in the stochastic self-propagating star formation (SSPSF) process.

The enormous energy released by the massive stars through stellar winds and supernovae creates a large pressure on the surrounding interstellar medium (ISM). The effects are maximized if/when several stellar associations act together, and supergiant structures, on kpc-scales, may be formed. The dense, neutral ISM is displaced by a low-density gas, producing a superbubble surrounded by a dense, ringlike supershell. The shock-compressed shells may induce further star formation, eventually releasing more energy. The process will continue until the superbubble breaks through to a low-density region, blowing a hole in the galactic plane and throwing material into a halo.

5.1   O stars and B–G supergiants
The classical study by Feast *et al.* (1960) lists 155 objects, 50 in the SMC and 105 in the LMC, and notes that the early A-type stars are the visually brightest. Most of the stars are 'normal' supergiants. Others are still more luminous hypergiants, some

Table 5.1. *O3 stars in the Magellanic Clouds*

	O3V Def.	O3V Poss.	O3III Def.	O3III Poss.	O3If*/WN-A	Total
LMC						
30 Doradus		26	3		5	34
30 Dor periphery		3	2		1	6
N11	1	2	2			5
N180			1	1		2
Single	1	3	3		1	8
SMC						
Single			2	1	1	4

The 30 Doradus periphery includes 30Dor B, C and N159A.

with very strong emission lines. The O stars are relatively rare in the list. Some of the most luminous ones have been shown to be dense clusters of stars (see Sect. 5.7 and Sect. 5.8). Particularly well known among the latter is the central object, R136, in the 30 Doradus cluster (Chap. 10), which for a while was believed to be a supermassive star by many astronomers, and has been resolved into a grouping of very massive hot stars. 28 components are now known within a radius of $4''.6$ and there are at least 56 others as luminous as the R136 components ($-8.1 < M_V < -5.2$). Many of them have been suspected of being very massive, over 100 $M_\odot$, with a maximum of 250 $M_\odot$ for the brightest unresolved component, a1, in R136 (Walborn 1986). However, this star has, like many others in the LMC, been shown to be a multiple object. It now appears unlikely that supermassive stars exist in the Clouds.

An important group of young luminous stars are the O3 stars, the hottest and most massive components of an extreme Population I. The number known in the Clouds is given in Table 5.1 (Walborn 1993).

Several important catalogues of luminous stars in the Clouds exist. References to a number of those compiled between 1960 and 1975 may be found in Martin *et al.*(1976), who reclassified 1471 OB stars and B–G supergiants within a radius of $4°$ from the LMC centre and studied the distribution of a total of 1569 luminous stars. They found that the extreme LMC Population I was principally found in eight well-defined large regions (see Chap. 6).

The distribution of Cepheids and supergiants in the LMC as a function of their ages has been investigated by Isserstedt (1984). He used the data compiled by Gaposchkin (1970) for the Cepheids and his own observations and those compiled by Rosseau *et al.*(1978) for the supergiants. Rapid changes in the distributions of these stars over the last 90 Myr were visible when the material was divided into groups of 10 Myr range. This result spoke in favour of a SSPSF process. As, however, strong fluctuations with time and positions in the LMC were seen, a single burst of star formation over the whole LMC could not be traced. This does not, however, exclude the fact that the superassociations are of about the same age. The star formation processes may have progressed differently in them.

Massey *et al.*(1995) have updated existing catalogues of luminous LMC (Fitzpatrick and Garmany 1990) and SMC (Garmany *et al.* 1987) members with new spectral types, and have derived HR diagrams of 1584 LMC and 512 SMC stars. The true field stars, i.e. stars located more than $2'$ (30 pc) from the boundary of an OB association, or the extreme field stars, more than $20'$ from a border, show the same distribution in the HR diagram as the association stars. Field stars as massive as 85 $M_\odot$ exist. Consequently, the field produces stars as massive as those produced in the associations. They are, however, proportionally less common in the field than in the large OB complexes. The slope, $\Gamma = -x$ (used in Chap. 4) of the IMF of the field stars is very steep, $\Gamma = -4.1 \pm 0.2$ (LMC) and $\Gamma = -3.7 \pm 0.5$ (SMC) compared to $\Gamma = -1.3 \pm 0.3$ for the Magellanic Cloud associations.* A similar difference exists in the Galaxy, with $\Gamma = -3.4 \pm 1.3$ and -1.5 ± 0.2 for the galactic field population and the total galactic sample, respectively. There are no differences in the IMFs of associations in the LMC, the SMC and the Galaxy.

Available low-metallicity evolutionary tracks reproduce well the distribution of the higher mass stars in both Clouds and also the width of the main sequence, showing that the 'main-sequence widening' problem† has disappeared with better data and models. A discrepancy between the theoretical ZAMS and the blue edge of the lower-luminosity main sequence appears to be due to the theoretical low-metallicity model, which places the ZAMS too far to the blue.

The rarity of the youngest (< 2 Myr) stars, noted also by Isserstedt (1984), remains. The main-sequence objects of this age are too faint for the present surveys. Deeper surveys will probably eliminate this deficit. The same applies to the lack of older stars near the main sequence.

5.2 The most luminous stars

Many of the most luminous stars are variables and many of them show strong emission lines. Several natural groupings can be identified among them (Stahl 1989): B[e] supergiants, Ofpe/WN9 stars, P Cyg type stars, and also 'Normal' stars. The B[e] supergiants appear to form a separate group. The S Dor variables belong spectroscopically to either of the three other groups, depending upon the phase. P Cyg line profiles can be found in stars scattered over the whole HR diagram and the group is thus ill defined. Here, 'P Cyg' refers to B and A supergiants with strong P Cyg profiles in many lines.

5.2.1 B[e] supergiants

The B[e] supergiants show a near-infrared excess which distinguishes them from other luminous emission-line stars in the Clouds. In a $(J-H)/(H-K)$ diagram they form a group well separated from the others (Gummersbach et al. 1995). The wavelength dependence ascribes the excess to hot circumstellar dust, about $1000°$ K. Specific conditions are required with regard to temperature and density for the formation of the dust grains. Their visual spectra show the Balmer lines in extremely strong emission as well as narrow emission lines of low-excitation species like [FeII], FeII, [OI], He I and Na I. They have a far-ultraviolet spectrum dominated by broad absorption lines, characteristic for stellar winds (the blue edge has a velocity of $\sim$

---

* Massey et al. discuss the meaning of an IMF for field stars, meaning 'stars that were not born together'. The period of age covered by the present field stars is about 10 Myr, and they are most likely all members of superassociations and thus parts of reasonably well defined units.
† The early models did not reach far enough towards the red during core-H-burning.

82 The youngest field population

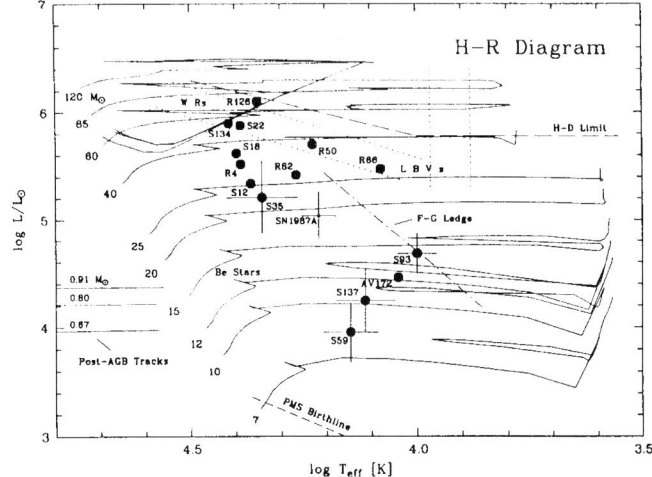

Fig. 5.1. HR diagram with all known B[e] stars in the Magellanic Clouds (Gummersbach et al. 1995). H-D = upper luminosity limit; F-G = the density contrast ledge for luminous LMC stars (see Fitzpatrick and Garmany 1990). LBV and dotted lines identify the luminous-blue-variables strip. The evolutionary tracks are for Z = 0.008.

-2000 km s^{-1}). Line forming regions with very different physical conditions must exist around these stars. CO first-overtone emission has been detected in the near-IR (2.2 μm) spectra of several B[e] stars (McGregor et al. 1988). This emission requires a cool environment of high density shielded from the UV-radiation of the star. The density in the CO producing region was estimated to be $N_H \geq 10^{10}$cm^{-3} at $T_{ex} \approx$ 3000–5000 K.

The circumstellar medium of the B[e] supergiants must consist of at least two components: a high-velocity stellar wind and a dense cool region with low velocity, possibly a disk. In the latter, molecules and dust have formed and TiO emission at 6159 Å has been observed (see Zickgraf 1993). Significant polarization has been detected in most of the objects (Magalhaes 1993), showing that the envelopes are non-spherically symmetric. The observed polarizations do not correlate with the near-IR excess, due to free–free radiation from the envelopes, but significantly with the far-IR excess due to dust in the outer parts of the envelopes. The evolutionary state of the B[e] stars – 15 are known in the Clouds, 11 in the LMC and 4 in the SMC – is not known. If it is assumed that all massive O stars have to pass a B[e] phase during their post-MS evolution, a lower limit for the duration of the phase is 10^5 yr, and the total mass loss during this phase may amount to a few M$_\odot$ (Zickgraf 1993).

Gummersbach et al. (1995) have shown that the B[e] phenomenon is not restricted to supergiants but extends over a large region of the HR diagram, down to luminosities of about $log L = 4 \log$ L$_\odot$. i.e. to stellar masses of about 10 M$_\odot$ (Fig. 5.1).

5.2.2 The Of/WN stars

The Ofpe/WN9 stars have strong emission in NIII and HeII and also in NII and HeI, and are morphologically between the Of and WNL stars as well as between

5.2 The most luminous stars

the Of and the P Cyg stars. The WN classification suggests enhanced N abundance. Overabundance of He has also been suggested (McGregor et al. 1988) because the HeI line at 2.058 μm is stronger than Brackett-γ. The group has ten members in the LMC, in three subgroups (Walborn 1989), none in the SMC, and no close galactic counterpart.

5.2.3 The luminous variables

The luminous blue variables (LBVs) are hot unstable stars and are the visually most luminous stars in the Universe. They are characterized by irregular photometric variations of more than one magnitude (in visual light) on time-scales of years to decades. In the visual minimum phase they are probably early B supergiants, and in outburst mostly A-type P Cyg stars. They are surrounded by cool, slowly expanding envelopes. It is believed that the most massive stars, above ≈ 40 M$_\odot$, become LBVs for a short time ($\approx 10^4$ yr), suffering episodic mass loss. They then evolve to the left in the HR diagram to become WR stars and spend the majority of their He-burning lives ($\approx 10^5$ yr) as such (Maeder and Conti 1994). The irregular variations are a consequence of the variable mass-loss rate.

S Dor, or Hubble–Sandage, variables belong to the LBVs. They occupy at quiescence an inclined instability strip in the HR diagram in the range $-9 \geq M_{bol} \geq -11$, $14\,000 \leq T_{eff} \leq 35\,000$ K, with the most luminous being the hottest (Fig. 5.2). During outbursts they expel dense envelopes which at maximum have velocities around 100–200 km s^{-1}. The stars move at virtually constant M_{bol} in the HR diagram to occupy at maximum a vertical strip at $T_{eff} \approx 8\,000$ K. This indicates the existence of an amplitude–luminosity relation. Wolf (1989) has derived

$$M_{B(max)} = -(7.72 \pm 0.25) - (1.33 \pm 0.15)\Delta B,$$

where B is the blue magnitude and ΔB is the amplitude (max–min).

Best studied of the S Dor variables are S Doradus, R71 ($\equiv$ HDE 269006), and R127. R71 is the least and R127 the most luminous of the three (see Fig. 5.2). R71 is a B2.5Iep star during minimum. The minimum spectrum of S Doradus is not well known. Its last deep minimum was in the 1960's.

R127 ($\equiv$ HDE 269858) is the hottest of the S Dor type stars known. It has been classified OIafpe but during outbursts it has shown a B-type and later an A-type P Cyg spectrum. In 1989 it was the visually brightest star in the LMC (Stahl 1989). The expansion velocity and the mass loss rate of the stellar wind were then ~ 100 km s^{-1} and 5×10^{-5} M$_\odot$ yr^{-1}, respectively. Its spectrum showed strong, extended nebular lines suggesting the presence of an expanding circumstellar shell. Coronographic images in Hα + [NII] displayed a nebula with the dimensions 1.9×2.2 pc and a bipolar morphology, similar to the well-known nebula around AG Car (Clampin et al. 1993). The ionized gas mass is estimated as ~ 3.1 M$_\odot$.

Inverse P Cyg absorption has been observed in the spectra of S Doradus and R127. In S Doradus, and probably also in R127, it appeared during the fading of the star, in S Doradus after the brightest phase ever observed for it (Stahl 1993). Combined with the fact that the inverse is only seen in the faint lines, Stahl suggested that it is caused by the inner parts of the wind falling back after the outburst, whereas the outer parts, where the stronger lines are formed, are outside the escape velocity and

84 *The youngest field population*

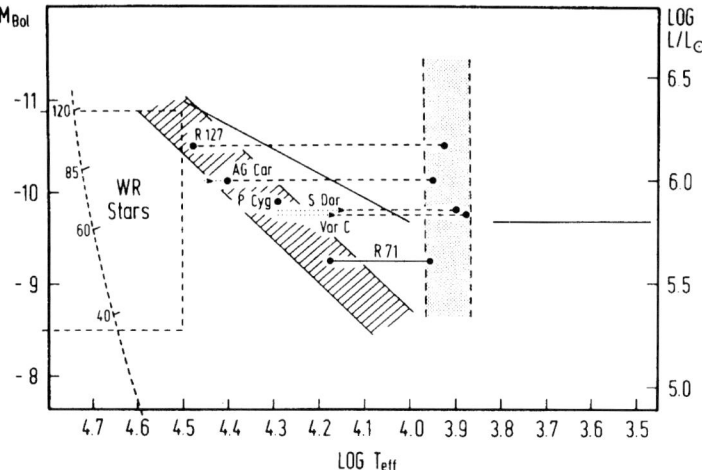

Fig. 5.2. HR diagram showing the positions of the best studied S Dor variables in the LMC, the Galaxy, and M 33 (var C) and of P Cyg. The S Dor instability strip is the inclined hatched area. The vertical shaded area is the opaque wind limit, the area reach by the stars at maximum. The ZAMS is the broken line going through the WR region (Wolf 1989).

continue to show the expansion of the wind. A problem of this model is that the infall continued for over two years in the case of S Doradus.

R110 ($\equiv$ HDE 269662 $\equiv$ S116) is the coolest LBV in the Clouds. It was classified B9:Ieq in 1960 by Feast *et al.*, in 1985 it showed an A-type spectrum and in 1989 an early F-type absorption spectrum (Stahl 1989). Later, it was observed in a cool state with a spectral type around F8–G0 (Stahl 1993). Contrary to all other S Dor variables observed so far it shows no signs of enhanced mass loss during the outburst, nor of any other spectral peculiarities.

Sk−69°142a ($\equiv$ HDE 269582), a suspected S Dor variable classified Ofpe/WN9, and S111 ($\equiv$ HDE 269599), another probable S Dor variable, are in Shapley II. This constellation, about 40′ west of 30 Doradus, is rich in red supergiants and contains several P Cyg stars and some peculiar early-type stars. This grouping may indicate a connection between the red supergiants and the S Dor variables (Lortet and Testor 1988). Also R127 belongs to a group of several early type supergiants in a star-rich but gas-poor region south of 30 Doradus.

A recent addition to the LBVs is the star R143, the first of this kind identified in the 30 Doradus cluster and the sixth to be identified in the LMC (Parker *et al.* 1993). During the past 40 years it has changed from an F6 to an F8 supergiant, then blueward, possibly as far as to O9.5, and is now moving back redward, currently being a late B supergiant.

R143 is relatively heavily reddened, $E_{B-V} \approx 0.65$. At least four curved nebulous filaments extend from the star and end either on knots of nebulosity or faint stars (Feast 1961). Parker *et al.* interpret the knots as parts of an incomplete shell with a radius of about 3.5 pc. R143 appears to have evolved at a constant luminosity of $M_{bol} = -10$ mag. Its original mass must have been around 60 $M_\odot$. Its current mass is estimated as between 56 and 46 $M_\odot$; the lower value is the mass at which a 60 $M_\odot$

star becomes a WR star. The star must be close to 4 Myr old, i.e. it represents a star formation event earlier than that resulting in the R136 cluster (see Chap. 10).

The SMC star R40 was classified B8Ie in 1960, with $V = 10.73$. In 1992 it had $V = 10.3$ and showed microvariations of a typical time scale of 120^d and an amplitude of 0.1 mag (Szeifert et al. 1993). Its mass is estimated as 13 $M_\odot$. During minimum it is in the inclined instability strip among the lower-luminosity LBVs ($M_{bol} = -9.1$), comparable to R110 in the LMC. It is similar to R110 also with regard to its low mass loss rate, a few times 10^{-6}–10^{-5} yr^{-1}. It appears to be in the post red-supergiant evolutionary phase. Because of the lower metallicity of the SMC, R40 may be a key object for investigating the metallicity dependence of the LBV phenomenon.

5.3 Wolf–Rayet stars

The number of WR stars known in the Clouds is now believed to be essentially complete; there are about 110 in the LMC and 9 in the SMC. The distribution in subtypes differs from that in the Galaxy; the ratio of WN:WC is 1 in the Galaxy, 4 in the LMC and 8 in the SMC (Breysacher 1988, Morgan et al. 1991). Of interest in understanding the evolution of massive stars is the ratio of red supergiants/WR stars: it is about 1 in the solar neighbourhood, 9 in the LMC and 24 in the SMC (Azzopardi et al 1988). The latter number will increase to 47 if the Sanduleak (1989) list of late-type supergiants is used.

A comparison of the WR stars in the Clouds and in the Galaxy ought to clarify how the formation of WR stars depends upon metallicity. As just noted, the most massive WR stars are expected to have evolved from the O stars via the LBV phase. Less massive stars (10–15 $M_\odot$) are expected to evolve from the main-sequence to the red-supergiant (RSG) phase and then turn back and spend their last productive periods as blue supergiants before becoming supernovae of SN 1987A type.

The mass above which a star becomes a WR star instead of a RSG depends upon the mass-loss rate, which, in turn, is coupled to the initial metallicity of the star. Recent models suggest that a mass of 60–85 $M_\odot$ is needed for a low-metallicity galaxy like the SMC, whereas only 25–30 $M_\odot$ is needed in conditions such as those in our Galaxy (Maeder and Meynet 1994).

The relative numbers of WR to main-sequence stars should be that of the relative He- and H-burning lifetimes which, using modern theories, is expected to be 1/10. In order to obtain a complete material Massey et al. (1995) use only field stars. The numbers and the derived ratios are given in Table 5.2.

The mass at which the ratio 1/10 is reached is $\approx 30\ M_\odot$ in the LMC and probably higher, $\approx 50\ M_\odot$, in the SMC (with only one field WR star the result is very uncertain).

5.4 The red supergiants

The existence of very red luminous stars in the Magellanic Clouds has been known since the early 1950s. Shapley and Nail (1953) deduced their probable existence in the LMC from star counts and statistical allowance for foreground objects. Shapley (1955) identified 'red stars later than M0' in the neighbourhood of 30 Doradus. These stars were sometimes referred to as M-type supergiants though their spectral types were unknown. The first proof of the existence of M supergiants was obtained when Westerlund (1960) carried out an objective-prism survey of the LMC in the near

Table 5.2. *Ratio of field WR stars to field main sequence stars*

Mass	MS stars	ratio WR/MS
LMC(No. of WR Stars =38)		
$> 85\ M_\odot$	2	19.0
> 60	11	3.5
> 40	83	0.45
> 25	740	0.05
SMC(No. of WR stars = 1)		
> 85	1	1
> 60	3	0.3
> 40	20	0.05
> 25	210	0.005

Table 5.3. *B/R ratios in the SMC, the LMC, the outer Galaxy, the solar neighbourhood (SN), and the inner Galaxy*

| | M_{bol} | SMC | LMC | Outer Gal. | SN | Inner Gal. |
Z		.002	.006	.013	.02	.03
Field	≤ -7.5	4	10	14	28	48:
Assoc.	≤ -7.5	4	10	14	30	89:
Clusters	< -2.5	2.5	6.7	7.7		20

The data for the field and association stars are from Humphreys and McElroy (1984), those for clusters from Meylan and Maeder (1982).The limiting magnitude used for the clusters is M_V, not M_{bol}.

infrared and showed that luminous red stars with TiO bands existed and could be classified in the Case system. M supergiants were also identified in several luminous clusters (Westerlund 1961b).

About 800 late-type supergiants are catalogued in the LMC (Rebeirot *et al.* 1983), close to 400 in the SMC (Sanduleak 1989). A major difference between the two populations is that the spectral types of the latter population, K to early M, are systematically earlier than those of the former. This is a consequence of the lower metallicity in the SMC, also seen in the spectral types of the red giants (see Chap. 7).

The ratio of blue to red supergiants is a major characteristic of the luminous star population in galaxies. For a given luminosity range it increases steeply with metallicity. The ratios (B/R) for (O + B + A) supergiants to red supergiants are given in Table 5.3 for the SMC, the LMC and the Galaxy according to Langer and Maeder (1995).

The ratios in Table 5.3 may be compared with the results for the two young SMC clusters NGC 330 and NGC 346. The CMDs for NGC 330 (Carney *et al.* 1995) and NGC 346 (Massey *et al.* 1989b) give B/R = 0.6 and 2.5, respectively. It is to be noted that the blue supergiants in NGC 330 are all of type B and the M_{bol} of the red supergiants ≤ -5. The selection of objects is thus different from that behind

Table 5.3. NGC 346 is also more an association than a cluster and shows clear signs of sequential star formation (see Sect. 5.10).

There may be inhomogeneities in the fields and groupings of the SMC and the LMC that need to be taken into consideration in the analysis of the evolution of the youngest population in these galaxies. It is desirable to have more detailed information about the B/R ratio over the Clouds, in particular to see how it varies inside each superassociation and between the superassociations. This does not appear to be available yet. The distribution of the OB stars and the early-type supergiants shows clear age effects inside and between the superassociations in the LMC (Sect. 6.1), but similar information about the red supergiants is not available. Some years ago the B/R ratio across the LMC was examined by Cowley *et al.* (1979) using available catalogues. The ratio was found to vary in an east–westerly direction with the easternmost regions having B/R $\sim$ 1.3, a central portion $\sim$ 1.8 and the western $\sim$ 2.3. With the criteria derived from Table 5.3, this suggests that the young population in the western part of the LMC should be more metal-rich than that in the east. This result must be considered in the light of the east–west asymmetry seen (i) in the far-IR to 6.3 cm brightness ratio, which is lower in the eastern giant HII regions, and (ii) in the diffuse X-ray emission which shows a plasma temperature ten times higher in the east than in the west (Sect. 12.1.2). As the composition of the 'blue supergiant material' is different in the two parts, there may be no real contradiction between the B/R and FIR/6.3 cm ratios. Supergiants of type B7–A9 dominate among the supergiants in the north-west, where only stars later than B 2.5 are found. In the eastern superassociations (IV + V, Martin *et al.* 1976) 57% of the stars are of type OB and 20% of B7–G, whereas in the western ones (I + II) the percentages are 48 and 32, respectively.

5.5 Supergiant stars with circumstellar shells

Among Magellanic Cloud stars with thick dust shells, emitting much (most) of their energy in the infrared, are young as well as evolved stars. A few years ago late-type supergiants with dust shells did not appear to exist in the Clouds. However, in 1986, Elias *et al.* reported the detection of two such sources in the LMC, *IRAS* 0453-6825 and *IRAS* 0536-6949. Their energy distributions resembled those of the most luminous galactic OH/IR stars and their M_{bol} were estimated as < -9 mag. They did show little variability over a two-year period. In the first source a M7.5 star was identified with a radial velocity confirming its LMC membership. In the second source the dust shell was so dense that no star could be observed.

In the SMC thin dust shells were known to exist around AGB stars. The *IRAS* satellite detected a number of sources, and Whitelock *et al.* (1989) monitored five of them in the JHK and L for 800^d and obtained optical spectra of two of the three with optical counterparts. One appeared to be a carbon-rich, high-luminosity ($M_{bol} = -6.6$) object (see Table 5.4). It may be in an evolutionary stage between a red giant and a planetary nebula or it may be an interacting binary. Another object, with $M_{bol} = -6.0$, is a long-period ($\sim 800^d$), large-amplitude ($K > 1$) variable, probably near the top of the AGB. Only one of the stars, 00483-7347, may be a core-burning supergiant. It could, however, be at the upper luminosity limit for the AGB stars. 01432-7455, $\equiv$ R50, may be a B[e] supergiant (Zickgraf *et al.* 1986).

Table 5.4. *Data for giant and supergiant stars with dust shells*

Star no. IRAS or R	M_{bol} mag	ZAMS mass $M_\odot$	Remarks	Ref.
LMC				
04553-6825	−9.5	30	variable, 930 d, $\delta K \sim 0.3$ mag,	1, 2
05346-6949	−9.1	30	no variations detected	1, 2
R 71	−8.8	25	optical outburst in early 1970s	2
R 126	−10.5	70–80	dust condensed ~ 300 radii from star	2
R 150	−9.4	30	CO emission at 2.3 μm	2
SMC				
00350-7436	−6.6	> 3	not variable	3
00483-7347	−7.2	> 6	decline 0.5^m in 800^d	3
00554-7351	−6.0	∼2.5	variable, 517^d	3
01074-7140	−6.4		variable, 517^d, HV 12956	3
01432-7455	−8.2	25–30	R 50	3

References: 1, Elias *et al.* 1986; 2, Roche *et al.* 1993; 3, Whitelock *et al.* 1989. Zickgraf *et al.* (1986) give a higher mass and a higher M_{bol} for 01432-7455.

Spectra at 8–13 μm have been obtained by Roche *et al.* (1993) of the two dusty red supergiants detected by Elias *et al.*. Their thick dust shells have deep silicon absorption features; they are probably late-type M supergiants undergoing very heavy mass loss. The apparent depth of the silicate feature in the OH/IR star *IRAS* 04553-6825 appears to have varied over a five-year interval. Spectra were also secured of the peculiar B[e] hypergiant R126, the G supergiant R150, and the B[e] star R71. R126 displayed a strong circumstellar dust shell, but with a very muted silicate emission feature. R150 and R71 showed prominent silicate emission bands. In both objects there is evidence that the emission increases away from the star, suggesting that the bulk of the dust was produced at a previous evolutionary stage in which the mass-loss rate was much higher. These stars are evidently more evolved than the two M supergiants and on a blueward track with cooling dust shells. The silicate features indicate that the material lost from the stars is oxygen-rich, C/O < 1, in agreement with stellar evolution models.

The best-known case of a LMC star showing a previously ejected shell is SN 1987A (see Sect. 9.4). It illuminates a remnant envelope believed to have been ejected by the precursor star when it underwent a red supergiant phase about 10 000 yr before the explosion (see Fransson *et al.* 1989).

5.6 Supergiant stars in small HII regions

Most of the luminous early-type stars in the LMC are located outside their parent nebulae, indicating that their strong stellar winds have transformed the initial cocoons into cavities. It has been suggested that the lower metallicities in the Magellanic Clouds, as compared with the Galaxy, cause the O-type stars in the former to have appreciably lower terminal wind velocities (Kudritski *et al.* 1988). With the metal

5.6 Supergiant stars in small HII regions

abundances $Z_{LMC} = 0.28\ Z_{Gal}$, $Z_{SMC} = 0.1\ Z_{Gal}$ the values derived for an O5 V star are 2435, 2900 and 3350 km s^{-1} for the SMC, the LMC and the Galaxy, respectively. The corresponding mass losses are 0.72, 1.35, and 2.12 × 10^{-6} M$_\odot$ yr^{-1}. It is therefore of interest to study the influence of individual Cloud stars on the surrounding medium.

5.6.1 The central star in the bubble N120A

The supergiant Bl141 is associated with a small bubble, N120A, in the HII region N120 (see Sect. 8.3.3). The dynamical age of N120A is estimated as ~ 6 ×10^4 yr (Laval et al. 1992). The bubble expands with a velocity of 50 km s^{-1} into an ISM with a density of $n \approx 10$ cm^{-3}. The radius of the bubble is 4.5 pc. If the bubble is produced by a stellar wind, a power of $\approx 7 \times 10^{36}$ erg s^{-1} is required.

Bl141 has been classified B1–B3Ib from its optical spectrum but it may be a peculiar or double star. The ultraviolet spectrum suggests a type earlier than B1; there are strong P Cygni profiles in the emission C IV 1548, 1550 and Si IV 1398, 1402 resonance doublets. A classification of O9.5–B0I is supported by fitting the continuum flux, corrected for reddening with $E_{B-V} = 0.17$, to a Kurucz' model of normal metallicity with $T_{eff} = 30\,000$ K and gravity log $g = 3.5$. With this classification $M_{bol} = -9.7$ is derived. The visual spectrum may then indicate that a secondary component exists (Laval et al. 1994).

Also the terminal velocity of 1800 ± 100 km s^{-1}, derived from fitting theoretical profiles to the P Cygni profiles of the UV CIV and SiIV doublets, confirms the classification of Bl141 as a supergiant earlier than B1. Its mass loss is estimated as $\geq 3 \times 10^{-6}$ M$_\odot$ yr^{-1}. The winds from Bl141 have created and maintained the bubble N120A. The star must be ~ 1 Myr old, the bubble is only ~ 6 × 10^4 yr old. The stellar wind must have been injected into the bubble for the latter period but not much longer. It appears likely that Bl141 left the main sequence to become a supergiant about 10^5 yr ago.

5.6.2 The star DD13 in the nebula N159A

The LMC giant HII region N159, with a diameter of ~ 60 pc, is the southernmost part in the string of nebulosities which extends south from 30 Doradus. It is near the northernmost peak of the strong CO emission mapped by Cohen et al. (1988) in that region. It is a very active region: it contains the X-ray source LMC X-1, two IR protostars (Jones et al. 1986), a most peculiar 'Blob', the dense nebula N159A (Heydari-Malayeri and Testor 1982, 1985), and a well-studied molecular cloud (see Chap. 8).

The main body of N159A covers 13 × 9 pc in a ring-like structure with the star DD13 (Dufour and Duval 1975) in its centre. There are extensions in the north and in the east. Two faint stars are located in the outer parts of the ring and one in the northern extension. The Q-method (UBV) leads to a probable spectral type for DD13 of O5 or earlier (Heydari-Malayeri and Testor 1982). The reddening varies over N159A; a maximum of $A_V \approx 2.5$ mag is found at DD13 (see also Sect. 5.9; the Jones et al. source # 45 corresponds to DD13). At three points in the nebula $A_V = 1.84$ mag is observed. The latter points coincide with the zones of highest excitation and suggest that part of the dust is associated with the ionized gas.

The spectral type of DD13 has been determined to O3–O6, its colour excess to $E_{B-V} = 0.64$ and its luminosity to $M_V = -6.93$ (Conti and Fitzpatrick 1991). These data lead, in connection with the Lyman continuum photon injection rate, $N_{Ly} = 5.6 \times 10^{49}$ photons s^{-1}, to the suggestion that DD13 actually consists of $\sim$ 2–4 young, early O stars still enshrouded by their natal dust cloud. It may be a younger example of the type of tight cluster represented by a number of now resolved 'stars', e.g. Sk−66°41, composed of six or more components (Heydari-Malayeri et al. 1988) and HDE 269828 (Heydari-Malayeri et al. 1993) ≡ Sk−69°209a.

5.6.3   The 'blobs' in N159 and N11A

The N159 'Blob' is slightly elongated E–W with the dimensions 2.2 × 1.4 pc. Heydari-Malayeri and Testor (1982) resolved it into two components separated by about 1″.5, i.e. $\sim$ 0.4 pc. It is highly excited; the [OIII]/Hβ ratio is as high as 8. The Hα/Hβ ratio is also high, $\sim$ 7.5. This may be explained by an interstellar absorption of A_V = 2.40 mag. The mass of the Blob is estimated to be $\sim$ 50 M$_\odot$. It may be excited by a ZAMS star of $T_{eff} \sim$ 44 000 K ($\sim$ 40 M$_\odot$) with a cluster of B and A stars playing a minor role in the excitation (Heydari-Malayeri and Testor 1985). It is, however, difficult to see how a large number of stars can be concentrated in such a small volume.

A similar blob is seen in N11A (see Sect. 8.3.1). Its diameter is 12″, i.e. $\sim$ 2.9 pc. The interstellar extinction towards N11A is normal for the LMC, $A_{H\beta}$ = 0.6 mag. Its excitation is high, [OIII] (5007 + 4957)/Hβ = 6.8, and its source of excitation may be the same as for the N159 'Blob'. As for it, an intense continuum source is seen in visual light (Figs. 2 and 3 in Heydari-Malayeri and Testor 1985). Its mass is estimated to $\sim$ 250 M$_\odot$.

5.6.4   Compact luminous groups

Hutchings and Thomson (1988) have used optical and UV spectroscopy to study stars in six compact, luminous groups in the LMC and the SMC which may be related to the small HII regions just discussed. The groups have a high concentration of nebulosity or starlight within areas < 30″ on each side; some stars can be resolved. Only one cluster, L56 in the SMC, has neither IR nor Hα emission; it is the only cluster with a normal HR diagram. All the stars were found to be of O and B type with little reddening and no signs of stellar winds but frequently with spectral peculiarities. Most spectra have strong, broad Lyα absorption. All the stars have masses > 10 M$_\odot$. Most of them have $M_{bol} \leq -6$ and fall on the main sequence, close to the locus for core hydrogen exhaustion. Several stars, in particular in the compact LMC region N130, lie off the main sequence and are suggested to be massive pre-main-sequence stars. The groups are all very young and may be considered as additional examples of star forming clusters.

5.7   Multiple stars in the LMC

Several among the supposedly most luminous stars in the Clouds have been shown to be multiple; among these are several in the central cluster in 30 Doradus (see Chap. 10). A few have been mentioned in the preceding section as possibly involved in the 'Blobs'. Some other interesting cases will be described here. The resolution of these objects into multiple ones contributes to eliminating the vision of supermassive stars.

5.7.1 The multiple star HDE 269828

HDE 269828 in LH90,≡ NGC 2044, has been classified O8, WR+OB, and O8:+WN5–6, etc., but it has also been noted as a close visual triple system with approximately equal components (Heydari-Malayeri et al. 1993). High-resolution imagery has revealed that it is a rather rich cluster of stars, including the Wolf–Rayet binary Brey65, WN7+O (Breysacher 1981). Within $\sim 12'' \times 12''$ (2.9 pc $\times$ 2.9 pc) there are about 30 stars, corresponding to a stellar density of $\sim$ 1 star pc^{-3}. This is a rather dense cluster. It contains another WR star, WN4+OB, and at least two Of stars. Massive star formation must have occurred in this region of the LMC at least 5 Myr ago. There is another dense cluster in the region, just to the east of Sk−69°212, an evolved O star cluster containing a star with WR features.

5.7.2 The multiple stars Sk−66°41 and Sk−69°253

Sk−66°41 is associated with the HII region N11C and has, as a single star, been classified O5 V with $V = 11.72$, $B − V = −0.12$, $U − B = −0.92$ mag (Heydari-Malayeri et al. 1987). Its $M_V = −7.3$ gives, with a bolometric correction $BC = −3.9$, $M_{bol} = −11.2$, indicating that it is probably a multiple star. CCD imagery has resolved it into at least six components (Heydari-Malayeri et al. 1988). The two most luminous are responsible for the O5 V type.The image of the most luminous of the two is slightly elongated, indicating that it in turn consists of more than one component.

Sk−69°253, ≡ HDE 269936, lies in the LMC nebula N158, between the components A and B. It has been classified O9.5:I–B0.7–1I (Heydari-Malayeri et al. 1989) and measured to $V = 11.23$. High-resolution imagery shows that it consists of 14 components. All the resolved components in the two clusters may be O-type members. The most massive stars then have masses of $\sim$ 50 and 70 M$_\odot$, respectively. The absence of WR and Of stars in these massive clusters as well as in HDE 269676 (Heydari-Malayeri and Hutsemékers 1991b) suggests a young age for these objects.

5.8 Multiple stars in the SMC

Sk157 is in the stellar association NGC 465 in the SMC and was resolved by Westerlund (1961b) into five components. The two most luminous components had $V = 12.59$ and 12.80, respectively. High-resolution imagery by Heydari-Malayeri et al. (1989) resolved Sk157 into 12 components.

The large HII region N66, 115 pc $\times$ 130 pc, in the northern part of the SMC Bar is ionized by the young association NGC 346. Its age is $\sim$ 2.5 Myr. Massey et al. (1989b) determined its stellar content as $\sim$ 800 stars brighter than $V = 19$. They noted that its brightest member, classified O5.5If by them and O4III(f) by Walborn and Blades (1986), was a blended multiple image. Heydari-Malayeri and Hutsemékers (1991a) have shown that this star consists of at least three components. Component (a) may itself be double. With $A_V = 0.25$, $m − M = 18.9$, and $BC = 3.93$ mag (valid for O4I), its luminosity as a single star becomes $M_{bol} = − 10.5$. With T_{eff} and gravity g from Kudritzki et al. (1989), its mass is 58 M$_\odot$. Using Maeder's models (1990) for SMC metallicity a mass of $\sim$ 85 M$_\odot$ is found.

5.9 Infrared protostars

Star formation occurs in massive molecular clouds. Optically visible HII regions reside near the cloud edge; their exciting stars have emerged from the cloud in which they were formed. Deeper in the clouds are the compact HII regions in some of which OH maser emission has been detected. These objects are supposed to represent an early stage of an OB star. The objects discussed in Sect. 5.7 and Sect. 5.8, stars previously classified as single but now shown to be multiple objects, and the 'Blobs', belong to these age groups. They are among the youngest objects known in the Clouds, all younger than ~ 6 Myr.

Still deeper within the molecular cloud is an extremely young population, the so-called 'infrared protostars'. They provide information about the current star formation and may delineate details about the star formation process.

The three young generations, the visible HII region, the compact HII region, and the protostar, are often found together within a few parsecs in the Galaxy and, as will be seen, also in the Magellanic Clouds. A typical region of this kind, and the one most studied, is the HII region N159 in the LMC. It is known to contain FIR and molecular emission as well as OH and H_2O maser emission (see Chap. 8).

Logically, the first infrared protostar was detected in N159 (Gatley et al. 1981). N159-P1 is a compact, very red ($J - H = 2.48$, $H - K = 2.10$, $K - L' = 2.7$) object. Its extreme colours may be appreciated by comparing them with those of a red-giant star in the same region with $J - H = 1.80$, $H - K = 0.71$; this star is redder than most LMC giant and supergiant stars and suffers an extinction of $A_V \approx 10$ mag, most of it caused by dust inside N159. In the $J - H$, $H - K$ diagram the protostar lies about one mag to the red of the reddening line for galactic and Cloud supergiants. Its intrinsic colour corresponds to a black-body temperature of ~ 850 K. A lower limit for its luminosity is $L \geq 2 \times 10^4$ $L_\odot$, indicating that the 'protostar' is most likely a cluster of objects.

The second protostar in N159, N159-P2, lies in a dark lane on the edge of N159A, about $20''$ ESE of a FIR peak. Two star-like sources are seen in the nebula, close to the FIR peak. Their reddenings amount to $A_V \sim 3.0$ and 1.5, respectively, judging from their E_{J-K}. The first source (# 45) is identical to the OB star DD13, discussed above. These two stars, each with $L \approx 2\text{–}3 \times 10^6$ $L_\odot$, are sufficient to provide the major fraction of the luminosity radiated in the FIR peak. The protostar, with $L \approx 2 \times 10^4$ $L_\odot$, is not a major luminosity source in N159A.

The first infrared protostar in the SMC was observed by Gatley et al. (1982) in the HII region N76B. Its properties are similar to those of the infrared protostar in the LMC (Table 5.5). IR spectra of the two stars show none of the features expected in the case of late-type giant stars, CO absorption and $B\gamma$ emission.

The stellar population in the N76B region shows a large number of red and yellow supergiants with masses $12 < M/M_\odot < 16$ and ages ≤ 8 Myr (Seal and Hyland 1983). As the protostar is ≤ 1 Myr, star formation may have been continuous there over the last 8 Myr.

The multi-aperture observations of the two protostars reveal the presence of extended emission bluer than the stars. There is likely to be reflection nebulosity in their vicinity.

5.10 The stellar associations

Table 5.5. *Protostars in the Magellanic Clouds*

Star	K mag	$J-H$ mag	$H-K$ mag	$K-L'$ mag	Beam arcsec	Ref.
N159-P1	12.11	2.48	2.10	3.5	2.7	1
	11.71	1.86	2.02	7		3
	11.90	2.07	1.91	10		2
N159-P2	11.78	0.95	1.90	7		3
N76B-K2	12.59	1.27	1.81	3.5	3.1	2
	12.51	1.09	1.62	5.0		2
N160A	11.88	0.33	0.85	7.5		4
N105A	13.02	0.58	1.49	7.5		4

The two stars in N160 A and N105A are young objects but probably not protostars in the same sense as the other three objects.
References: (1) Gatley *et al.* 1981; (2) Gatley *et al.* 1982; (3) Jones *et al.* 1986; (4) Epchtein *et al.* 1984.

5.10 The stellar associations

Galactic OB associations are rather ill-defined and it is often difficult to identify their members, in particular the less luminous ones, with full certainty. The problem is apparently easier in the Magellanic Clouds. Luminous blue stars stand out well when they are seen from outside a galaxy, and they may be considered to form an association when there is one clear density concentration. It may, on the other hand, be difficult to define the boundaries of the associations in the Clouds as most of them, probably all, are members of superassociations, in the same way as galactic associations are members of spiral arms. Luminous field stars, OB stars and supergiants, may then be counted as association members as their ages are more or less identical to those of the associations. This has to be avoided for the sake of understanding the sequential star formation that may play an important role in the evolution of the superassociations. It is also essential to avoid classifying luminous blue clusters as stellar associations; some of the former may be sufficiently open for this to occur. The problem of defining an OB association and separating it from an OB aggregate has been discussed by Kontizas *et al.* (1994). The latter are larger and have internal subclusterings; both are, however, members of superassociations.

The most complete survey of stellar associations in the LMC is the one by Lucke and Hodge (1970). Their catalogue contains 122 objects with diameters between 1' and 20' (~ 15 and ~ 300 pc) and between 2 and 225 stars brighter than $V = 14.7$. Some of the larger associations may be aggregates according to the definition above.

In the SMC there are few true associations. Recognized ones are the two groups of associations in the Wing, the sequence of NGC 456–NGC 460–NGC 465, called a constellation by Shapley (1956), and those in SGS SMC 1 (Meaburn 1980). Also NGC 346, in the northern end of the Bar, is considered to be an association. Hodge (1985) lists 70 associations of which he considers 16 as doubtful. About 35 of the catalogued objects are in the crowded parts of the Bar and are difficult to delineate; the identification of their members is also difficult. The well-studied young cluster NGC 330 is part of an association. The metallicity estimates for NGC 330 are between $-0.4 > $ [Fe/H] > -1; the association members may then be expected to show similar metal abundances (see Chap. 11).

The limit for Hodge's counts is $V = 14.2$, i.e. $M_V \sim -4.7$ mag. The mean diameter of the SMC associations is about the same as for the LMC associations, ~ 70 pc (distances of the Clouds 63 and 52 kpc). In the Galaxy the corresponding mean is ~ 125 pc. The average number of stars more luminous than $M_V \leq -4$ mag is estimated as 11 in an SMC, 18 in an LMC and 7 in a galactic association. It is difficult to judge the importance of these values. The search for associations in the two Clouds must be considered identical enough as far as plate material and techniques are concerned, whereas the conditions for surveys in the Galaxy are vastly different.

An important structural difference between the LMC and the SMC is that the LMC Bar is predominantly red, the SMC Bar blue. This difference is underlined by the fact that the former has only ~ 15 associations in the Bar region, i.e. about 10%, whereas in the SMC almost 50% of its associations are in the Bar. This difference has to be understood. The SMC Bar does contain a large amount of red stars, but not enough to dominate in integrated light. The lower metallicity in the SMC shows up in the lack of M giants; its late giants appear mainly as K stars. In the LMC the number of M giants is large, in particular in the Bar region..

As in the Galaxy, associations in both Clouds contain 'nuclei' of clusters. Hodge and Lucke (1970) list 16 such cases in the LMC; in the SMC a similar percentage may be seen. The number of associations with 'nuclei' may increase appreciably if the 'multiple stars' are considered.

Nearly 90% of the LMC associations coincide with one or more HII regions, as do most of the well-known SMC associations. Detailed studies show a clear relation between ages of associations and their content of HII. The youngest are embedded in nebulosity; the more evolved have a shell of nebulosity or have lost so much gas that no emission is visible.

The associations in Table 5.6 generally have ages ≤ 10 Myr. Kontizas et al. (1994) investigated 15 stellar associations centred on five LH associations and found their ages to be between 10 and 30 Myr with disruption times of about the same order, 2–20 Myr. The associations should thus be dissolved soon.

IMFs for the LMC associations LH9, LH10, LH58, LH117 and LH118 and for NGC 346 in the SMC, have been determined by Massey et al. (1995) using the data referred to in Table 5.6 but with new bolometric corrections, new colour/log T_{eff} transformations and new lower metallicity evolutionary tracks. The derived IMF slopes are similar to what is observed in galactic OB associations: Cyg OB2 has $\Gamma = -1.0$ (Massey and Thomson 1991) and Tr 14/16 has $\Gamma = -1.3$ (Massey and Johnson 1993).

Several of the associations in Table 5.6 show relatively large age spreads. In LH117 and LH118 most of the 50 stars with masses ≥ 10 M$_\odot$ are on or near the ZAMS and have ages of about 4 Myr. There are, however, two red supergiants and one B2I star of lower mass than several of the unevolved stars. They must have formed 6–10 Myr earlier than the majority of the association members. A similar situation appears in the rich stellar association NGC 346 (Massey et al. 1989b, 1995). The IMF has a slope of -1.3 and the upper-mass limit is not much lower than in the Galaxy. Also this association has evolved members (five red and two blue supergiants) of considerably

5.10 The stellar associations

Table 5.6. Well-studied OB associations in the Magellanic Clouds

Association LH,NGC	Photometry spectroscopy	E_{B-V} mag	Age Myr	$\Gamma(IMF)$	Ref.
LMC					
LH9	UBV, sp	0.17	5	-1.4 ± 0.2	1, 10
LH10	UBV, sp	0.05	2–3	-1.1 ± 0.1	1, 10
LH13	UBV	0.19			2
LH58	BV				3
	UBV	0.05		-1.4 ± 0.2	4, 10
	UBV	0.09		-2.5 ± 0.3	5
	BV	0.15	4–9	-1.2 ± 0.1	6
LH76	UBV	0.16			2
LH90	BV,sp	0.4	3–8		9
LH105	UBV	0.45			2
LH117	UBV,sp	0.10	1–12	-1.6 ± 0.2	7, 10
LH118	UBV,sp	0.08	1–12	-1.6 ± 0.2	7, 10
SMC					
NGC 346	UBV,sp	0.14	≤ 5	-1.3 ± 0.1	8, 10
NGC 456-460-465	BV		10		11
NGC 602	BV	~ 0	10		12
	Sp(*IUE*)		5		13

References: 1, Parker *et al.* 1992; 2, DeGioia-Eastwood *et al.* 1993; 3, Lortet and Testor 1988; 4, Garmany *et al.* 1994; 5, Hill *et al.* 1994b,c; 6, Will *et al.* 1995; 7, Massey *et al.* 1989a; 8, Massey *et al.* 1989b; 9, Testor *et al.* 1993; 10, Massey *et al.* 1995; 11, Westerlund 1961b; 12, Westerlund 1964d; 13, Hutchings *et al.* 1991.
LH9 and LH10 are in the HII region N11: see Sect. 8.3.1.
LH90: The reddening varies strongly and reaches $E_{B-V} \sim 0.8$ in its western part. There are two age groups, 3–4 Myr and 7–8 Myr, respectively. The association contains 7 WR stars.
LH117 and LH118: The large age spread is discussed in the text.
LH58: Hill *et al.*(1994b) found a dispersion of 0.092 mag in E_{B-V} due to differential reddening. Garmany *et al.* (1994) give a low E_{B-V}, but note that there is a sprinkling of about 30 stars with appreciably higher reddening values. The $\Gamma(IMF)$ given for LH9, LH10, LH58, LH117, LH118, and NGC 346 are those for 'variable reddening' (ref. 10).
NGC 456-460-465 has been discussed as an outstanding prototype of sequential star formation on a scale of 10 Myr and ~ 500 pc (Testor and Lortet 1987) but no recent age determinations are available.
No recently determined CM diagrams of NGC 602 appear to exist. This is regrettable as NGC 602 is part of a superassociation and of SGS SMC 1 (Meaburn 1980). It is suggested in ref.13 that some of the stars are pre-main-sequence; this is not confirmed by the photometry in ref.12.

lower mass than the unevolved members. In NGC 346 these stars form a spatially distinct subgroup, indicating that sequential star formation has taken place.

The association with the flattest slope, $\Gamma = -1.1 \pm 0.1$, in Table 5.6 is LH10. It is close to LH9, an association with $\Gamma = -1.4 \pm 0.2$. Both are situated in the HII region N11. LH10 is surrounded by nebulosity, LH9 has little but a shell structure centred on it. The difference in Γ indicates that LH10 contains a higher ratio of higher-mass to lower- mass stars than LH9. LH10 contains at least three, possibly six O3 stars. The most luminous of the O3 stars has been resolved into six components, the most luminous of which has $V = 13.07$ and a mass of ~ 100 $M_\odot$, whereas the other two remained unresolved in the HST observations (Walborn et al. 1995). The least luminous of the two has a mass of ~ 60 $M_\odot$, the lowest possible mass for an O3 star.

LH10 also contains at least five ZAMS O stars.* Only 30 Doradus surpasses this region in stellar content and activity. The reddening of LH10 is greater than that of LH9; $E_{B-V} = 0.17$ and 0.05, respectively. LH9 may have very little intrinsic reddening even if signs of an uneven distribution of dust can be seen. (In the HR diagrams in Fig. 5.3 the cases of variable reddening are shown.) In the area of LH9 are two K supergiants, in the area of LH10 there is one. The first two are definitely members of the LMC; whether any of the stars are members of the associations remains to be proven. LH9 contains the multiple star HD 32228 with the WC5+O9.5II component Brey9 (Heydari-Malayeri and Testor 1983; Parker et al. 1992; Schertl et al. 1995; Walborn et al. 1995; see, further, Sect. 8.3.1). Schertl et al. find some evidence for the outer parts of LH9 being the youngest with an upper age of about 5 Myr. The WC5 component in its centre has an age in excess of 5 Myr.

The stars in NGC 1962-65-66-70, $\equiv$ LH58 ($5^h26^m.8$, $-68°50'$; 1975) at the NW edge of Shapley's constellation II also cover a relative large age range. The association, or at least its northern part, is embedded in a bright nebula, N144, $\equiv$ DEM199, with a shell of filaments, $13' \times 12'$. There are 22 O-type stars in the association (Garmany et al. 1994), and several evolved supergiants. The earliest spectral type is O3–4V. The stars in the association have ages between 4 and 9 Myr. There are indications of a propagating star formation from the south towards the north. The southern half contains the most luminous stars, three A–MIa stars, and three WR stars, Brey32 $\equiv$ HD 36521 (WC4+O6), Brey34 $\equiv$ HDE 269546 (B3I+WN3), and Brey33 (WN3+abs). Brey34 is a multiple system (Lortet and Testor 1988). One of the supergiants, HDE 269551, is suspected to be a VV Cephei star (M+OB), a P Cyg type star, or a Be double star. Will et al. (1995) have resolved it into three components, F8Ia, B6Ia, and B0V, of which the last one may not be physically connected with the other two. In the northern part, N144A contains all the hottest unevolved stars; this part of the nebula is about five times as bright in [O III] as N144B.

Sequential star formation may be traced over the whole constellation. The region is rich in red supergiants but contains no early O stars, no WN6–7 stars, and no bright nebulosities outside N144, where the only WC star is found. There are in the field also several emission-line hot supergiants, some of which are of S Dor type (see Sect. 2.3). The turn-off ages for the clusters and older associations in the region go from 5.5 to 3.6 Myr (Westerlund 1961a), with the southernmost clusters being the oldest. (It appears unlikely that modern techniques would change these values much.)

14 OB associations in the LMC have been investigated by Hill et al. (1994a,b,c). Using UBV CCD photometry they derived colour–magnitude diagrams, measured the interstellar reddening and determined luminosity functions and initial mass functions for them. The LFs of the 14 associations are quite similar with an average slope of

---

* ZAMS O stars are recognized by having their HeII λ 4686 absorption lines substantially stronger than any of the other HeII absorption lines.

5.10 The stellar associations

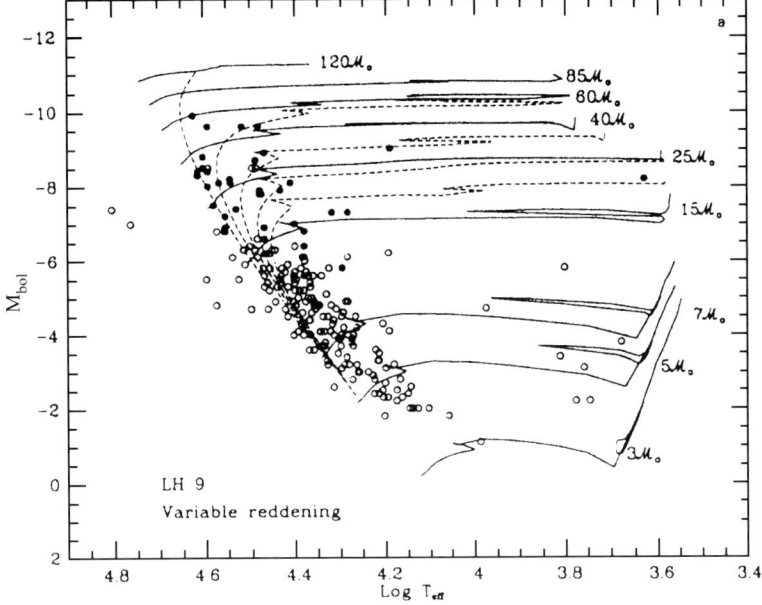

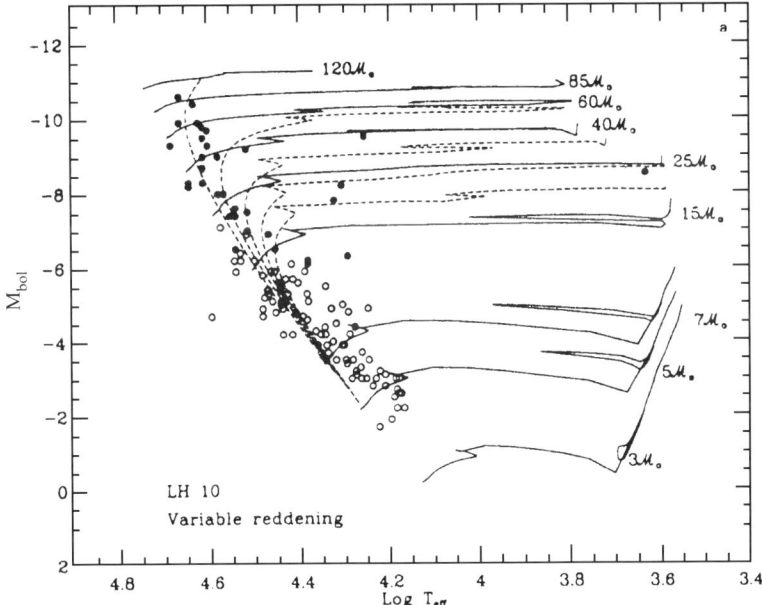

Fig. 5.3. HR diagrams for LH9 and LH10. Dots identify stars with spectroscopy, open circles stars with photometry, only. The evolutionary tracks are for $Z = 0.008$; dashed lines are isochrones at 2 Myr intervals. (From Massey et al. 1995.)

$S = 0.30 \pm 0.06$. Their IMFs span similar ranges about a common mean for the LMC and the SMC; the average slope is $\Gamma = -2.0 \pm 0.5$ for $M > 9\ M_\odot$. This indicates that metal abundance does not have a strong effect on the IMF, at least for the metallicity range observed in the Clouds. The observed scatter is probably due to the large uncertainties in the calculations of the MFs.

The considerably steeper slope found by Hill *et al.* for the IMF of the Cloud associations than by Massey *et al.* (see Table 5.6) is probably due to the use by the former of only photometry in the treatment of the hotter stars. It has been noted that the IMF from photometry alone tends to have a steeper slope than when the temperatures for the OB stars are derived from spectral types (Parker and Garmany 1993), and this has been confirmed by Massey *et al.*(1995).

6

The superassociations and supergiant shells

The concentrations of luminous, blue stars in the Magellanic Clouds have attracted much attention. Shapley (1956) noted that the large gaseous nebulae in the LMC are frequently associated with groups of stars but also that some of the larger star groups are free of conspicuous nebulosity. As these stellar aggregations were too large to be called clusters or associations in the sense used in our Galaxy he called them 'Constellations'. He estimated their diameters to be between 250 and 600 pc and their content of blue supergiants, with a red magnitude brighter than 14.0, to be between 14 and 32. A few red supergiant stars were seen in each of them. In the region of 30 Doradus, Shapley (1955) identified a number of red stars by comparing blue (B) and infrared (I) plates and concluded that the very red stars were of spectral class M0 or later. Only two of the 21 most luminous of these stars are in the vicinity of the core of 30 Doradus.

In the SMC Shapley found only one object rich enough to be called a constellation in the sense used for the LMC. It is the aggregate comprising NGC 456, NGC460 and NGC 465 in the Wing area.

Shapley's designation is still used to identify the five most conspicuous stellar aggregates in the LMC. Improved techniques have extended and redefined them and led to the identification of more such formations. The 'Constellations' are now parts of large 'superassociations', which may coincide with or contain 'supergiant shells' (SGS). The youngest objects in the Clouds are found in them (See Chap. 5). Our understanding of large-scale stellar evolution will improve if we can find irrefutable answers to questions such as 'How have these superaggregates formed?' 'Have they evolved by the SSPSF process?' Our understanding of the evolution of the Magellanic Clouds will gain by finding the answer to the question 'Are the superaggregates all of about the same age so that a global burst of star formation may have occurred in the relatively recent past?' Undoubtedly, supergiant shells play a dominating role in controlling the evolution of galaxies up to the size of the LMC (de Boer and Nash 1982).

The appearance of a superassociation may be summarized roughly as follows. In a region with relatively weak HI gas there are stellar groups, some embedded in HII regions, others without obvious nebulosity. As a rule the latter are slightly older than the former; they are also less reddened. In most superassociations the youngest stellar associations and HII regions form a peripheral structure which may be more or less ringlike. If this structure is sufficiently well defined it is called a supergiant shell (SGS). Often there is a shell of neutral hydrogen on the outside of the ionized shell. In some cases a decrease in age of the stars and stellar groups can be seen from the centre of the superassociation towards the ring. This can be understood in terms of a SSPSF process. Some young clusters and isolated HII regions may remain far inside the shell, indicating that during the outwards propagation some minor clouds were left behind the propagating front and the star formation in them was delayed for one reason or another.

6.1   Superassociations in the LMC

In a photometric study of clusters, associations and field stars in four regions of the LMC Westerlund (1961a) noted that all the objects were of more or less the same age independent of their positions. He suggested that they were all parts of giant stellar aggregates, superassociations. Westerlund and Smith (1964) noted that all the Wolf–Rayet stars (identified at that time) were within such superassociations.

Martin et al. (1976) have defined eight giant concentrations of OB2 stars, WR stars, and B–G supergiants within a radius of $4°$ from the LMC centre (Fig. 6.1). The largest ones correspond to Shapley's Constellations, strongly extended. Inside these regions the ratios of early- to late-type stars are different, giving evidence of a variable rate of star formation in the LMC, or of a variable rate of the SSPSF inside each region. On the average, OB2 stars are relatively ill-correlated to neutral hydrogen. On the other hand, the WR stars are strongly associated with HI and HII. The OB2 group is, however, not homogeneous. It contains very hot stars, O7 and earlier, likely to ionize all nearby hydrogen completely, and cooler stars, O8–B2. The correlation between ionized and neutral gas seems to be the most representative of the bright part of the luminosity function.

The distribution of the stars in superassociations nos. I–VI shows clear age effects but of a different nature. In no. I, the youngest stars (OB2) are in the north-western part, in no. II they are in the eastern part (see Fig. 6.1). The separation between the OB2 stars and the later-type supergiants is surprisingly complete, indicating that star formation has continued longer in the western and eastern parts, respectively. SGS LMC 1 makes up the northernmost part of no.I, and the active nebular region N11 (Sect. 8.3.1) occupies a central position. In superassociation no.V ($\equiv$ SGS LMC 4, see below) the supergiants are concentrated in the central parts, and the OB2 stars populate the peripheral regions. The impression of shell-like structures in the distribution of the hot stars themselves is supported by the structure of the LMC in UV light (Fig. 12.2, see Smith et al. 1987).

Superassociation VI is in a low-density HI field with a higher-density region just south of it. The associations LH119–121 (possibly also LH122) and the emission arc DEM328 are in its southern part. In its northern extreme LH112 and LH116 are found as well as a number of HII regions, dominated by DEM309 and DEM310. Between these two groups, several B2.5–A9 supergiants are seen as well as a few OB2 stars. The overall impression is that star formation has progressed outwards and is now going on in the two extremes just described.

There thus exist in the LMC two types of superassociation: (1) one which has sufficient hydrogen gas along the periphery so that an ionized and, in most cases, an outer neutral shell exist, and (2) one where most of the gas is used up, or lost, so that the boundary is defined by the stellar population itself.

6.1 Superassociations in the LMC

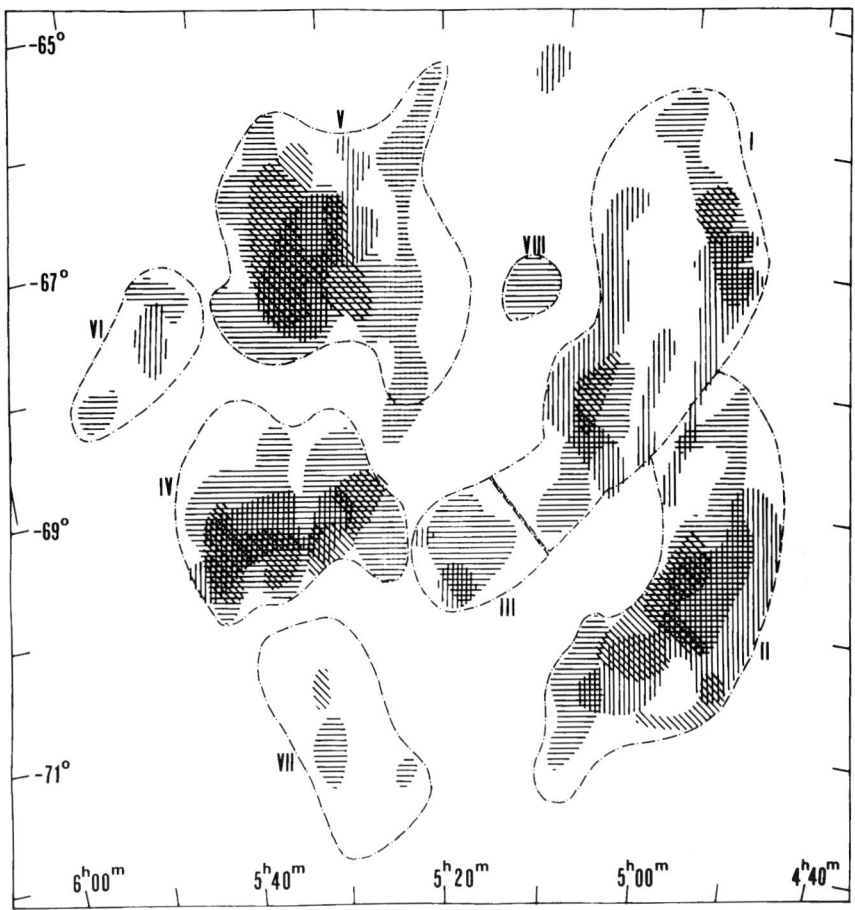

Fig. 6.1. The superassociations in the LMC according to Martin et al.(1976). The most important are: IV, containing 30 Doradus and SGS LMC 2 and SGS LMC 3; V, corresponding to SGS LMC 4 and 5 and containing Shapley Constellation III; III, including Shapley Constellation V; and VIII, ≡ Shapley Constellation IV. OB2 star concentrations are identified by horizontal; B2.5–B6 by diagonal and B7–A9 stars by vertical hatches.

The distribution of the two types of superassociation over the LMC is such that type (1) exists mainly in the eastern part, type (2) over the whole Cloud but preferentially in the west. This adds to the east–western asymmetry of the LMC that is pronounced in the distribution of dust, hot plasma, etc. (see Chap. 12).

Meaburn (1980) used the appearance of nebular filaments to identify nine possible supergiant shells in the LMC and one in the SMC. Typical for the supergiant shells is that they are radiatively ionized on the inside by the numerous OB stars in the superassociations whereas the outer shells may be neutral. Kennicutt et al. (1995a) have determined Hα fluxes for the proposed SGSs in the LMC and the SMC and gathered some other properties which are reproduced in Table 6.1.

Of the nine supergiant shells in the LMC, the positions of nos. 2, 3, 4, and 5 are referred to in Fig. 6.1. Of the others, SGS LMC 1 lies, as mentioned above, in the

Table 6.1. Properties of supergiant shells in the LMC and the SMC

SGS	Radius pc	L(Hα)	E_{B-V} mag	Blue stars	Q(Hα)	Q(UV)	$M(H^+)$ 10^5 $M_\odot$
LMC 1	350	tot 1.4 diff 1.2	0.04	49	0.96 0.90	2.61	1.3
LMC 2	450	tot 20.5 diff 15.9	0.24	125	14.56 11.26	30.87	1.3
LMC 3	500	tot 26.1 diff 5.9	0.11	138	18.57 4.18	14.62	7.8
LMC 4	700	tot 32.9 diff 13.9	0.05	> 394	23.36 9.93	45.21	6.5
LMC 5	400	tot 4.8 diff 3.1	0.12	25	3.40 2.21	2.61	2.6
LMC 6	300	tot 5.0 diff 2.5	0.21	25	3.57 1.81	4.36	0.9
LMC 7	400	tot 13.3 dif 7.8	0.15	135	9.45 5.57	6.26	6.4
LMC 8	450	tot 4.7 diff 2.4	0.10	125	3.30 1.73	6.30	2.1
LMC 9	450	tot 9.9 diff 2.9	0.12	44	7.08 2.06	2.40	2.1
SMC 1	300	tot 1.4 diff 0.9	0.08	( )	1.06 0.64		1.3

The Hα luminosities in col. 3, expressed in 10^{38} ergs s^{-1}, are reddening corrected; Fitzpatrick's (1986) extinction law is used on the E_{B-V} in col.4 (from Lucke (1974) for the LMC and a mean E_{B-V} for the SMC (Table 2.1)).
'Tot' and 'Diff' in col. 3 refer to total and diffuse luminosities. The latter are derived by subtracting the contribution of HII regions from the total $L(H\alpha)$.
$Q(H\alpha) = 7.1 \times 10^{11} L(H\alpha)$ s^{-1}.
$Q(H\alpha)$ and $Q(UV)$ are expressed in 10^{50} s^{-1}.
Vacuum UV fluxes for the LH associations are from Smith et al. (1987, 1990).
Distances used are LMC 50 kpc, SMC 60 kpc.

northern end of superassociation I; nos. 6, 7, 8 are parts of superassociation II; and SGS LMC 9 is a part of superassociation VII.

6.1.1 The supergiant shell SGS LMC 4

The most typical of the supergiant shells is SGS LMC 4, connected with Shapley Constellation III (Westerlund 1964c; Westerlund and Mathewson 1966; Meaburn 1981), and the major part of superassociation no. V. It consists of an HI shell with a maximum diameter of $\sim$ 1900 pc in the plane of the LMC, a ring of HII regions with a diameter of $\sim$ 1400 kpc, and, in its southern part, Shapley Constellation III. Its HI gas has three components (Dopita et al. 1985c): one in the plane with $V_{LSR} = 285$ km s^{-1}, one above and one below it; the ejected gas has an expansion velocity of 36 km s^{-1}. The gas may reach a height of 2.5 kpc before returning. It may then induce a renewed process of star formation.

The HI column density is extremely low in the centre, $5^h 30^m.5$, $-66°45'$, reaching only 5×10^{19} cm^{-2}. This implies that virtually no dust exists there. The HI column

6.1 Superassociations in the LMC

density peaks in the north-west with 2.4×10^{21} cm^{-2}. This is in a region with four supernova remnants and where the shell may be in collision with SGS LMC 5. The locally generated HI motions are as high as 55–70 km s^{-1}.

The ages of the supergiant stars in the region indicate that star formation started about 15 Myr ago near the centre of the shell and propagated outwards at a constant speed of 36 km s^{-1}.

Reid et al.(1987) have discussed the evolution in SGS LMC 4 on the basis of V, I photometry of stars in the southern part of the shell, dominated by the rich stellar arc, LH77. The luminosity function for the MS stars in their field showed a break at $M_V = -3$ mag. This was interpreted as an age of about 20 Myr for the involved stars. This age is high in comparison with that of most clusters and supergiants in the field. The central association in SGS LMC 4, LH72 ($\equiv$ N55), has an age of about 4 Myr, and almost all the other clusters and associations, mainly in its shell region, are of this age (Braunsfurth and Feitzinger 1983). Among the exceptions is LH65 with an age of over 10 Myr.

If SGS LMC 4 has resulted from a symmetrical expansion from its origin, the part of the shell moving towards the centre of the LMC would be the first section to become stationary (Westerlund and Mathewson 1966). With an expansion velocity of 36 km s^{-1} the front would have reached the southern stellar arc LH77 in about 10 Myr but the time may well have been longer. Cohen et al. (1988) suggested a characteristic kinetic time scale of expansion of $\approx$ 26 Myr. The difference in age between the stars in the LH77 arc, $\approx$ 20 Myr, the supergiants spread over the whole area, 1–15 Myr, and the majority of the clusters and associations, $\approx$ 4 Myr, may be interpreted in the following way. The expansion of the shell began at least 26 Myr ago, probably soon after the last perigalacticon of the LMC. The original energy input must have been extremely strong so that the expansion continued rather undisturbed. A number of clouds of various sizes were left behind in a very weak intercloud medium. These clouds functioned as star forming sites, but star production began in them at times that were determined not only by their distances from the origin of the expansion but also by other factors. This is evident as LH72, in spite of being close to the origin of expansion, is only 4 Myr old. The mass of each cloud must have played a significant role (see Dopita 1987 for a discussion of the phases of the ISM.) for the star formation if the rich arc LH77 and its immediate neighbourhood contain the oldest stars of this generation. Some of the young star clusters are surrounded by 'bubbles' of HII, which in turn are still embedded in HI clouds (e.g. N57, N58 $\equiv$ LH75, LH76, LH78). The SSPSF model needs some modification to cover this course of events also. 'Left-overs' in the expansion process are also seen in the 30 Doradus complex which may give a view of an earlier phase of SGS LMC 4 (see Sect. 10.2).

Feitzinger et al.(1987) acknowledged the SSPSF as having been active in the Constellation III area in the way suggested by Dopita et al.(1985c) and considered it as an isolated space–time cell. (Nevertheless, they also make the star forming shell in Constellation III part of two of their long ridges of star-formation loci in the LMC (see Sect.12.1).)

Detection of soft X-ray emission, at 1/4 keV, from SGS LMC 4 was claimed by Singh et al.(1987b), who suggested that SGC LMC 4 was the first X-ray superbubble detected in an external galaxy. It is, however, likely that the X-ray surface brightness seen by Singh et al. is a residual effect of the strong X-ray point sources in the region.

Diffuse X-ray emission originating in SGS LMC 4 has been detected in a pointed *ROSAT* PSPC observation (Bomans *et al.* 1994) covering its northern part. The enhancement of the diffuse emission is strong *inside* the supergiant shell (defined by the Hα arcs) and also *inside* SGS LMC 5, to the NW of SGS LMC 4. Confirmation of the X-ray emission coming from the cavity is obtained by comparing the X-ray emission with the *IRAS* 100 μm far-infrared (FIR) map: the X-ray emission comes from the region of low FIR emission, i.e. from a region with low HI density and very little dust. The results strongly suggest that it originates in a volume of hot gas, devoid of HI, where the dust grains are effectively destroyed. The spectral properties of the X-ray emitting gas were obtained by modelling the observed pulse height distribution by a thermal plasma emission spectrum. The best-fitting spectrum yielded a temperature of $(2.4 \pm 0.2) \times 10^6$ K, an HI column density of $n_H = 0.7^{+0.3}_{-0.2} \times 10^{21}$ cm^{-2} and a mean *unabsorbed* X-ray flux of $F_x = (9.0 \pm 1.0) \times 10^{-15}$ erg cm^{-2} s^{-1} arcmin^{-2} for the energy range $E = 0.1$–2.4 keV. (The n_H value is consistent with the average foreground density from HI data (McGee and Milton 1966).) If the derived value is typical for the whole cavity of SGS LMC 4, its total X-ray luminosity is $L_x(0.1$–2.4 keV$) = 1.4 \times 10^{37}$ erg s^{-1}. The X-ray emitting volume may be considered cylindrical; it may be estimated as $V = \pi \times (600\text{pc})^2 \times 1200$ pc.* The electron density, n_e, can be obtained from

$$L_x = n_e^2 V \Lambda_R(T),$$

where $\Lambda_R = 4.0 \times 10^{-24}$ erg cm^3 s^{-1} is the emissivity of hot gas at the derived temperature and LMC abundance. With the derived $n_e = 8.0 \times 10^{-3}$ cm^{-3} the thermal pressure of the hot gas inside SGS LMC 4 is found to be $\sim 2.0 \times 10^4$ K cm^{-3}. This is much more than the pressure of the warm ionized medium in the LMC of $\sim 4 \times 10^3$ K cm^{-3} (de Boer and Nash 1982), indicating that SGS LMC 4 is still expanding into the ambient ISM of the LMC. A thick HI layer should prevent it from breaking through into the halo regime; HI shells are seen expanding upwards and downwards from the plane (see above).

The electron density and pressure inside SGS LMC 4 are comparable to the conditions inside SGS LMC 2 (Wang and Helfand 1991b, see below). This result is inconsistent with the higher age of SGS LMC 4 indicated by its much larger size. The roughly equal expansion velocities of the two SGSs and the somewhat lower temperature of the gas inside SGS LMC 4 may be the results of internal structural differences in the ambient ISM and in the energy sources.

The surface brightness distribution inside SGS LMC 4 is indeed not homogeneous. Density fluctuations must exist in the hot plasma as seen in the bright filamentary ridges of enhanced X-ray emission. They may be caused by SN blast waves hitting interstellar clouds. A significant number of SNRs may then remain undetected in the LMC because of their location inside hot, low-density cavities. The SN rate in the LMC would then be higher than one SN per 500 yr (Chu and Kennicutt 1988b).

Using the Australia Telescope Compact Array, Dickey *et al.* (1994) have surveyed 21 cm absorption towards the LMC. Six continuum sources were seen in the direction of SGS LMC 4. Absorption by cold gas was seen at the rest velocity of the LMC, ~ 285 km s^{-1} (Dopita *et al.* 1985c). At the position of the source, J0526-658, at about

---

* The possible hot halo around the LMC is assumed to make a negligible contribution to the X-ray luminosity compared to the gas inside the supergiant shell.

6.1 Superassociations in the LMC

the centre of the supergiant shell, a shallow absorption component was also seen at 304 km s^{-1}. This corresponds to the HI emission-line velocity of $300 \leq RV_{LSR} \leq 325$ km s^{-1} and indicates that cool gas also participates in the outmoving motion.

Domgörgen et al. (1995) have derived velocities and column densities of absorbing gas clouds from high-dispersion *IUE* spectra of stars in the southern half of SGS LMC 4. They have combined these absorption line data with a reanalysis of the Rohlfs et al. (1984) HI 21 cm profiles to derive the location of the various gas components and of the stars. Low column-density gas, having velocities near 325 km s^{-1}, is seen in emission in the southern part of SGS LMC 4 but not in absorption. This gas must be receding from the LMC main body with a velocity of ~ 40 km s^{-1}. Its mass is estimated to be $M \leq 4 \times 10^4$ M$_\odot$. The total mass of the gas in the inner part of the shell is $\sim 4.7 \times 10^5$ M$_\odot$. As noted also by Dopita et al., this is a fragment rather than a complete shell.

Gas having velocities ~ 260 km s^{-1} is seen in emission and in absorption along several lines of sight. Including broad components there is a smooth range of velocities from ~ 250 km s^{-1} up to the system velocity, 285 km s^{-1}. The mass of this gas is 3–4 times as large as that of the receding fragment.

Domgörgen et al. also discuss various possibilities for the formation of SGS LMC 4, stellar winds and supernovae of a central association, collision with a high-velocity cloud and SSPSF. They conclude, in agreement with other authors, that the SSPSF is most likely responsible for its creation, and place the breakup on the distant side.

6.1.2 The supergiant shell SGS LMC 2

SGS LMC 2 is the most important region in the LMC as far as star forming activities are concerned. It is centred at $\sim 5^h 42^m$, $-69°30'$ (1950), on the SE side of 30 Doradus. Optically, it shows some similarities to SGS LMC 4. It is delimited by HII regions and filaments (see Fig. 10.1), but it is smaller than SGS LMC 4 and has more structure inside the shell. The stellar associations inside the shell produce enough photons to ionize the inner edge of the shell. They show evidence of a typical SSPSF event; the propagation has occurred from the NE toward the SW (see Table 6.2). It has, physically, stopped when the gas hit the dust clouds going N–S from 30 Doradus and led to enhanced star formation there. The ISM structure, displayed as HI, FIR, HII and X-ray emission, may be interpreted as shells of various types and motions, and supports the idea that star formation has propagated from NE towards SW. Throughout the region the associations line up parallel to this direction; this is also true of the groupings just west of 30 Doradus, and may be more significant than has been realized so far.

The huge molecular cloud embedding SGS LMC 2 has a north–south extension about twice that of the supergiant shell as seen in the size of the CO cloud (Cohen et al. 1988). The whole complex can be understood if this cloud swept through the LMC following the collision with the SMC ~ 0.2 Gyr ago and arrived at its present position about 50 Myr ago (Fujimoto and Noguchi 1990). Its leading edge has been stopped in the 30 Doradus position, possibly as a result of a collision between SGS LMC 2 and SGS LMC 3. The expansion of the shell towards the west has been stopped by heavy dust lanes, causing a ridge of HII regions to form south of 30 Doradus. The ridge contains dust, hydrogen, CO (Booth 1993) and plasma emitting diffuse X-rays (see Sect. 9.1.2) and shows multiple signs of on-going star formation. N158C, N159 and

Table 6.2. *Age estimates of clusters in SGS LMC 2*

Cluster NGC	Distance (pc) from NGC 2102	NGC 2100	Age Myr	Number of red supergiants	HII region Age, M yr
2092	260	60	12	3	No
2100	260	0	8.5	18	No
2081	200	220	6	2	Ring, 6.8
2074	260	350	5	0	Embedded, 6.5

The four clusters lie nearly on a straight line going NE–SW; most stellar groupings in the SGS LMC 2 region are extended and lined-up in this direction.
The age estimates for the clusters are from Westerlund (1961a) and for the HII regions from Dottori and Bica (1981).

N160 are outstanding among the star forming giant HII regions in the ridge. N159 (see Fig. 6.2) was the first identified CO emission source in the Clouds (Huggins *et al.* 1975). It contains the first detected Type I OH maser (Caswell and Haynes 1981) and the first H_2O maser (Scalise and Braz 1981). The first infrared protostars in the LMC were discovered in N159 (see Sect. 5.9).

Also, N160 has objects almost as young as the protostars in N159. Epchtein *et al.* (1984) discovered a compact object in N160A near an H_2O maser (Whiteoak *et al.* 1983). The nebula N160A is a strong radio continuum source. CO emission has been detected near it (Israel *et al.* 1982).

Wang and Helfand (1991b), using the IPC on the *Einstein* Observatory, discovered a bright extended X-ray emission region which is enclosed by SGS LMC 2. The diffuse X-rays appear to come from hot gas in a superbubble. The luminosity, in the 0.16–3.5 keV band, is $\sim 2 \times 10^{37}$ erg s^{-1} with a gas temperature of $\sim 5 \times 10^6$ K. The X-ray emission appears as a ring with a radius of $\sim 15'$. It correlates well with a cold ISM cavity seen in HI and *IRAS* maps, centred at $05^h 43^m$, $-69°35'$ (1950) (near NGC 2102) and with a diameter of $\sim 30'$. The ring is much smaller than the region bounded by the optical filaments.

It is of fundamental importance for understanding the whole 30 Doradus region, including SGS LMC 2, SGS LMC 3 and the filaments to the NE of 30 Doradus, to determine the spatial distribution of the various velocity components seen in HI, HII, CO, and in the interstellar and stellar lines in stellar spectra. There is some evidence for the 300 km s^{-1} component of neutral and ionized gas being in front of the ionizing stars and the 243 km s^{-1} component being on the far side.

Dickey *et al.* (1994) have suggested that the low-velocity HI component L (Luks and Rohlfs 1992) is on *the far side* of the LMC and that SGS LMC 2 is even further away, maybe as much as 1–2 kpc, on the back side of the L-component. Their suggestion is based on a 21 cm absorption survey of the LMC where six background continuum point sources were observed near 30 Doradus and the two HII regions N159 and N160. Three of the background sources are intrinsic to the LMC. The absorption line velocity components referred to the L-component differ about ~ -17 km s^{-1} from the velocity of the disk (D-) component. They have, however, a velocity gradient of ~ -9 km s^{-1} kpc^{-1} along the line of sight. This is interpreted as

6.1 Superassociations in the LMC

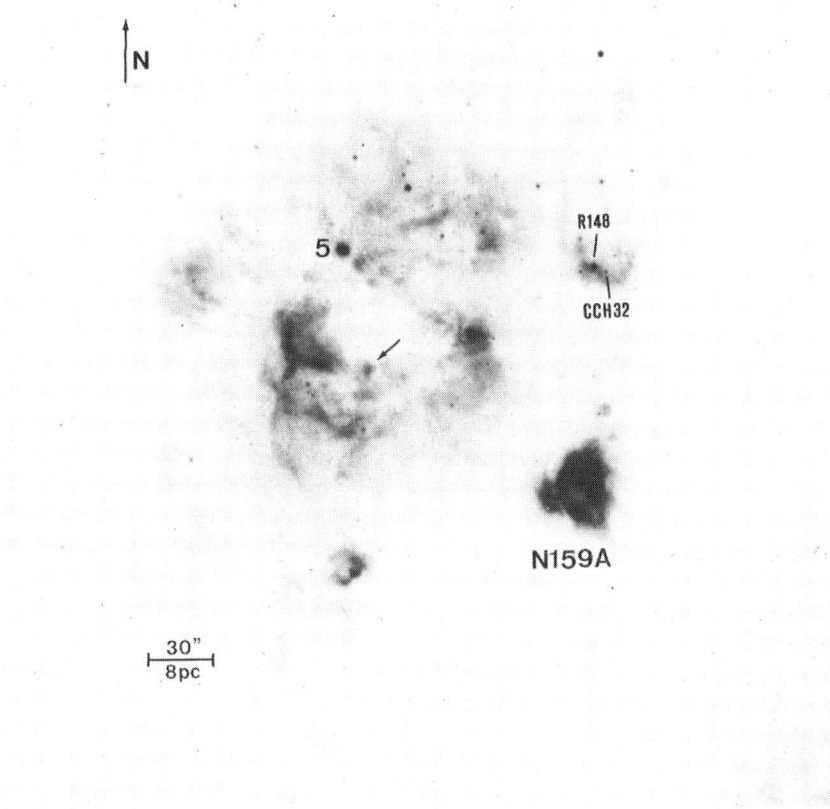

Fig. 6.2. The giant HII region N159. Identified are the HII region N159A, the X-ray source LMC X-1 ≡ CCH 32, the 'Blob' ≡ 5 and the infrared protostar P-1 (at the arrow). The second protostar, P-2, not marked, is in the NE edge of N159A. (From Heydari-Malayeri and Testor 1982.)

placing the L-component behind the D-component. Luks and Rohlfs (1992) discussed this possibility but discarded it as they saw no absorption effects. The CO velocities (Cohen et al. 1988) agree with the velocities in the HIL component, only one CO cloud has a velocity corresponding to that of the HI disk.

Support for a position of SGS LMC 2 behind the disk is also found in the fact that both the L- and the D-components are seen in absorption toward the sources intrinsic to the LMC. If the L-component is behind the D-component, their observed velocities indicate that they will eventually collide. This position as well as the possible future merger is in agreement with the Fujimoto–Noguchi model referred to above.

The dynamics of SGS LMC 2 have been studied by Caulet et al. (1982). From 16 Hα photographic Fabry–Perot interferograms 967 radial velocities were obtained over its ionized filaments and individual HII regions. The velocity pattern derived, lowest velocities (230–250 km s^{-1}) in the eastern and southern filaments, highest (270–285 km s^{-1}) in the central and NW regions, corresponds to that derived by McGee and Milton (1966) for neutral hydrogen. The observed velocity field can be accounted for by the expansion of a portion of a spherical shell of radius 475 pc, with a velocity of 30 km s^{-1}, into an ambient neutral medium of 245 km s^{-1}. Its age is $\sim$ 10 Myr. The supergiant shell may have resulted from the combined action of stellar winds and supernova explosions.

Similar conclusions have been drawn by Wang and Helfand (1991b). They suggest that SGS LMC 2 has been created by the break out of a superbubble from the dense galactic plane of the LMC. The superbubble is near this plane; it is energized by stellar winds and supernova explosions. The blown-out superbubble expands rapidly in the low-density region on the far side of the LMC. Its boundary is marked by absorption of soft X-rays in the dense HI shell it creates and by rim-brightened optical filaments at the interface between the hot interior and the cold swept-up medium. They further speculate that the massive ridge of gas at the eastern edge of the LMC, in which this bubble is, was formed as a result of the interaction of the LMC disk with the gaseous halo of the Milky Way. The high pressure and density characteristic of this interaction region should then have triggered the massive star formation responsible for creating the superbubble SGS LMC 2.

So far, the only model giving a likely explanation for the existence of the huge amounts of gas and dust in the SGS LMC 2/ 30 Doradus region of the LMC is the one proposed by Fujimoto and Noguchi (1990).

6.1.3 The other supergiant shells in the LMC

The expansion of SGS LMC 2 and SGS LMC 4, as well as of the other SGSs in the LMC, has been questioned by Meaburn *et al.* (1987). Some support for this is found in the case of SGS LMC 1 in Hunter's data (1994).

SGS LMC 1 is a complete shell composed of filaments of rather low surface brightness. It is partly incomplete on the western side. In the southern part some HII regions are seen. The stellar association LH15 is responsible for the ionization. It is similar in structure to SGS SMC 1.

SGS LMC 3 is a large shell, possibly in collision with SGS LMC 2 at 30 Doradus (see above). Chu *et al.* (1994) have obtained *IUE* absorption line spectra at five positions in the shell and detected blue-shifted absorption in low-ionization species as well as C IV and Si IV absorption at some positions.

There are a fair number of filamentary structures extending NE and SW of SGS LMC 3, which has been referred to as 'supergiant features' (Meaburn *et al.* 1987).

SGS LMC 5 is the second smallest SGS; it is probably in collision with SGS LMC 4 (see above). It has a high HI column density and a relatively large CO concentration. In the 'collision area' X-ray emission is seen (Bomans *et al.* 1994).

The remaining shells, SGS LMC 6, LMC 7, LMC 8, and LMC 9, have not been studied in any detail. They appear to be rather similar in size and luminosity with no particularly outstanding filaments or HII regions.

We may conclude from our brief summary of the superassociations that they include the supergiant shells and practically all of the youngest population in the LMC. There are, however, a number of 'run-away' stars and clusters *outside* the LMC main plane, which have to be fitted into the picture. They all appear to move in the way seen in the expansions of SGS LMC 4 and SGS LMC 2 perpendicular to the plane.

6.1.4 Stars and clusters outside the LMC main plane

SGS LMC 2, the 30 Doradus nebula, the HI 'L-component' (Luks and Rohlfs 1992) and some HI sheets (Meaburn et al. 1987) appear to be outside the main HI plane of the LMC.

Evidence exists that several young stars are also outside the major HI plane. McGee (1964) identified 18 early-type supergiants with velocities and reddening, indicating that they are outside the LMC plane. Twelve of them group around a velocity of 35 km s^{-1} larger than the HI velocity in their neighbourhood, six around a 34 km s^{-1} smaller velocity. These velocities are similar to the expansion velocity of SGS LMC 4 (see above) and may be typical for the motion of young objects forced out of the plane.

Also several clusters and associations may be in front of or behind the HI main plane. By comparing their ultraviolet observations with Hα and HI surveys and using colour excesses Page and Carruthers (1981) found four stellar groupings behind the HI clouds and 34 in front of them.

Fragmentation of the LMC in the line of sight has been observed by several astronomers (see Sect. 8.7). Thus, Songaila et al. (1986) found that their CaK absorption spectra showed features at four general velocity regions: strong absorption near zero (LSR) velocity and greater than $+180$ km s^{-1} and weaker features near $+60$ km s^{-1} and near $+20$ km s^{-1}. The zero velocity material is gas in the galactic disk. The strongly absorbing high-velocity gas, generally with several components between 180 and 280 km s^{-1}, is in the LMC. Along most lines of sight the CaK velocity width is similar to, or larger than, that of the hydrogen, suggesting that most of the observed stars lie *behind* the bulk of the gas. Also the weak features were considered to belong to the LMC, which is thus extended and fragmented along the line of sight. These fragments belong to the younger population in the LMC and may be connected with the sheet of neutral hydrogen mentioned above.

We may ask to what extent the older populations in the LMC have been exposed to similar fragmentation in their early ages. The distribution of the intermediate-age populations over the LMC is nowadays such that conclusions about the intial conditions are difficult to draw. There is, however, no reason against assuming that they originated in bursts of star formation and that their evolution continued in a way similar to that of the youngest generation. As intermediate-age stars and clusters are well dispersed over the whole LMC, star formation must have started more widely. A major centre of star formation must have been in the Bar region. More recent interactions between the SMC, the LMC and the Galaxy have disturbed the original structures severely.

6.2 Superassociations in the SMC

The only constellation identified in the SMC by Shapley is the one with NGC 456, NGC 460, and NGC 465 as major concentrations. It will be identified below as SMC SGS 2. Another possible superassociation further out in the Wing is the one with the stellar groups NGC 602a, b and c as major concentrations (Westerlund 1964d). This aggregate was later described as a supergiant shell, SGS SMC 1 (Meaburn 1980) and given about twice the size of the NGC 602 complex with a diameter of ~ 600 pc. On its western side are the HII regions N88 and N89 and the clusters L101, L103, and L104.

At least two superassociations are likely to exist in the SMC Bar. The surface distribution of O–B2 stars (Azzopardi and Vigneau 1977) shows one major concentration in the northern part of the Bar and three minor in the central and southern parts. The separation between the northern and the other aggregates is rather complete and is also seen in the distribution of clusters. Shell structures are evident in the Hα + [NII] charts presented by Davies *et al.* (1976); the question is if the giant shells are part of a supergiant shell. The structures seen in their Plate XVIII may be compared with those in their Plate III, of SGS LMC 4. If we bear in mind that the SMC may be more extended in depth than the LMC, a SGS SMC 3, centred $\sim 0^h 54^m.5, -72°34'$ and with a diameter of ~ 1200 pc, may be identified. It is also possible that a fourth supergiant shell exists in the southern end, SGS SMC 4, see their Plate XVII.

There is every reason to consider the youngest population in the two Clouds as predominantly located in superassociations, frequently inside supergiant shells. A large number of studies have led to the belief that sequential stellar evolution exists and many propose that large-scale SSPSF dominates the formation of the youngest population in the LMC. There appears to be little reason to exclude the SMC from this kind of process.

7

The intermediate-age and oldest field populations

For a complete picture of the history of star formation in the Magellanic Clouds, knowledge about the age distribution as well as the surface distribution of the intermediate-age and the oldest stars is needed. The evolved red giants are the most luminous in these generations and have therefore attracted most interest and contributed much to our understanding of the stellar evolution. It is now possible to reach stars as faint as $V = 24$ mag, so that main-sequence field stars much older than 3–4 Gyr may be observed in the Clouds. Their importance for solving star formation problems will increase as observations of still fainter stars have become available (using the *HST*). A brief summary of what is known today is therefore appropriate.

7.1 Main-sequence stars

A large number of colour–magnitude diagrams reaching main-sequence stars below $M_V = 2$ mag have been presented for regions in many parts of the LMC. The bulk of stars appear to have formed about 3–4 Gyr ago (Butcher 1977, Stryker 1984, Hardy *et al.* 1984, Hodge 1987). Some doubts have been expressed about the reliability of the photographic photometry on which these results were based, as well as on the age calibrators used, but the overall results have been confirmed (Vallenari *et al.* 1994c). The main-sequence differential luminosity functions for seven LMC fields are compared in Fig. 7.1. The fields identified in the diagram are: Ardeberg *et al.* (1985), ~ 1 kpc S of the Bar; Butcher (1977), 3.5 kpc N of the Bar; Hardy *et al.* (1984), in the Bar; Hodge (1987) 4.5 kpc SW of the Bar; Stryker (1984), 8 kpc NW of the Bar; HS96, SL196, NGC 1787 and NW (comparison field), 3.5 kpc NW of the Bar (NW region); and LW55 (Westerlund *et al.* 1995a, SW region), 3.8 kpc SW of the Bar.

The slopes of the LFs five of the fields also agree well with the general luminosity function (ϕ, Salpeter function). Exceptions are the two fields in the SW, 'LW55' and 'Hodge' which are appreciably steeper, indicating a lack of the younger population. This is in line with the east–west asymmetry in the LMC (see Sect. 12.1). The Bertelli *et al.* (1992) fields at NGC 1783 and NGC 1866 are in full agreement with those in the 'NW region'.

A further analysis of the main sequences in the NW and SW regions of the LMC shows that the former contains a young component, 0.2–0.6 Gyr old, which is completely missing in the SW region. The latter contains a well-identified old generation, $\geq$ 7–10 Gyr old, which may also be traced in the NW region. The

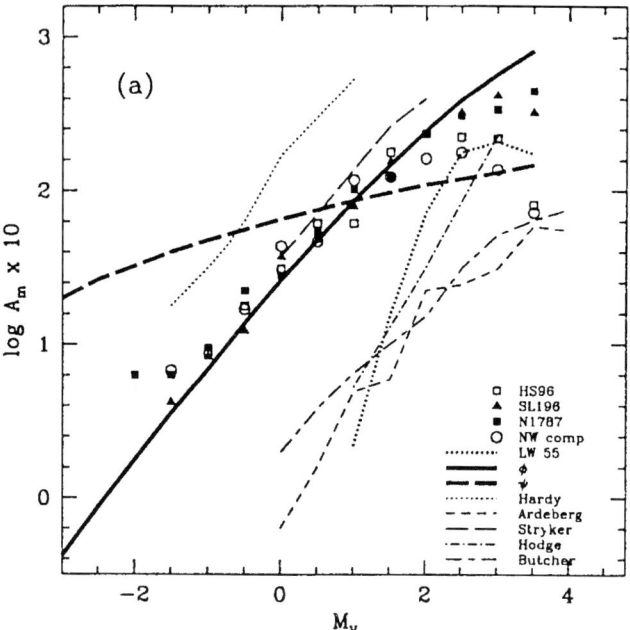

Fig. 7.1. Comparison of the main-sequence differential luminosity functions for seven fields in the LMC. ψ is the Salpeter initial luminosity function and ϕ the corresponding general luminosity function, both normalized to the field around SL196.

star formation behind this generation must have been weak as compared with that occurring during the period 4–0.2 Gyr ago. Vallenari et al. (1994c) found, however, that in their southern field 'LMC 56' the bulk of star formation started 7–8 Gyr ago; it is not clear if younger populations were identified.

In the SMC most studies of the field stars have dealt with the youngest populations, the variables and the red-giant stars. For these stars as well as for all other objects in the SMC a major uncertainty in interpreting the data is the possible, and unknown, extension in depth of their population (as well as the true location of any cluster along the line of sight). Some idea about the age distribution of the old main-sequence stars may nevertheless be obtained. Colour–magnitude diagrams exist for fields around a number of clusters situated between 3.7 kpc west of the centre to 4.2 kpc east of it: L1, 3.7 kpc W of the SMC centre (Olszewski et al. 1987); NGC 121 (Stryker et al. 1985) and K3 (Rich et al. 1984), 2 kpc W of the SMC centre; NGC 152, 1.6 kpc W of the SMC centre (Melcher and Richtler 1989); NGC 330 in the Bar (Carney et al. 1985); NGC 411, at the northern end of the Bar (Da Costa and Mould 1986); and L113, 4.2 kpc E of the SMC centre (Mould et al. 1984). The field population has in all cases [Fe/H] between -1.3 and -1.4 except in the case of the NGC 330, where the values found in the literature, [Fe/H] ~ -0.7 (see Chap. 11), probably refer to young association stars.

The field around L1 contains stars of the same age (and metallicity) as the cluster, i.e. around 10–12 Gyr old. There are no main-sequence stars above the turn-off, at $V = 22.3$ mag.

In the field near NGC 121, 2 kpc W of the SMC centre, the main sequence is seen at and below the turnoff at $R \approx 22.2$ mag. The bulk of the field population has an age of 8–14 Gyr.

The field around K3 has a weak main sequence reaching $R \approx 20.2$ mag. The ages of the youngest stars may be estimated to ~ 1 Gyr. A relatively rich population older than 10 Gyr appears to exist. These results are confirmed by the shape of the main sequence in a field 300 pc NE of K3 (Hawkins and Brück 1982), where the main-sequence turn-off is at $V \simeq 21$ mag with a number of blue stars scattered above this point.

The field around NGC 152 shows a mixture of populations. The main sequence is weak and reaches $V \simeq 16.5$ mag, corresponding to an age ≥ 0.1 Gyr. The majority of the blue stars have $V \geq 20.5$ mag.

In the field around NGC 330 two populations are identified, one very old, ≥ 10 Gyr, and one relatively young, $\simeq 18$ Myr, but older than the cluster itself (12 Myr). NGC 330 is possibly the youngest member of a stellar association (see Sect. 5.9), represented by the observed field stars.

Around NGC 411 the field stars cover a large range in age. The youngest are < 0.2 Gyr old; the majority are 7 Gyr old or older.

The bulk of field stars around L113 are older than the cluster, probably around 10 Gyr old. A few main-sequence stars are brighter than $R = 22$ mag, some of them are probably younger than the cluster itself. They are probably members of the SMC Wing and do not represent the general field.

In order to obtain as complete a coverage as possible some photographic studies may be considered even if they are not as deep and well defined as those just described. Brück and Marşōglu (1978) observed two fields north of NGC 458, one at the top of the Bar, the other 0.9 kpc further east, reaching $V \approx 21$ mag. In the former field a well-populated main sequence is seen, with a turn-off at about $V = 17.5$ mag and a top at about $V = 15$ mag. In the latter no main-sequence stars exist brighter than $V \approx 20$ mag. The stellar distribution in this area appears similar to that around NGC 152, on the opposite side of the Bar.

In the outlying areas studied by Hatzidimitriou and Hawkins (1989) (the SW area is between 2.5 and 4.1 kpc SW, the NW area between 2.2 and 4.5 kpc NE of the SMC centre) the bright end of the main sequence reaches $R \simeq 18.0$ mag in the innermost parts but fades away rapidly and is below the limiting magnitude ($R = 20$ mag) about 4 kpc from the centre. This corresponds to the minimum ages of the youngest population in these areas of 0.3 and $\simeq 1$–2 Gyr, respectively.

7.2 Red-giant stars

Red-giant stars have been identified in both Clouds and are now known to exist in tens of thousands. In the LMC they are easily recognized as M stars. Few such stars are seen in the SMC, where the lower metallicity makes the red giants appear mainly as K stars. Extremely red stars with $B - V \approx 3$–4 mag were first recorded in and near the SMC cluster NGC 419 (Arp 1958). They were later identified as carbon stars and several thousands are now known.

The main branches in the evolution of *low-mass* stars, i.e. stars with MS masses ≤ 2.3 $M_\odot$, are best identified in HR or CM diagrams of globular clusters (see Figs. 4.3

and 4.5). Major features are: the main sequence (MS), the turn-off point (TO), the subgiant branch (SGB), the red-giant branch (RGB), the horizontal branch (HB) and the asymptotic giant branch (AGB).

The RGB life of a star begins when the He core produced by the H burning collapses and the star leaves the main sequence. About half of the gravitational energy of this collapse becomes available as heat. The star develops a H-burning shell and a deep convective envelope as it becomes cooler and more luminous. Eventually He starts burning in the minute core. If the core has electron degeneracy it is highly conductive and the burning may start explosively. The RGB period ends with a 'He flash'.

The low-mass stars reach a maximum luminosity of $\sim$ 2500 L$_\odot$ ($M_{bol} \approx -3.8$ mag), at the top of the RGB. They reduce their luminosities to $\sim$ 60 L$_\odot$ ($M_{bol} \approx 0$ mag) at the beginning of the core He-burning since the degeneracy is lifted by the core flash. Population I stars with initial masses of 1 M$_\odot$ remain on the RGB as so-called clump stars (RGC), whereas Population II stars with initial masses of 0.8 M$_\odot$ reach positions along the horizontal branch depending upon the thickness of the remaining hydrogen envelope.

Intermediate-mass stars do not produce a degenerate core and they do not climb the RGB. In the HR diagram they evolve nearly horizontally from the main-sequence to the red-giant region. From there on they move in loops (towards the blue) during central He burning. They spend a minor part of the time in the loops as Cepheids; most of the time is spent at the high-temperature end of the loop.

After central helium exhaustion the low- as well as the intermediate-mass stars evolve upwards to and on the AGB; the very blue HB stars may evolve directly towards the white dwarf stage. The stars increase in luminosity on the early AGB (E-AGB) during a phase of thick helium shell burning. This is followed by a double-shell-burning phase in which helium is mainly burnt during shell flashes and thermal pulses (TP). A degenerate core of C/O is produced. Its growth is responsible for the increase in luminosity along the AGB. The maximum luminosity of an AGB star is derived from the 'Paczynski relation':

$$L/L_\odot = 59000 \times (M_{core}/M_\odot - 0.50).$$

The maximum is set by the requirement $M_{core} \leq$ the Chandrasekhar limit of 1.4 M$_\odot$, thus $L < 53\,000$ L$_\odot$ ($M_{bol} \approx -7.2$ mag).

The AGB evolution is normally terminated by the loss of the outer envelope due to stellar winds and planetary nebula (PN) formation. The remnant core will evolve horizontally in the HR diagram towards higher temperatures. When the star is hot enough to ionize the surrounding material, a PN is seen.

Overshooting and early mass loss will change the picture. The former will increase the stellar cores during phases where central burning occurs under convective conditions. An important result is that the limit between low- and intermediate-mass stars (i.e. the limiting mass for non-degenerate helium ignition) is shifted from $\sim$ 2.3 M$_\odot$ to 1.7 M$_\odot$. Also the upper limit for nondegenerate carbon ignition, at $\sim$ 9 M$_\odot$, is shifted to $\sim$ 6 M$_\odot$. The range for the production of an AGB star from an intermediate-age star is thus reduced. Mass loss will modify the picture further (see Sect. 7.2.6).

7.2 Red-giant stars

7.2.1 Surveys

The first systematic search for M and carbon (C) type stars in the Magellanic Clouds was carried out by Westerlund (1960, 1964a), who used the Schmidt telescope at Uppsala Southern Station, Mount Stromlo Observatory, to survey the central parts of the LMC. With a dispersion of 2200 Å per mm at the atmospheric A-band, stars brighter than $I \approx 13.5$ mag could be classified with the aid of the TiO and CN bands in the (photographic) near infrared (7000–8600 Å). 225 stars were classified as M0 or later in the Case system (see e.g. Mavridis 1967) and 9 as carbon stars. The central region of the LMC was surveyed by Blanco and McCarthy (1975) using objective-prism spectra with a dispersion of 6700 Å mm^{-1} at the atmospheric A-band. They identified 803 late M stars (M6.5 and later) in a field of 8.75 deg^2. The dispersion was too low to permit the identification of carbon stars. These surveys were extended later on to cover most of the LMC (Westerlund et al. 1978, C stars; Westerlund et al. 1981, M stars). A confirmation of their classification of M stars was obtained when Humphreys (1979) carried out slit spectroscopy of a number of M supergiants and giants in the LMC. Sanduleak and Philip (1977d) used the TiO bands at $\lambda 4955$ and $\lambda 5168$ to identify M stars in the LMC. They also identified C stars from the C_2 bands in the visual-green spectral region. Most of the early M stars identified in these surveys are supergiants (see Sect. 5.4).

Both Clouds have been sampled for carbon stars by Blanco et al. (1980), using the near infrared spectral region. On the basis of the sampled areas the total number of carbon stars was estimated to be about 11 000 in the LMC and about 2900 in the SMC (Blanco and McCarthy 1983). Identification charts for 849 C stars in the LMC sample have been presented (Blanco and McCarthy 1990). A more complete survey of the central part of the SMC was carried out by Rebeirot et al. (1993). Their C star catalogue lists 1707 stars with positions, apparent magnitudes, colours, and C_2 band strengths. A survey of the outer parts of the SMC has been carried out by Morgan and Hatzidimitriou (1995), adding 1185 carbon stars to those previously known. They covered a total area of ~ 220 deg^2 including the InterCloud region. On the basis of the two most recent surveys the total number of carbon stars in the SMC is estimated to be $\simeq 3060$. Their surface distribution is identical to that of the red horizontal-branch/clump stars (see below).

Few S stars have so far been found in the fields of the Clouds. Blanco et al. (1981) discovered an S star in the SMC field and Richer and Frogel (1980) identified an SC star in the LMC. During the investigation of the AGB stars in clusters, several S stars have been identified. They define the transit stage between M and C types during the evolution along the AGB in clusters of ages corresponding the SWB type V and IV (see Westerlund et al. 1991).

The red variables in the field have been investigated rather extensively. Wood et al. (1985) found in a study of long-period variables in and to the north of the LMC Bar that (i) the bulk of the stars had MS masses $M_{MS} \leq 1.6$ M$_\odot$ and ages $\geq$ 1 Gyr, and (ii) that there was a distinct group of stars with $M_{MS} \approx 5$–6 M$_\odot$ and age ≈ 60 Myr. There are also Cepheids of this mass in the LMC Bar, as well as a number of clusters, suggesting that a burst of star formation occurred there about 60 Myr ago. Any luminous AGB stars corresponding to these Cepheids would have to have luminosities in the range $-6.5 > M_{bol} > -7.3$ and would then be seen as red variables, bright early M giants, or early M supergiants (Becker 1982).

7.2.2 The RGB stars

The RGB stars represent the oldest population in the Clouds. Its contribution is rather insignificant in the LMC Bar area,* though it has been traced (Hardy et al. 1984). Wood et al. (1985) have suggested that a group of short-period oxygen-rich variables in and to the north of the LMC Bar are analogous to the Population II Mira variables (such as in the globular cluster 47 Tucanae) in the Galaxy. They would belong to a very old, ≥ 10 Gyr, moderately metal-poor population.

In the NE part of the LMC, around NGC 2257, the field population appears to be significantly younger than the cluster itself (Stryker 1984). Stryker suggests that the outer LMC regions may have been inactive for a few Gyr after the initial collapse of the CLoud, and then formed the observed field stars.

In the Shapley Constellation III area the metallicity of the red-giant stars has been determined to be [M/H] $\sim -1.1 \pm 0.2$ with the aid of the $V-I$ colour of the RGB at $M_I = -3.0 \pm 0.25$ mag (Reid et al. 1987). This is slightly more metal-rich than the estimate, [M/H] ~ -1.4, obtained from RR Lyrae stars near the LMC Bar (Butler et al. 1982). The $V-I$ width of the RGB is, however, appreciable, so it is possible that an older, more metal poor generation, also exists in this field.

In the SW part of the LMC, 4.5 kpc from its centre, the field is dominated by an intermediate-age population, 2–3 Gyr old, with no trace of younger stars (Hodge 1987). It is also suggested that no trace of a true Population II can be seen, but this conclusion appears premature due to the large scatter in the diagram.

In another field, 'LW 55', in the SW, 3.7 kpc from the LMC centre, a weak RGB has been traced with a population similar to that in old galactic globular clusters (Westerlund et al. 1995a). Its red star luminosity function (LF) is shown in Fig. 7.2 and is compared with that of other fields and of the galactic globular M 3. The agreement with the M 3 population is good around the HB range ($M_V \sim 0.5$ mag). A difference not visible in the LF is that the HB stars in M 3 are blue, in the LW 55 field red. The latter population is younger and less metal-poor than that of M 3. The dominating red population in the other fields is appreciably younger and more mixed. The RGC stars contribute strongly and the ages of the involved stars may be estimated to between ~ 3 and 0.5 Gyr.

In the SMC the oldest field population appears to be about 10 Gyr. In its outer parts, around L1 in the west and L113 in the east, this generation dominates (see Sect. 7.1). This is in contrast to the situation in the LMC where the intermediate-age population also dominates the field in the outer regions.

7.2.3 The HB stars

Horizontal-branch stars may be red (RHB), blue (BHB) or RR Lyrae variables. The ratio of RHB/BHB stars is strongly variable and depends upon metallicity and age (see Sect. 4.4.1 for clusters).

7.2.3.1 RR Lyrae variables

The first RR Lyrae variables in the Clouds were discovered by Thackeray (1951) in NGC 121. As they were of the 19th magnitude they were first considered as possibly lying beyond the SMC, but it was soon understood that the distance scale

---
* Population II stars should reach $V \sim 15$ mag in the Clouds and be relatively easily identified.

7.2 Red-giant stars

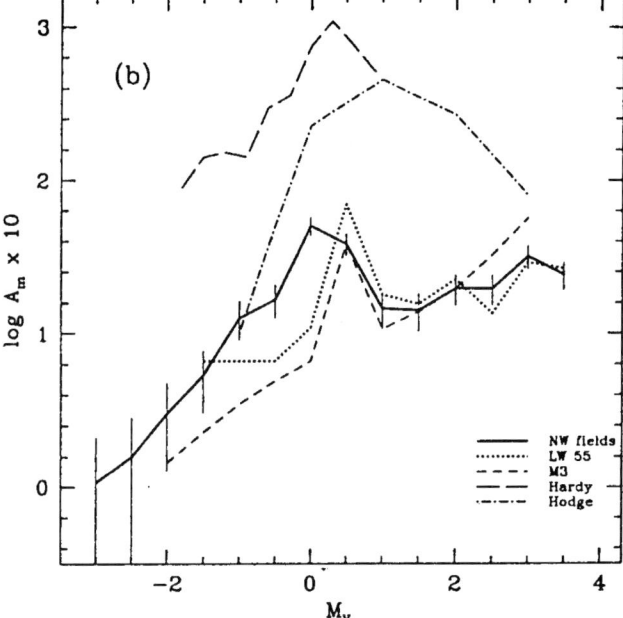

Fig. 7.2. Red-giant luminosity functions for the LMC fields: Fields 'NW' and 'LW 55' (in the SW) are from Westerlund et al.(1995a), 'Hodge' (in the SW) from Hodge (1987) and 'Hardy' (in Bar region) from Hardy et al.(1984). The LF for the galactic cluster M 3 is normalized to the LW 55 field.

was wrong. (The distance scale in use in 1951 was wrong by a factor of ~ 2; Baade's announcement of a correction of $1^m.5$ in the modulus of M 31 was presented at the IAU General Assembly in Rome in 1952.) As more RR Lyraes were discovered in the Clouds, in NGC 1978, 1466 (Thackeray and Wesselink 1953, 1955) and NGC 2257 (Alexander 1960), they became the first proof that a population similar to Population II in the Galaxy existed there. There are now seven globular clusters with RR Lyraes known in the LMC (Table 4.2).

Our knowledge about the field RR Lyrae variables in the LMC and the SMC is still limited, as there are no complete surveys of the Clouds for these stars. They may have a smooth surface distribution in both Clouds with slight increases towards the central regions (see Graham 1984). In the SMC, the RR Lyraes appear to concentrate towards the west of the main body.

Most of the field RR Lyraes studied have been found in the neighbourhood of globular clusters. Well-studied regions are a 50' diameter field around NGC 2257 (Nemec et al. 1985) and the 1° × 1.3° fields around NGC 121 and NGC 1783 (Graham 1975, 1977). The latter two fields may be incomplete for RRc stars.

Nemec et al. identified 47 variables in the field around NGC 2257. Period and light curves were determined for 24 RR Lyrae stars. Mean periods were $\langle P_{ab} \rangle = 0^d.564$ (13 stars) and $\langle P_{cd} \rangle = 0^d.363$ (10 stars), characteristic of Oosterhoff group I systems. The identification of two large-amplitude, relatively short-period ab-type RR Lyraes suggests that the field around NGC 2257 may contain old stars of higher metallicities than the cluster. A wide range of metal abundances may exist in this remote field.

RR Lyrae stars near NGC 2210 have, from their period–amplitude relation, a metallicity of [Fe/H] = −1.8 (Hazen and Nemec 1992). This value is slightly larger than that found for the eight certain LMC globular clusters (Suntzeff et al. 1992) for which ⟨[Fe/H]⟩ = −1.9 (Sect. 4.4.1).

Kinman et al. (1991) found that the cluster-type variables in the LMC halo (flat or spherical) have metallicities similar to those in the (outer) galactic halo; the latter RR Lyrae stars have ⟨[Fe/H]⟩ = −1.65.

The bulk of the field stars in the LMC halo may thus have formed *after* the LMC globular clusters. The LMC halo would then resemble the outer regions of giant ellipticals where the globular clusters are more metal-poor than the dominant stellar population.

The period distribution for the RR Lyraes in the LMC resembles that found at high latitudes in the Galaxy, whereas in the SMC it is more characteristic for dwarf spheroidal galaxies. RR Lyraes of type ab with large amplitudes and periods ≤ 0.45 days, relatively common in the solar neighbourhood, are rare or absent in both Clouds. Two such stars have been found in the field of NGC 2257 (see above).

The SMC contains a small group of apparently normal RR Lyraes which have a mean apparent magnitude about 1.5 mag brighter than the average value. No such stars are yet known in the LMC (Graham 1975, 1977). These overluminous stars have a metallicity of [Fe/H] =− 0.4 and an age of 0.25 Gyr. They belong, consequently, to the intermediate-age generation. The normal RR Lyraes in the Clouds have −1.4 ≥[Fe/H] ≥ −2.0 and ages ≥ 10–12 Gyr.

The observed surface densities of RR Lyraes in the LMC are 88 deg^{-2} in the NGC 2257 field and 75 in the NGC 1783 field (Nemec et al. 1985). The corresponding value for the SMC is 80. If the distributions are uniform the total numbers should be a few times 10^4 in each galaxy.

Frogel (1984) used the statistics for the RR Lyrae stars in the Magellanic Clouds to estimate the contribution of the oldest generation to the total mass. He derived a value of about 6% – comparable with the ratio of spheroidal to disk mass for the Milky Way within the solar circle.

7.2.4 The RGC stars

The RGC stars are more massive and younger than the HB stars. In the HR diagrams of fields and clusters in the Clouds they appear as a group on the low-temperature side of the RGBs, level with the HBs. Various views on the dependence of the RGB on metallicity and age have been presented. Mateo and Hodge (1985) have argued that the M_V (clump) is a constant if the cluster is older than ∼ 0.3 Gyr but otherwise age-dependent. Olszewski et al. (1987) have confirmed the constancy for stars up to ∼ 10 Gyr (see also Sect. 4.4.2).

The RGC is probably metallicity dependent in the same way as the RGB at the level of the HB. For galactic globular clusters Sandage (1982) derived the relation

$$(B-V)_{0,g} = 0.180 \,[\text{Fe/H}] + 1.10,$$

and Zinn and West (1984) obtained

$$[\text{Fe/H}] = -5.00 + 4.30 \,(B-V)_{0,g},$$

where $(B-V)_{0,g}$ is the intrinsic colour of the RGB at the level of the HB.

7.2 Red-giant stars

Important for the use of the RGC stars as tracers of the structure of the LMC and the SMC is the knowledge of the size of the clump in the HR diagram for a coeval population. Hatzidimitriou and Hawkins (1989) found from galactic globular clusters a mean $M_V = 0.75 \pm 0.06$ mag; the same value was derived from their SMC field clump stars (all stars are assumed to be older than ~ 0.5 Gyr). The range is estimated to be 0.6 mag (from their R mag). Any spread over this size is interpreted as a spread of the stars in depth. It should be observed, however, that the spread in M_V is appreciable if stars younger than ~ 0.5 Gyr are involved. The RGC magnitude difference between a cluster 0.3 Gyr old and a metal-rich galactic globular cluster (M 71) is ~ 1.5 mag (Mateo and Hodge 1985, Fig. 7). The colour range, from the metal-poor M 92 to M 71, is ~ 0.3 mag. Consequently, the size of the clump will, to some extent, depend upon the composition of the field.

Hatzidimitriou (1991) uses the colour difference between the median colour of the RHB or RGC and the RGB at the level of the RHB as an age indicator for clusters and field populations. The difference is clearly independent of reddening and distance but is also insensitive to metallicity, at least for the range from solar abundances to [Fe/H] $= -1.7$. Its application to a few fields in the SMC gives ages varying from 3–4 Gyr near NGC 411, 1.6 kpc from the SMC centre, to 10–12 Gyr, 2–4 kpc from it. The older field population near NGC 2257 in the LMC obtains an age of 7–9 Gyr, in agreement with the determination by Stryker et al. (1985).

7.2.5 The AGB stars

Most stars in the mass range ~ 0.8 to ~ 7–9 M$_\odot$ become AGB stars. Cepheids are the immediate precursors of the more massive AGB stars ($M > 4$ M$_\odot$). It should be possible to infer the expected number of luminous AGB stars from the period distribution of the Cepheids.* In an attempt to do this Becker (1982) concluded that several of the Blanco et al. (1980) fields, containing Cepheids, should have had bright AGB stars. They were not found. A reason for this could be that their AGB surveys did not detect early M giants; the most massive AGB stars could possibly be of this type.

There are two main types of AGB stars, M stars on the E-AGB and C stars on the TP-, or upper, AGB, with S-type stars as a transit phase. Decisive for the type is the C/O ratio in the outer layers of a star. With both elements present CO is formed. If there is still an excess of O, TiO and other oxygen-rich molecules may form, and bands typical for M stars appear in the spectra. When carbon is overabundant, molecules such as C_2 and CN form and the result is a carbon star. During the stellar evolution a period may come when C/O is ~ 1.0. Then, bands of ZrO, LaO, and YO form if these metals are available in the stellar outer layers, and an S star results. Enrichment of these elements in these layers is possible by convection from the stellar interior where such nuclei are formed by the 's-process'. The enrichment of carbon in the outer layers of a star may occur through a dredge-up process in the stellar convective envelope. These processes are, according to current theories, possible when a star is on the AGB and the energy is mainly produced in the H-burning shell with a small contribution from the inner He-burning shell. The latter shell may periodically sink into higher-temperature regions, increasing the He-shell energy output and leading

---
* Stars less massive than ~ 4 M$_\odot$ fail to intersect the instability strip during the post-helium flash blue loops at the metallicity in the Clouds.

to a 'thermal pulse'. Following each such event the outer envelope may reach into the region where He has been transformed into C and bring up C-enriched matter and probably also s-process elements (e.g. Ba, Zr). Thus, successive pulses lead to an increase in the C/O ratio in the outer layers and an M giant may be transformed into a C star via MS, S, CS star phases. If these processes of convective mixing occur on the RGB, the dredge-up of C is supposed to be less efficient, and still more so the formation of s-process elements.

On the upper AGB, where the stars undergo thermal pulsations, they may also experience the shorter-term atmospheric pulsations that lead to Mira-type variability.

The astrophysical processes involved in these processes are mixing to the stellar photosphere of products of interior nucleosynthesis, mass loss from the stellar surface, and return to the interstellar medium (ISM) of processed material. Detailed observations are needed to determine the importance of mass loss for the evolution and to test the surface composition predictions.

Comparisons of theory and observations of the AGB stars in the Magellanic Clouds led immediately to the important 'carbon star mystery'. Many low-luminosity carbon stars ($M_{bol} > -4$ mag) were observed that had not been predicted, and few of the numerous high-luminosity carbon stars ($M_{bol} < -5.5$ mag) predicted were observed.

The luminosity of a star on the AGB is a direct function of the mass of the degenerate core. The maximum luminosity is set either by the removal of the outer envelope and the formation of a PN, or by the core mass reaching the Chandrasekhar limit (see above). The luminosity of the tip of the AGB is a monotonically increasing function of stellar mass (see e.g. Renzini and Voli 1981). All AGB stars with $M_{bol} < -6$ mag have descended from MS stars with $M \geq 3$ M$_\odot$.

The AGB stars are of intermediate age. The Magellanic Clouds are, in contrast to our Galaxy, rich in clusters of ages between 4 and 0.1 Gyr, and AGB stars are found in many of them (see catalogues by Frogel et al. 1990 and Westerlund et al. 1991). It is evident that younger clusters contain brighter AGB stars. Carbon stars are seen in clusters with ages between 0.1 and 3–4 Gyr; in the SMC there are possible C stars even in older (~ 10 Gyr) clusters.

7.2.6 AGB stars in the LMC

In a survey for M giants in the western part of the LMC Bar, Frogel and Blanco (1983, 1990) have shown the existence of two distinct groups of E-AGB stars (Fig. 7.3). Stars in the most luminous of the two groups are similar to stars in LMC clusters of ages a few times 10^8 yr. The fainter branch corresponds to older clusters, of SWB type V–VI, i.e. of ages a few times 10^9 yr. The bright AGB giants have systematically stronger CO indices than the fainter ones of the same colour. The rate of star formation in the older episode was estimated to be about ten times the rate in the younger one. However, as the older episode may cover a period ten times as long as that of the younger one, the rates of star formation may well have been the same. Between the two episodes the rate of star formation must have dropped significantly. This supports the view that star formation in the Magellanic Clouds has been dominated by bursts throughout their history.

On the basis of theoretical calculations Frantsman (1988) and Frantsman and Shmield (1995) have suggested that the more luminous of the two groups of M giants observed in the Bar West field consists of TP-AGB stars. However, the TP-AGB

7.2 Red-giant stars

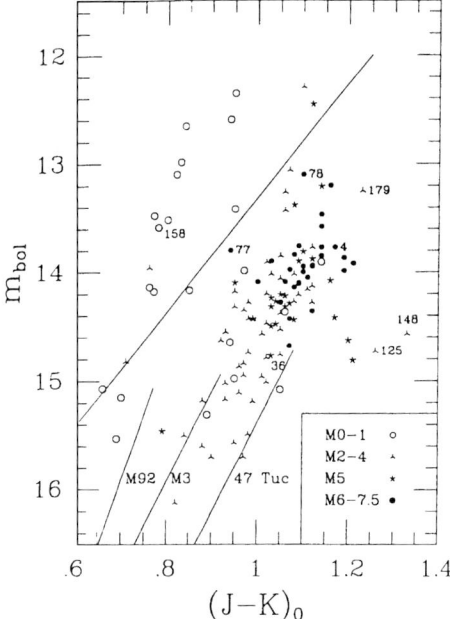

Fig. 7.3. Colour–magnitude diagram for M giants in the LMC Bar West field (Frogel and Blanco 1990). The sloping straight line separates AGB stars of SWB cluster types I–III from later types (right side). Numbered objects are extreme in colours or CO content. The giant branches for the three galactic clusters are drawn assuming a LMC distance of 45.7 kpc.

branch in their $K_0/(J-K)_0$ diagram occupies part of the position taken by carbon stars in the Clouds (the carbon stars reach much redder colours), and *not* by the M giants. Their E-AGB branch has the position of the SWB cluster type VII AGB branch (with SMC metallicity and ages ≥ 10 Gyr), whereas the region occupied by the E-AGB branches of the younger SWB cluster types is empty (see Westerlund *et al.* 1991). With the information available it is difficult to find an explanation for this lack of E-AGB stars of ages ≤ 3 Gyr in the theoretical diagram. There is, anyhow, no reason to reject the suggestion of two age groups of AGB stars in the Clouds.

Extensive photometric and spectroscopic studies of AGB stars, identified by having colours $(V-I)_0 > 1.50$ mag, in a 15 deg² field in the northern part of the LMC ($\sim 5^h - \sim 5^h 45^m$; north of $-68°20'$) have been carried out Reid and Mould (1984, 1985, 1990), Mould and Reid (1987), and Reid *et al.* (1987, 1990). A bolometric luminosity function was derived. The deficiency of luminous AGB stars of all spectral types, i.e. not only of carbon stars, noted in earlier spectroscopic studies, was confirmed. Significant variations in the LF over the field were noted. Particularly strong was the effect of red supergiants, non-AGB stars, in the quadrant containing Shapley Constellation III.

A comparison with the Frogel and Blanco (1983) data for the western part of the LMC Bar gave a good fit and a reason to believe that at least two discrete epochs of star formation have also occurred in this part of the LMC.

Models for the star formation in the region were calculated, considering constant and declining star formation rates (SFR). All led to a confirmation of the deficiency of luminous AGB stars. The existence of 240 Cepheids in the field (Payne-Gaposchkin (1971) with an age of ~ 0.1 Gyr shows that star formation has been active there over the past few hundred Myr. Their distribution in the field corresponds to that of the more luminous AGB stars, $M_{bol} \leq -5.5$ mag, which is to be expected as they will eventually evolve to the AGB limit.

An explanation of the absence of luminous AGB stars (mainly luminous carbon stars) may be found in the mass-loss rates for red giants. A high mass-loss rate may either prematurely remove a star from the AGB or hide it in a 'cocoon'. The hidden stars would have strong infrared excesses. A search at 2 μm by Frogel and Richer (1983) in a field in the LMC Bar (west) failed to show any such objects. For the detection of cool stars, such as the OH/IR stars, the *IRAS* FIR data are needed (sect. 7.2.7).

Most of the stars with $-5 < M_{bol} < -5.5$ and several of those with $-4.5 < M_{bol} < -5.0$ show evidence of dredge-up s-process elements. 10 % of the sample were C stars, several objects were of type MS, one of type SC, all representing transition stages in the M–S–C sequence.

The C/M ratio for stars in the range $-7 < M_{bol} < -4$ varied over the field. The ratio may be metallicity dependent. A lower metallicity increases the fraction of C stars and shifts the M stars towards higher temperatures, thus increasing the ratio.* The ratio also depends upon the age of the population. If it is old, ~ 10 Gyr, the stars will have lost their envelopes before they have reached the critical core mass for thermal pulsing. The Reid and Mould data seem to indicate that a pronounced metallicity gradient exists in the LMC such that the Bar is less enriched than the outer regions. This appears, however, unlikely and is contradicted by studies of other objects. A likely explanation of the observations is, instead, that the regions studied have different star formation histories.

The best-fitting model to the Reid and Mould complete sample has a smoothly distributed population of stars with ages up to a few Gyr, together with a patchy young population from a recent burst of star formation. The former contains C and S stars in the range $-4.5 > M_{bol} > -5.5$. The young population consists mainly of luminous early M-type stars which are more likely to be core-helium burning than AGB stars. Spectroscopy of the 20 most luminous red stars ($-7 < M_{bol} < -6.4$) has confirmed this. There are, however, AGB stars in the luminosity range $-7 < M_{bol} < -5.5$ which show signs of the third dredge-up, namely the long-period variables (Wood *et al.* 1983) with pulsational masses in the range 2–7 M$_\odot$ (15 Miras are known in the field in question).

A search for hidden luminous AGB stars in a particularly crowded area in Shapley Constellation III was carried out among stars with $V - I > 1.6$ mag and $M_{bol} < -4$ mag. Two populations were found: a low-luminosity population, $M_{bol} > -5$ mag, of highly evolved late-type stars with signs of the third dredge-up and carbon stars at the upper limit, and a higher- luminosity population reaching $M_{bol} \simeq -7.5$ mag and containing no S, C or barium stars The mean velocity of the latter component was close to that of the nearside of the expanding HI shell of SGS LMC 4 (Sect. 6.1.1). It may represent a population slightly older than that now being formed in the

---

* Abundance differences explain the increase in C/M in going from the galactic bulge through the galactic disk to the LMC and on to the SMC.

7.2 Red-giant stars

shell proper. The older population had a mean velocity 21 km s^{-1} lower; in good agreement with the HI disk velocity.

The following explanations to the paucity of luminous AGB stars are suggested by Mould and Reid (1987):

(i) Envelope burning (Renzini and Voli 1981) could convert carbon produced in the third dredge-up to nitrogen. There is, however, no sign of *s*-processed material in the atmospheres of the luminous stars present.

(ii) Star formation may have ceased in this part of the LMC with a resulting deficiency of post-main-sequence stars of masses > 2 M$_\odot$ (to an upper limit of ~ 9 M$_\odot$). However, there are 180 Cepheids with periods between 3 and 15 days in the searched field so there should be more than 100 AGB-TP stars with $-7 < M_{bol} < -6$. Only 12 are known. Similarly, there are 9 such Cepheids in the Constellation III area but only 2 Miras and they are too luminous to be AGB stars. On the other hand, the bulk of star formation in Constellation III began ≈ 20 Myr ago, so the turn-off mass is ≥ 9 M$_\odot$.

(iii) Convective overshooting may reduce the upper limit for AGB stars to $M \sim 5$ M$_\odot$ (Bertelli *et al.* 1985). Then, the missing stars, with higher initial masses, may have become supernovae after having ignited their carbon.

(iv) Mass loss may be underestimated in the standard theory and may actually truncate AGB evolution at $M_{bol} \sim -5.8$. Thus, it is necessary to assume that the Mira phase occurs briefly after thermal pulses are ignited. This would lead to a superwind phase, lasting $\sim 10^5$ yr. About ten undetected OH/IR stars should then exist in the surveyed 15 deg^2 field.

The possibility exists that a significant fraction of the stars evolving on the upper AGB, $M_{bol} < -6$, may be hidden from optical detection by being embedded in dusty circumstellar envelopes. Attempts to find such objects have been made (see the following section) but so far have not led to the detection of a sufficient number of obscured stars. Mass loss appears, therefore, to be the most obvious mechanism for the upper cut-off of the AGB branch.

7.2.7 Optically obscured AGB stars in the LMC

The northern 15 deg^2 region has been searched by Reid *et al.*(1990) for optically obscured AGB stars using *IRAS* data and near-IR photographic plates. Of 156 *IRAS* sources detected in more than one passband, 63 have [12–25] colours, consistent with their being either stellar photospheres or circumstellar dust shells. 17 were identified with foreground stars in our Galaxy; a further 17 are associated with red LMC supergiants. At most 5 of the remaining sources are likely to be optically visible AGB stars while the rest have no obvious optical counterparts. The number of high-luminosity 'cocoon' stars is not sufficient to explain the observed deficit of several hundred luminous ($M_{bol} < -6$ mag) AGB stars between the predictions of standard models of AGB evolution and the observed luminosity function. Most of the unidentified objects may be dusty AGB stars, evolving through a phase of enhanced mass loss towards becoming PNe.These sources may have low M_{bol}, ~ -5 mag, and have been triggered at low luminosities so that the AGB evolution gets truncated, leading to the scarcity of AGB stars with $M_{bol} < -6$ mag.

A further analysis of *IRAS* data added 21 point sources, not in the *IRAS* point source catalogue. *JHK* photometry confirmed that nine had the expected flux distributions for dust enshrouded AGB stars (Reid 1991). The faintest had $M_{bol} = -5.1$ mag. The duration of this phase was estimated at 10^5 yr.

7.2.8 AGB stars in the SMC

In the SMC the most luminous AGB stars are red variables of spectral type MS (or S, depending upon classification criteria used). C stars of lower luminosity and higher temperature than in the LMC have been identified, indicating that C stars form at lower masses and higher temperatures in this more metal-poor galaxy (Westerlund et al. 1991, 1995b). Li has been found to be overabundant in luminous AGB stars ($M_{bol} = -6$ to -7) in the SMC (Smith and Lambert 1989). Enhanced Li may be common among luminous long-period variables in the LMC.

A photographic survey for the identification of AGB stars was carried out by Reid and Mould (1990) in a field of 0.82 deg^2 just east of the SMC Bar. V and I magnitudes were eventually derived for more than 61 000 stars. The resulting colour–magnitude diagram (I vs $V - I$) shows a giant branch with a position similar to that of the metal-poor galactic globular cluster M 92, well to the blue of that for their northern LMC field (Reid et al. 1987), but in slope more like that of the LMC field.

A moderate number of stars were found above the SMC giant branch. They are intermediate-age AGB stars and their Cepheid progenitors as well as more recently formed stars which have evolved beyond the main-sequence turn-off. Some, about 3 mag above the giant branch at $V - I = 1.6$ mag, are young red supergiants. Others, about 1 mag above the AGB, may be young and intermediate-age giants, showing a continuing star formation.

Spectra were taken of most AGB candidates with $M_{bol} \leq -3.25$ mag in three small fields.* Of 28 SMC stars with $M_{bol} < -5.5$ mag only four (are cool enough to) show TiO absorption. One of them is a MS star at -5.85 mag, the other three show no S type features. They may have reached the top of the AGB without changes in surface composition. The most luminous C stars in the field are at $M_{bol} = -5.7$ mag. 22 of the luminous AGB candidates are of type K4 or K5 (little evidence of TiO bands).

There are few red stars in the SMC with $M_{bol} < -6.0$ mag. As also predicted for the LMC, many should have been found. Again, underestimation of the effects of mass loss may be the explanation of the discrepancy.

The luminosity functions of the AGB stars in the present SMC field and in the northern LMC field, discussed above (Sect. 7.2.6), are very similar, suggesting similar star formation histories.

7.3 The carbon stars

Carbon stars are found in intermediate-age clusters in the two Clouds and are numerous in the field. Their luminosities reach from $M_{bol} \sim -6.5$ to -3 in the LMC and to -1.8 mag in the SMC (Westerlund et al. 1995b). It is not yet clear if the difference in the lower luminosity limit between the two Clouds is real or not, but, so far, searches for C stars in the LMC with $M_{bol} \geq -3$ mag have been in vain. The field carbon stars cover a larger range in luminosity than those identified in clusters. The most luminous field C stars in the LMC are J-type stars, possibly

---

* Galactic foreground stars were eliminated with the aid of their radial velocities. SMC stars are expected to have velocities between ~ 120 and ~ 200 km s^{-1}. Galactic halo giants may have similar velocities but only 1–2 are expected in this limited field.

CO-rich (Westerlund et al. 1991), CH stars (Sect. 7.4) and MS variables (Wood et al. 1983).

In the LMC C stars exist with luminosities placing them as a continuation of the two M giant branches discussed above (Westerlund 1987). The younger and most luminous C-star branch contains mainly J-type stars, i.e. ^{13}C rich, which, as in the case of the M giants, are more CO-rich (Cohen et al. 1981) than those on the fainter branch. At still higher luminosity, MS stars are observed (Wood et al. 1983, Lundgren 1988). The whole scenario agrees remarkably well with the evolutionary pattern modelled by Renzini and Voli (1981) for stars with masses between 3.3 and 5 M$_\odot$ and 'hot-bottom burning', making the C star phase short or non-existing.

In this scenario the older M giant branch continues towards higher luminosities as a rather wide sequence of C stars. It must be composed of a number of evolutionary paths indicating an appreciable spread in age.

In the SMC 70 % of the carbon stars fall in the magnitude range $16.5 \leq V \leq 17.7$ with a relatively symmetrical distribution (Rebeirot et al. 1993). A few stars have $M_{bol} \leq -6$ mag, most of them are very red. A rather large number of C stars have $M_{bol} \geq -3$ mag; the least luminous has $M_{bol} \approx -1.8$ mag. JHK photometry and medium-dispersion spectroscopy of about 30 of these stars (Westerlund et al. 1995) show that those with $M_{bol} < -2.7$ mag are on the post-flash side of the locus for the onset of the first helium flash in double-shell sources; for the fainter ones the choice of mechanism is still open. This is shown in Fig. 7.4, in which are drawn, from left to right, the loci for the onset of helium shell flashes in double-shell source stars as read from the high-temperature envelopes for the C stars in the HR diagrams for the SMC (S) and the LMC (L) (Westerlund et al. 1991) and the theoretical locus for galactic (G) disk composition (Scalo 1976). Plotted are also the starting points for thermal pulses in low-mass stars (0.70–0.55 M$_\odot$) of metallicities $Z = 0.0004$ [1], [2], 0.006 [3], and 0.02 [4], according to Castellani et al. (1991). The dashed line gives that locus for stars of 1.0–3.0 M$_\odot$ (Lattanzio 1986).

The faint C stars in the SMC overlap in the HR-diagram to some extent with the most luminous C stars in the galactic bulge.

7.4 The CH stars

The CH stars are typically metal-poor objects with a high carbon abundance. In the Galaxy they are members of an extreme halo population (a velocity dispersion perpendicular to the galactic plane of ≥ 114 km s^{-1} (Hartwick and Cowley 1985). The galactic globular cluster ω Centauri contains CH stars. This confirms that they are evolved giants. Several have been found to belong to long-period binary systems. It has been suggested that they have all obtained their high carbon abundances by mass exchange in binary systems (McClure 1984, 1989). Since their main features are easily seen on objective prism plates, they are important tracers of the galactic halo population.

A large number of CH stars have been discovered in the Magellanic Clouds. Hartwick and Cowley (1988) have identified ~ 200 CH-star candidates within 8° of the centre of the LMC. A greater number of the stars appear on the eastern side of the LMC. Spectroscopic confirmation and radial velocities have been obtained for 74 of the CH candidates (Cowley and Hartwick 1991, Hartwick and Cowley 1991).

Infrared (JHK) photometry has been secured by Feast and Whitelock (1992) for all the proposed CH stars in the LMC. Their luminosities fall in the range $-4 \geq M_{bol} \geq$

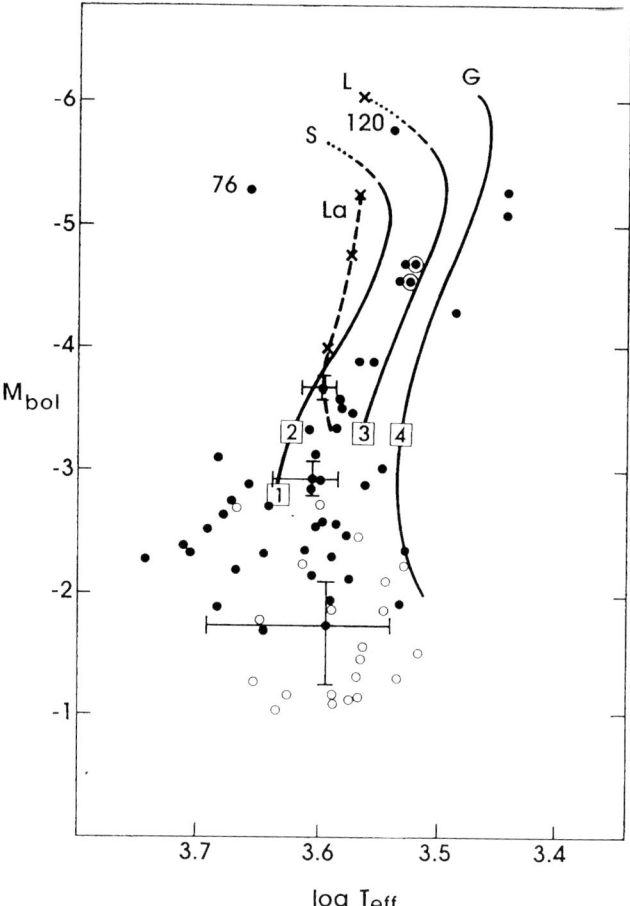

Fig. 7.4. The HR diagram for faint C stars in the SMC (dots) compared with the most luminous ones in the galactic bulge (circles). Error bars are drawn for selected SMC stars. For further description, see the text.

−6 mag or even brighter, placing them among the most massive C stars in the LMC. Most of the CH stars have $(J-K)_0$ between ∼ 1.0 and ∼ 1.5 mag; those with $M_{bol} \leq$ −5 mag are all bluer than the LMC cluster C stars with the same luminosities. Most of them are also more luminous than the most luminous C stars (intermediate-age objects) in the Local Group dwarf spheroidals. None of the CH stars detected so far in the LMC fall on the globular-cluster-type giant branch, $M_{bol} \geq -3.5$ mag, where the classical galactic CH stars and the oldest C stars in the dwarf spheroidals are found. Typical of the latter group are the Draco and UMi dwarfs with all C stars on this branch. The (intermediate-age) Fornax and Leo I and II dwarfs, on the other hand, contain only cooler, more luminous C stars, reaching $M_{bol} \sim -5$ mag.

The luminosities of the CH stars thus indicate that they are younger than most C stars in the intermediate-age dwarf spheroidals, i.e. younger than ∼ 3 Gyr, provided they result from single-star evolution. As they are also bluer (hotter) and more luminous than most LMC cluster C stars, they are, under this assumption, younger (∼ 0.1 Gyr) and more massive ($\geq$ 3 $M_\odot$) than the latter. This contradicts the age

7.5 Long-period variables

derived kinematically. It appears necessary to accept the kinematical age and masses of the order of ≥ 3 M$_\odot$ and to consider the possibility that the objects are binary mergers (and in that respect are similar to the galactic CH stars).

RIJHK photometry of the Hartwick and Cowley sample of CH stars in the LMC has been carried out by Suntzeff *et al.*(1993). They confirm that these stars are significantly more luminous than the red-giant branch tip for Population II stars and, therefore, younger than the latter. Their mean M_{bol} is -5.3 mag, the brightest star reaches -6.2 mag. As the CH stars are also younger than most other C stars in the LMC field, they may be associated with a population with an age around 100 Myr. They have enhanced *s*-elements, such as Ba and Sr, in agreement with their belonging to the luminous C-type class and having passed the third 'dredge-up' phase. The total number of CH stars in the LMC is estimated at 10–25% of the number of C stars.

The question about the age(s) of the CH stars in the LMC remains open.

7.5   Long-period variables

Little is known about the origin of the long-period variables (LPVs), and their masses and luminosities are not well determined. This is due to the fact that almost all galactic LPVs are field stars without any connection to groups with well-known properties. A few short-period LPVs are known members of old galactic clusters, and a few supergiant LPVs have been identified in very young galactic open clusters. No large-amplitude LPVs are known to belong to clusters with a turn-off mass in the range 1–9 M$_\odot$. In this range of stellar masses high-mass-loss-rate AGB stars are formed, such as Mira variables, OH/IR stars and dusty carbon stars.

The LPVs are among the intrinsically most luminous stars at near-infrared wavelengths. They are easy to recognize in deep R- and I-band surveys of the Clouds and other nearby galaxies. JHK photometry has shown that they obey a well-defined period–luminosity relation. Among them, the Mira-type variables have served to establish a distance scale independent of the Cepheid calibration.

A practical result of studying LPVs in Magellanic Cloud clusters would be the determination of the mass of stars giving rise to LPVs of a given period. A pulsation mass may also be determined for comparison with the turn-off mass. If these masses agree, the colour/T_{eff} relation, the link between observations and theoretical models, could be considered correct. For the more evolved LPVs (longer period, more dusty) the pulsation mass is expected to be lower than the turn-off mass, due to the high mass loss rate in these stars. The accumulated mass-loss would be derived directly.

Until about ten years ago most of the LPVs in the Clouds were those reported by Payne-Gaposchkin (1971) and Payne-Gaposchkin and Gaposchkin (1966). The objects reached in the Harvard surveys were the young, massive (~ 5 M$_\odot$) red giants on the upper AGB and the younger, more massive (≥ 9 M$_\odot$) supergiant variables.

Extensive searches for LPVs in the Clouds have now been carried out by Reid *et al.*(1988) and Hughes (1989). Together with previous surveys (Payne-Gaposchkin 1971, Lloyd Evans 1971, Butler 1971, Wood *et al.*1985) they make the coverage of the LMC for LPVs nearly global.

The LMC and SMC LPVs fall into two distinct groups, core-helium- (or carbon-) burning supergiants with small pulsation amplitudes in K (< 0.25 mag) and AGB stars with typical amplitudes of 0.5–1.0 mag (Wood *et al.*1983). The AGB stars extend up to $M_{bol} \approx -7.1$ mag, the AGB limit. Carbon star LPVs are confined

128 The intermediate-age and oldest field populations

to luminosities fainter than $M_{bol} \sim -6$ mag. Many of the oxygen-rich AGB stars show enhanced oxide bands of the s-process element Zr, indicating that dredge-up of material at helium shell flashes occur in these stars as well as in the carbon stars. The absence of carbon stars among the uppermost AGB stars is due to CNO cycling of dredged up carbon to nitrogen during quiescent AGB evolution. The AGB stars are, thus, sources of primary nitrogen as well as of carbon and s-process elements.

From the theories of stellar pulsation and evolution, Wood *et al.* deduced that AGB stars with initial masses ≥ 5 M$_\odot$ produce supernovae while the less massive stars produce planetary nebulae with nebular masses from ~ 0.1 to 2.1 M$_\odot$. The core-burning red supergiants appear highly overluminous for their pulsation mass; they must have lost up to half their mass (initial MS mass ~ 16–23 M$_\odot$) since the MS-phase.

The carbon stars and the oxygen-rich LPVs obey different PL-relations, with the carbon stars having longer periods at a given luminosity (Wood *et al.* 1985) . The PL distribution of the LPVs in the central LMC Bar suggests that the bulk of the stars in this region have MS masses $M_{MS} \leq 1.6$ M$_\odot$ and ages ≥ 1 Gyr. A distinct group in the region has $M_{MS} \sim 5$–6 M$_\odot$ and an age ~ 60 Myr. Also the PL distribution of the Cepheids in this region suggests a burst of star formation about 60 Myr ago. A group of short-period ($P < 250^d$) oxygen-rich variables, analogous to the Population II Miras in the Galaxy, probably belong to a very old (≥ 10 Gyr) moderately metal-poor population. A comparison with three similar Miras in the galactic globular cluster 47 Tucanae suggests a distance modulus to the LMC of 18.6 ± 0.25 mag, if 47 Tuc has a modulus of 13.34 mag (see Chap. 2).

Hughes and Wood (1990) have obtained JHK photometry for 459 of the LPVs found by Hughes and low- or medium-resolution spectra of a subset of these stars. 267 of the variables are Miras, defined as having large amplitudes ($\Delta_I \geq 0.9$ mag) and well-defined periods, and 117 SRa variables (lower amplitudes, well-defined periods). The remainder have uncertain periods or are non-periodic. The LPVs may be divided into old, OLPV ($P < 225^d$, age ≥ 5 Gyr), intermediate-age, ILPV ($225^d \leq P < 450^d$, age ~ 1–3 Gyr), and young, YLPV ($P \geq 450^d$, age ≤ 1 Gyr). The OLPVs are rather uniformly distributed over the LMC; the ILPVs are also spread throughout the LMC but show a higher concentration at the Bar than the OLPVs. The YLPVs occur preferentially in two regions, the Bar and around Shapley Constellation III. This is reminiscent of the distribution of Cepheids with periods of 3–5 days; the YLPVs have evolved from stars going through that phase. For all three age-groups populations of precursors are known with distributions over the LMC that are similar to those of the three groups.

The absolute bolometric mean magnitudes of the LPVs were determined by Hughes and Wood from the observed K and $(J - K)$ following Bessell *et al.* (1983). The following relations were derived between $\langle M_{bol} \rangle$ and P:

$$\langle M_{bol} \rangle = -4.39 - 2.91 (\log P - 2.4), \sigma = 0.32, \text{ for } P \leq 450^d;$$
$$\langle M_{bol} \rangle = -3.32 - 7.76 (\log P - 2.4), \sigma = 0.38, \text{ for } P > 450^d.$$

The luminosity functions show a deficiency in the number of LPVs between $M_{bol} = -5.5$ and -6.5 mag. This may be due to a decrease in star formation ~ 1.5 Gyr ago, followed by a burst of star formation ~ 50 Myr ago. It is essential, however, to

7.5 Long-period variables

understand the mass-loss rates for the Miras with $P > 425^d$ well before reaching a definite conclusion.

Most of the LPVs in the LMC are AGB stars; around one-third of them are carbon stars, and they tend to concentrate in the period range 250–400^d. The LPVs in the LMC have shorter lives as large-amplitude pulsators than the LPVs in the Galaxy. The lower metallicity in the LMC means that a large fraction of its evolved intermediate-age AGB stars become carbon stars with a smaller pulsation amplitude than the Mira M stars; $\sim 60\%$ of the LMC intermediate-age Miras are carbon stars as compared with $< 10\%$ in the Galaxy. The number of Miras observed in the LMC is an order of magnitude less than may be expected on the basis of current theories of AGB evolution. Some unknown mass-loss mechanism with a rapid onset probably terminates the Mira phase. There are also fewer large-amplitude LPVs per unit galactic stellar mass in the LMC than in the Galaxy.

There are relatively more short-period Miras in the LMC than in the Galaxy, possibly indicating that relatively more stars of ages ≥ 5 Gyr exist in the LMC than in the Galaxy, or a difference in the relative scale heights and/or central densities of the old and intermediate-age populations.

The short-period LPVs in the LMC have a high velocity dispersion, 33 km s^{-1}. Their kinematics must be dominated by random motions and not by rotation. They belong to a spheroid population. The height of this spheroid is constrained by their dispersion about the mean PL relation to be less than ~ 2.8 kpc. The distribution of the population represented by the OLPVs is highly flattened, with an axial ratio of $\sim$ 0.3–0.5, and its mass is $\sim 2\%$ of the LMC total mass (Hughes et al. 1991).

In the NGC 330 region of the SMC Sebo and Wood (1994) found 22 Cepheids, 20 LPVs and 9 other variables, including two eclipsing variables (one previously known), and four close binary stars, two of which show emission lines. None of the LPVs are cluster members; all appear to belong to the old field population of the SMC.

7.5.1 The Mira variables

As mentioned above, Miras are long-period variables ($P > 100$ days) with large amplitudes ($\Delta M_{bol} \geq 0.5$ mag). They may be found among the optically invisible OH/IR stars and their carbon-rich analogues. Those seen optically are of spectral types Me, Se or Ce. They must be close to the top of the AGB and they have high mass-loss rates, typically between 10^{-7} and 10^{-4} M$_\odot$ yr^{-1}. Their initial masses increase with the pulsation period.

Extensive surveys of Miras have been published (Reid et al. 1988, Feast et al. 1989, Hughes 1989, Hughes and Wood 1990). Some thick-shelled sources have been detected in the LMC as a result of investigations of *IRAS* sources (Reid et al. 1990, Wood et al. 1992, and Reid 1991). Of particular interest are the OH/IR sources, with periods in excess of 1000 days (see next section). They fit well on an extrapolation of the Mira PL relation, derived for Miras with $P < 420$ days.

The Magellanic Clouds contain both oxygen-rich (M- and S-type) and carbon-rich (C-type) Miras. The oxygen-rich Miras obey a period–luminosity–colour (PLC) relation (Feast et al. 1989):

$$m_{bol} = -4.32 \log P + 2.37 \overline{(J-K)}_0 + 21.63, \sigma = 0.13;$$
$$K_0 = -4.58 \log P + 2.00 \overline{(J-K)}_0 + 19.71, \sigma = 0.13.$$

The scatter, σ, can be accounted for by observational uncertainty so that the scale height must be less than 2 kpc.

The PLC relation implies that the instability strip in the HR diagram has a finite width. The period of an individual Mira would then be expected to change as it evolves through the strip. Little evolution is, however, seen, at least among the short-period Miras. Wood (1990) has suggested that the PLC relation exists because the T_{eff} of the giant branch at a given luminosity varies with metal abundance. Feast (1992) has pointed out that the tracks derived by Wood imply that the short-period Miras, at least, have already evolved away from the AGB due to depletion of their envelopes by mass loss. They are already post-AGB stars.

The number of Miras has been compared with the numbers of PNe in the galactic centre (Feast 1993). There are about 1000 PNe within 10° of the centre and about 1.4×10^4 Miras, giving a ratio of Miras/PNe of 14. The observable lifetime of the bulge PNe is ~ 8000 yr (Pottasch 1992) and of the Miras $> 10^5$; this gives a ratio of > 12. All, or almost all, Miras may then evolve into PNe. It is desirable to obtain the same comparison for the Magellanic Clouds.

7.6   OH/IR stars and *IRAS* sources

An important category of the AGB stars are the OH/IR stars. In the Galaxy they have a distribution similar to that of the PNe. Both belong to the 'old disk' population and come from the same main-sequence stars. Wood *et al.*(1992) have determined periods for a number of these objects in the Magellanic Clouds. A total of 54 sources was searched for 1612 MHz OH maser emission with the aid of the Parkes 64 m radio telescope. OH spectra were obtained with the Parkes digital correlator, providing a resolution of 1.5 km s^{-1}. The objects were mainly *IRAS* point sources (*IRAS* Point Source Catalog. 1985), the brightest in 25 μm flux, S_{25}, and with $0.0 < \log S_{25}/S_{12} < 0.5$ Jy and no obvious HII regions or large clusters. Six OH sources were detected; all in the LMC. Two of these sources also showed OH maser emission at 1665 MHz. The radial velocities of all objects confirm their membership of the LMC.

The areas of the *IRAS* sources were also scanned in the 2.2 μm K band to a limiting magnitude of $K \sim 11$mag. The six OH maser sources were easily seen. 19 of the detected objects have been monitored in *JHKL* over ~ 1700 days. The light curves split into two groups: seven with large K amplitudes, $\Delta K > 1$ mag, and strong, long-period pulsation, and the remainder with $\Delta K < 0.5$ mag. All the OH masers with luminosities consistent with AGB membership (i.e. $M_{bol} \geq -7.1$ mag) are large-amplitude variables. The brightest OH maser ($M_{bol} = -9.38$ mag) is the red core-helium-burning supergiant *IRAS* 04553−6825; it shows a ΔK of ~ 0.3 mag. None of the other sources brighter than the AGB limit show large-amplitude variability. Most show, however, some evidence of small-amplitude pulsation with periods of several years. They are presumably core-helium-burning supergiants. Some objects with AGB-like luminosities, including three SMC sources detected at K, have small, possibly negligible, amplitudes, and they may presently be leaving the AGB.

An interesting object is the probable supergiant *IRAS* 05389−6922 with $M_{bol} = -8.85$ mag, which showed a continuous rise in K over ~ 1600 days before beginning to decline. This indicates a period ≥ 3200 days, longer than any period known for galactic OH/IR stars.

7.6 OH/IR stars and IRAS sources

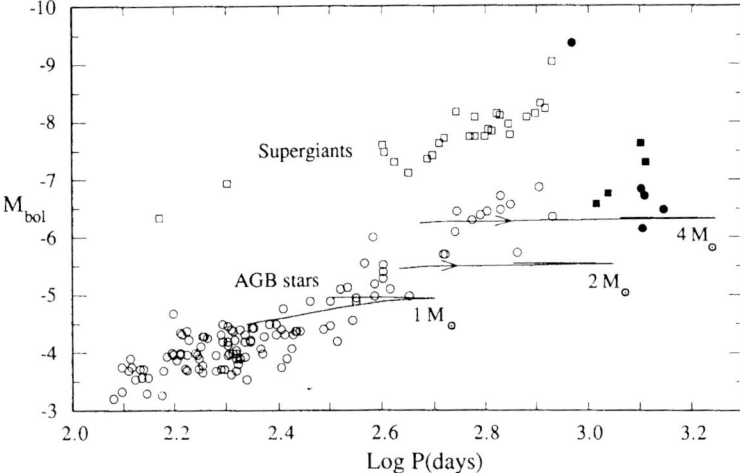

Fig. 7.5. The M_{bol}, log P diagram for LMC long-period variables. Open symbols identify objects found in optical and I searches, circles AGB and squares supergiants. Filled symbols are objects found in the OH, JHKL search by Wood et al. (1992). Evolutionary tracks for AGB stars with high mass loss (Wood 1990) are drawn.

In the SMC five *IRAS* sources have been detected and monitored by Whitelock et al. (1989).

The very red *IRAS* sources are a mixture of types. Some of the sources resemble those non-variable *IRAS* sources that are thought to be post-AGB stars or pre-planetary nebulae in the Galaxy. A few of them show no evidence of pulsation, giving further support to a post-AGB stage. Two vary, however, with small amplitudes characteristic for red supergiants; they have luminosities well above the AGB limit.

Another distinct group of *IRAS* sources consists of those dominated by near-IR (photospheric) emission and small amounts of dust beyond 10 μm. They are all supergiants which have yet to develop large mass-loss rates. There are optically identified LPV supergiants with similar flux distributions (Hughes and Wood 1990).

Pulsation periods were derived by Wood et al. for nine of the sources. They are in excess of 1000 days and fit rather well on an extrapolation of the PL relation for Miras with periods less than 420 days (Whitelock and Feast 1993). The sources are classified as AGB stars, core-helium-burning supergiants or preplanetary nebulae according to pulsation amplitude, luminosity and redness (see above). The stellar wind expansion velocities are only ~ 0.6 times those expected for comparable OH/IR stars in the Galaxy, a result explicable in terms of the lower metal abundance in the LMC. Mass-loss rates for the LMC OH/IR stars are estimated to be from a few times 10^{-5} to a few times 10^{-4} $M_\odot$ yr^{-1}. Low-metallicity (SMC) AGB stars have lower mass-loss rates than higher metallicity stars provided carbon star formation is not involved. Carbon star formation in low-metallicity systems appears to keep mass-loss rates high. No evidence has been found that AGB stars exist with luminosities exceeding the classical AGB limit of $M_{bol} = -7.1$ mag.

7.7 Planetary nebulae

The emission-object character of the PNe makes them ideal for attacking many problems in external galaxies. A large number of such objects has been identified in the Clouds. During the early years small HII regions were frequently misidentified as PNe; the spectral features used to identify them, strong Hβ and [OIII] 4959 and 5007 Å lines, are similar in the two classes. Typically *bright* PNe in the Magellanic Clouds are expected to be < 1", so that they are not resolved with medium-sized ground-based telescopes. Even with speckle interferometry their diameters are difficult to measure from the ground. Older, larger PNe have, however, been resolved. Jacoby (1980), in his deep survey resolved nine faint PNe and managed to estimate masses for two of them to 0.3 and 0.8 $M_\odot$, respectively. The PNe are well suited for imagery with the Faint Object Camera (FOC) and the Planetary Camera (PC) on the *HST*, and several PNe have already been studied in detail (see below).

7.7.1 Surveys

Sanduleak *et al.* (1978, SMP) surveyed both Clouds to a limiting flux of log $F_{H\beta} \sim -13.3$ erg cm^{-2} s^{-1} and found 102 PNe in the LMC and 28 in the SMC. Jacoby (1980) searched the central regions of the LMC and the SMC for very faint PNe. His catalogue contains 41 objects in the LMC and 27 in the SMC, of which seven and eight, respectively, were previously known. On the basis of the ratio of faint to bright PNe in these fields and the number of bright PNe in the surrounding fields he estimated the total number of PNe to 285 $\pm$ 78 in the SMC and 996 $\pm$ 253 in the LMC

Ten new PNe were identified in the SMC by Morgan and Good (1985). Azzopardi (1989), summarizing surveys and spectroscopic confirmations of PN candidates in the SMC, concluded that about 70 PNe are known.

The most recent survey of the LMC (Morgan 1994), undertaken with the UK 1.2 m Schmidt Telescope (UKST), lists 265 PNe. Their surface distribution is shown in Fig. 7.6.

The highest density is found in the Bar, despite the difficulty of identifying objects there. The density is moderate in a region roughly circular in shape and $\sim 6°$ in diameter and approximately concentric with the Bar. Finally, there is a low-density, outer region, also circular in shape with a diameter $\sim 13°$, but not concentric with the Bar.

The asymmetry of the PN system in the LMC is striking; 66% lie within the western half. Conversely, if the centre of the PN system is taken to be the centre of the inner, high-density region, then, of the 53 objects which are at radial distances $\geq 5°$, 70% lie in the eastern half of the LMC.

The outer boundary of the PN distribution matches neither that of the clusters, which is distinctly elliptical with an axial ratio of $b/a \sim 0.7$ (Kontizas *et al.* 1990b), nor that of the isopleth map of the outer regions of the LMC which also shows an overall ellipticity (Irwin 1991). The distribution of the brighter carbon stars, $I \leq 13.5$ mag (Westerlund *et al.* 1978), though incomplete in the Bar region, appears to match that of the PNe relatively well.

The centroid of the entire PN system is $5^h 23^m.7$, $-69° 1'$ (1950.0), to the north of the optical centre of the Bar. The centre of the outer envelope is $0°.7$ farther east and $0°.25$ farther north at $5^h 31^m.4$, $-68°47'$.

7.7 Planetary nebulae

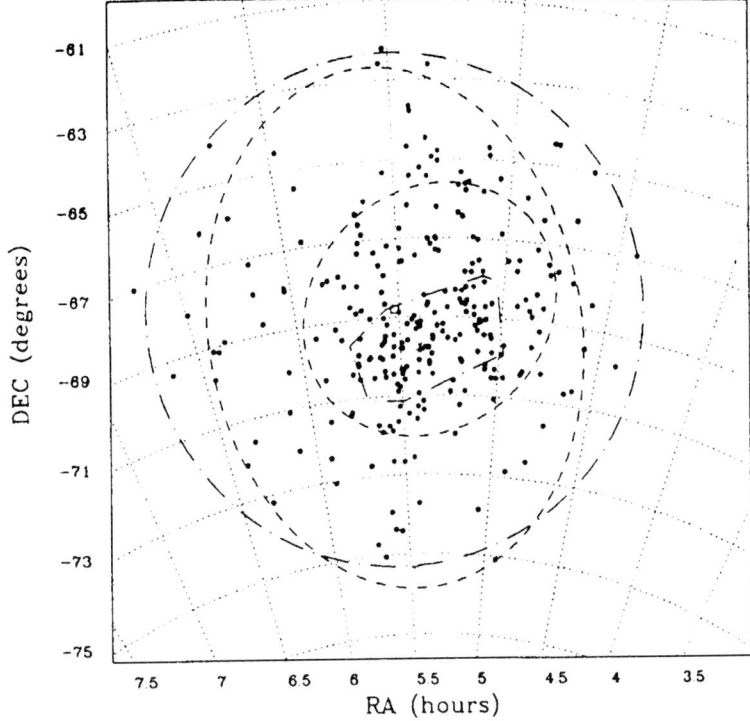

Fig. 7.6. The surface distribution of PNe in the LMC according to Morgan (1994). The dash–dot lines indicate the outer boundary of the PNe and outline the Bar. The dashed lines are elliptical representations of the cluster distribution in the LMC; the inner one marking the region where the density is high (Kontizas et al. 1990b).

In the SMC the centroid of the PNe is at $0^h 49^m 30^s$, $-73°20'$ (1950), close to the brightest part of the Bar (Dopita et al. 1985a). The major axis of the PNe system is aligned more or less with that of the Bar. The distribution of the 42 PNe involved gives the impression of a spheroidal population. There is no evidence of any organized rotation, nor is there in this material any evidence of the bimodal velocity distribution suggested by Feast (1968) and seen in several other components of the SMC (Sect. 12.2), most of which are, however, of a younger generation.

There is some evidence for a concentration of a younger, high-velocity sub-group of PNe in the NE corner of the SMC. These objects may be dense Peimbert type I PNe; they are likely to have relatively massive central stars and stellar precursors (see Sect. 7.7.3) and may represent a dynamically young region, maybe ≤ 0.1 Gyr. Judging from their mean velocity, $V_{GSR} = +0.2$ km s^{-1}, they belong to the H-component in the Martin et al. (1989) model (Sect. 12.2)..

7.7.2 The planetary nebula luminosity function

The bright end of the PN luminosity function (PNLF) has been shown to be an accurate standard candle for E and S0 galaxies (see Jacoby 1989, Ciardullo et al. 1989). However, the zero point is determined only with the aid of M 31 (Sb) and the dependence of the PNLF on the Hubble type is not clear. It is therefore important

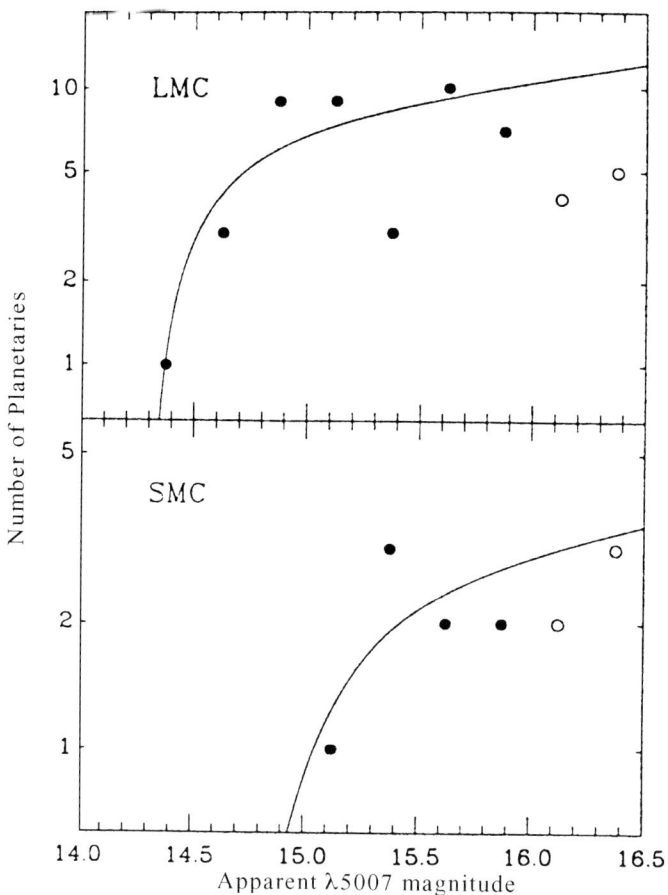

Fig. 7.7. The PN luminosity functions for the LMC(top) and the SMC binned into intervals of 0.25 mag. Filled circles represent objects above the completeness limit and the full-drawn lines are models (Ciardullo et al. 1989) fitted to them after adjustment for extinction and convolved with photometric and depth uncertainties.

to test the PNLF on the Magellanic Clouds to see (i) if the derived distances agree with previously derived ones and (ii) if the PNLF is independent of metallicity. Jacoby et al. (1990) have explored these issues by measuring the [OIII] $\lambda5007$ fluxes of the 102 brightest PNe in the LMC and the 31 in the SMC. The fluxes are transferred to a monochromatic magnitude, m_{5007}, using the relation

$$m_{5007} = -2.4 \log F_{5007} - 13.74.$$

An extinction correction at $\lambda5007$ of 0.36 ± 0.11 mag is applied to the LMC and 0.21 ± 0.04 mag to the SMC PNe, corresponding to $E_{B-V} = 0.10$ and 0.06 mag, respectively. The complete sample of bright LMC PNe is used in combination with the PN luminosity function (see Fig. 7.7) to derive a distance modulus of 18.44 ± 0.08 mag for the LMC and of $19.09^{+0.22}_{-0.30}$ mag for the SMC. The good agreement with other distance determinations (Chap. 2) is strong evidence that the PNLF method is independent of host galaxy Hubble type, colour, and metallicity.

7.7 Planetary nebulae

7.7.3 The evolution of the planetary nebulae

The key to understanding the late stages of stellar evolution is the determination of PN nebular abundances and the position of the central star in the HR diagram. It is evident that two classes of PNe exist in the Magellanic Clouds, one more massive and younger than the other. Their precursors are most likely AGB stars of different ages and masses, presumably all carbon stars. The relationship between the PNe and the known C stars in the fields of the two Clouds thus becomes important but is not yet well known. The overall distributions appear to agree, and in the SMC the kinematics of the PNe and the C stars also.

The main bulk of the PNe in the LMC have ages near 3.5 Gyr (Meatheringham et al. 1988), with some younger objects as young as 0.5–1.3 Gyr. The precursors of the PNe are thus to be found in the intermediate-age population though it is not excluded that older ones exist.

Among the PNe in the Clouds there is a subgroup with rather low [OII]/Hβ ratios and very high HeII/Hβ ratios. They are either dense Peimbert type I objects (Boroson and Liebert 1989) or bright density-limited objects and/or symbiotic stars (Dopita et al. 1985b). Type I PNe are believed to originate from the high-mass end of the AGB; their progenitors have masses $\geq 3 M_\odot$, and they are ≤ 0.1 Gyr old. Such stars have been observed in the SMC and represent the present end stage of the AGB (see Sect. 7.2.5).

The mass boundary between stars that evolve off the AGB to the PN stage and those which evolve to type I SNe is not yet well known. It is therefore important to identify the PNe that have evolved from the most luminous AGB stars. Such objects are rare because the rate of evolution through the PN stage increases when the PN nucleus increases. In the LMC, this has led to a sharp cutoff in the PNLF, corresponding to a core mass of 0.7 $M_\odot$. Dopita and Meatheringham (1991a, 1991b) have, by photoionization modelling of a sample of 82 PNe in the Clouds, constructed a HR diagram for the central stars and derived the chemical abundances and the nebular parameters. The majority of the central stars have masses between 0.55 and 0.7 $M_\odot$ and move under hydrogen burning towards higher temperatures. Optically thin objects are found scattered throughout the HR diagram; they tend to have a somewhat smaller mean mass. The Peimbert type I PNe have generally higher mass central stars, ranging up to 1.2 $M_\odot$ and very high efficient temperatures. They are observed during the fading part of their evolution, consistent with their expected more rapid evolution across the HR diagram.

The nebular mass of the optically thick objects is closely correlated with the nebular radius. PNe with nebular masses in excess of 1 $M_\odot$ are observed. The velocity of the expansion of the nebula is well correlated with the position of the central star in the HR diagram and is evidence for continual acceleration of the nebular shell during the transition towards higher T_{eff}.

There is clear evidence for a spread in the abundances of the α-process elements in both Clouds. This is probably a consequence of the spread in ages of the PN precursor stars. In the LMC the metallicity range is a factor of about 10, in the SMC about 3–4. For the LMC there is also a correlation between core mass and metallicity for these elements. This relationship may eventually permit derivation of an age/metallicity relationship for the Clouds.

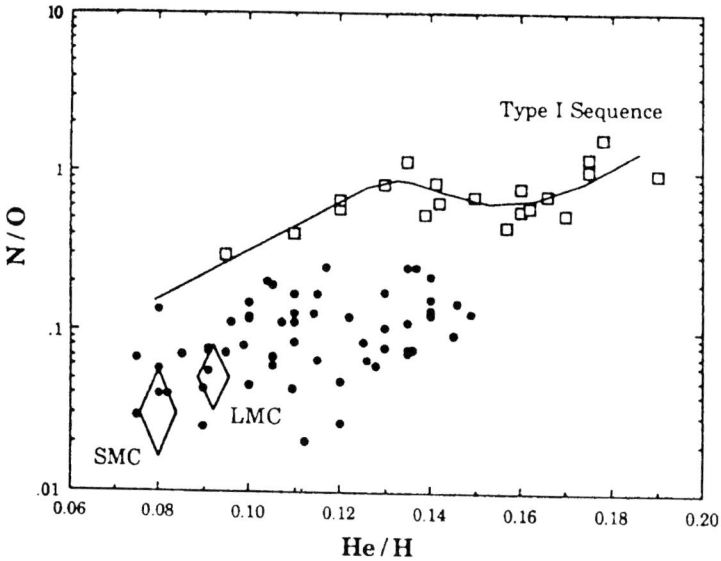

Fig. 7.8. The relation between N/O and He/H for PNe in the LMC and the SMC. Type I objects, defined as having N/O > 0.3, are plotted as open squares. The error diamonds represent the mean abundance ratios for the SMC and the LMC as derived from HII regions and supernova remnants (Dopita and Meatheringham 1991c).

Excluding the type I PNe, the mean abundances derived for the LMC and the SMC agree well with the mean abundances previously derived from observations of HII regions and evolved radiative supernova remnants.

The type I PNe lie on a distinct sequence in the N/O versus He/H diagram (Fig. 7.8). They are deficient in O with respect to the other α-process elements. This shows that they have undergone CN processing, dredge-up and ejection, as well as an appreciable O–N processing.

The uppermost end of the mass range of the precursor stars, which are able to evolve into the PN phase, may be represented in the LMC by some highly energetic type I PNe (Dopita et al. 1985b). They have higher expansion velocities than any of their galactic counterparts and probably show directed expansion flows rather than spherically symmetric expanding shells. They are highly luminous bipolar type I PNe with the central star undergoing continued energetic mass loss, which is, nevertheless, insufficient to supply the energy and the momentum of the nebula. Therefore, the bipolar morphology and outflow must have already been established when these objects became PNe. One of these objects is LMC N66, which will be dealt with more intensively in the next section.

In order to understand the distribution of SMC and LMC PNe in the HR diagram it appears necessary to accept that the PNe nuclei (PNN) eject a large amount of matter in the final shell flash event, causing them to leave the AGB as helium burners. The PNN fade until hydrogen reignition takes place. At this time, the rate of evolution across the HR diagram is slowed, increasing the probability that objects may be observed in this phase. In the He-burning tracks the fading of the more massive objects is considerably delayed, so that the type I objects may be observed while still relatively bright (Dopita 1993).

7.7 Planetary nebulae

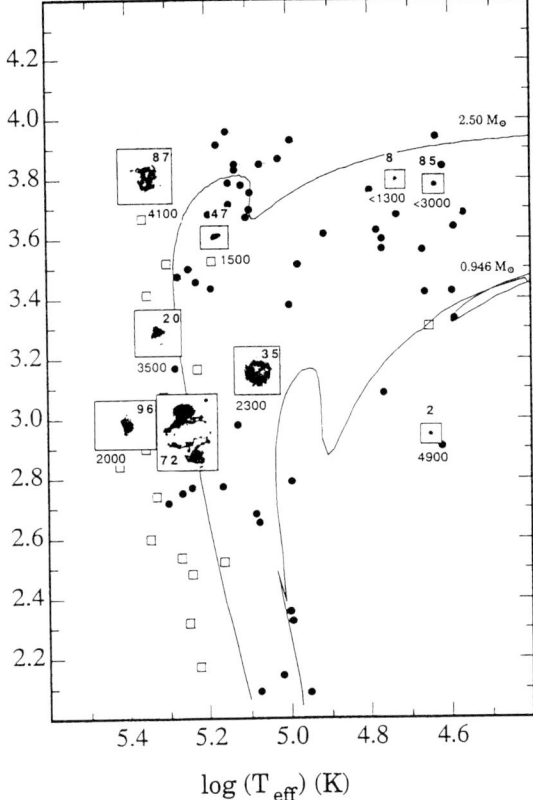

Fig. 7.9. The HR diagram of PNe in the LMC (Dopita *et al.* 1995). Type I PNe are shown as open squares. *HST* images of observed PNe (SMP numbers), all in the same scale, are superimposed at the points in the HR diagram indicated by the photoionization model, and their dynamical ages, in years, are given. The tracks, for He-burning stars, correspond to core masses of 0.57 (lower) and 0.68 $M_\odot$.

The HR diagram of the PNe (Fig. 7.9), if interpreted as He-burners, implies that the rate of star formation in the LMC has not been uniform over time. The faintest PNe are consistent with the lower track, for which the stellar age is ~ 15 Gyr. For $\log T_{eff} > 5.0$, there is a concentration of objects near the 1.5 $M_\odot$ track, implying a burst of star formation about 3–4 Gyr ago. This age agrees with the one, of 2.5–3.6 Gyr, derived from the interpretation of the velocity dispersion of the LMC PNe as achieved by orbital diffusive processes operating over the lifetime of the precursor stars. Finally, the large concentration of PNe near and above the 2.5 $M_\odot$ track implies that a burst of star formation took place within the last ~ 0.3 Gyr.

In a further interpretation of the HR diagram Dopita *et al.* (1995) find clear evidence for size evolution along the evolutionary tracks. The evidence for a corresponding increase in dynamical age is weaker. The evolutionary time scales are consistent with evolutionary ages derived from theory provided that some PNe are H-burning and others are He-burning. In the LMC the latter appear to outnumber the former approximately in the ratio 2:1.

Table 7.1. *Dimensions of bright planetary nebulae in the Clouds*

PN No.	Diameter arcsec	Diameter pc	$-\log F_\beta$ erg cm^{-2} s^{-1}	$M/M_\odot$	Expansion age, yr
SMC 2	0.20	0.064	12.69	0.06	940
SMC 15	0.15	0.048	12.44	0.05	1500
LMC 1	0.22	0.055	12.47	0.05	840
LMC 52	0.16	0.040	12.71	0.02	510
LMC 62	> 0.5	> 0.126	12.30	> 0.19	> 1400
LMC 83	0.32	0.081	12.65	0.07	360
LMC 97	0.23	0.058	12.82	0.03	480

7.7.4 Dimensions of planetary nebulae

Several attempts have been made to determine the dimensions of PNe in the Magellanic Clouds. Speckle interferometric angular diameters have been measured by Wood *et al.*(1986). The masses of the ionized gas in these nebulae were derived using the angular diameters and known Hβ fluxes. They range from ≤ 0.006 M$_\odot$ to ≥ 0.19 M$_\odot$. The PNe observed were, naturally, among the brightest in the Magellanic Clouds. Consequently, they are also small (diameter ≤ 0.13 pc), young (age ≤ 1500 yr), and dense. As they are probably only partly ionized, the derived masses are lower limits to the total masses.

The diameters agree rather well with what have been derived later from *HST* imagery. The ages appear to be underestimated (see below).

As mentioned above, several PNe have been resolved with the aid of the *HST*. Blades *et al.*(1992) used the FOC to obtain [OIII] λ 5007 and Hβ images of four PNe, SMC N2 and N5, and LMC N66 ($\equiv$ SMP83 $\equiv$ WS35) and N201 (see SMP). Structures as small as 0″.06 are easily discernible after deconvolution (Fig. 7.10).

SMC N2 is a slightly elliptical ringlike nebula, with its greatest elongation in the EW direction; its dimensions are 0.26 $\times$ 0.21 arcsec^2 in [OIII].

SMC N5 looks more like a circular ring with a diameter of 0″.26; the ring is significantly narrower than that of N2 with a width as small as 0″.06 in some places.

LMC N201 is the most luminous PN known in [OIII] and Hβ in the Magellanic Clouds or any other galaxy. Its nebular C/O ratio is < 1 and its N/O ratio is close to 0.5. SMC N2 and N5 have C/O ratios > 3 (Aller *et al.*1987). The interpretation of its structure is difficult due to the very bright central star; the deconvolved image is Gaussian in shape with a FWHM of 0″.21.

All three PNe, SMC N2, N5, and LMC N201, are optically thin in the hydrogen Lyman continuum, so that their nebular envelopes are completely ionized (Barlow *et al.*1986).

LMC N66 is a multi-polar nebula (Fig. 7.10) with, in [OIII], a weak central star. Its spectrum reveals that it belongs to Peimbert's type I class, showing both high excitation and strong emission lines of nitrogen (Meatheringham and Dopita 1991a). The FOC images show that Hβ is nine times weaker than [OIII] λ 5007 and reveal structures unprecedented for a PN with loops and knots. The loops are almost

7.7 Planetary nebulae

Table 7.2. *Dimensions of PNe from FOC observations*

Name	Size arcsec	Line	Structure
SMC N2	0.23 × 0.13	Hβ	ring
	0.26 × 0.21	[OIII]	ring
SMC N5	0.31	Hβ	FWHM
	0.26	[OIII]	ring
LMC N66	complex	Hβ	see text
	∼ 4	[OIII]	see text
LMC N201	0.21	Hβ	Gaussian
	saturated	[OIII]	

orthogonal to the axis defined by the main lobes. Lobe A has an extent of $1''.0$, lobe B of $1''.3$. From each of these lobes, secondary lobes extend more than $2''$ from the central star. Five knots are seen; they have dimensions between $0''.27$ and $0''.09$.

Two bright and three faint filaments are recognized in the [OIII] λ 5007 and Hα PC, *HST* images of LMC N66 (Dopita *et al*. 1993a). They are also seen in the Blades *et al*. FOC images in [OIII]. Approximately 45 % of the total [OIII] λ 5007 flux is produced by the two brightest filaments. The overall bipolar butterfly appearance of the nebula is similar to galactic type I PN such as NGC 6302 and NGC 6357. The nebula may consist of a partially equatorial ring (dimensions in Table 7.3), with an inclination to the line-of-sight of $63° \pm 5°$, and a bipolar extension. The existence of faint, radially directed streamers suggests that giant lobes may exist similar to those seen in H2-111 (Webster 1978) and NGC 2440 (Heap 1992). Such lobes ought to be large enough to be seen in ground-based imagery also. They are, however, not visible in the [OIII] images obtained by Jacoby *et al*.(1990). All the filaments are resolved with a characteristic FWHM of $\approx 0''.2$, corresponding to 1.4×10^{17} cm (distance 49 kpc).

Dopita *et al*. combined the *HST* results with optical and UV spectrophotometry, absolute flux measurements and dynamical and density information, to derive a fully self-consistent nebular model. LMC N66 is an extremely massive type I object with a central star of 170 000 K and a luminosity near 3×10^4 $L_\odot$. The core mass is in the range 1.0–1.2 $M_\odot$; the main-sequence mass must have been greater than ~ 6 $M_\odot$. The nebular abundances are higher than the average for the LMC and show signs of hot-bottom burning of C to N but little evidence of dredge-up of C–N processed material on the AGB and no evidence of O–N processing.

The four discrete velocity components seen in LMC N66 suggest line-of-sight expansion velocities of ± 46 and ± 68 km s^{-1} (Dopita *et al*. 1988). If the stronger, lower-velocity components correspond to the main lobes of N66, then the expansion time scale is ≤ 3000 yr for the centroids of the lobes and double for the lobe extremities (Blades *et al*. 1992). There is an obvious axis almost devoid of material at a PA of 45° through the central star (Fig. 7.10). It may represent the plane of a disk that is now almost dissipated.

In a similar analysis Dopita *et al*.(1993a) found a dynamical age between 2400 and 3400 yr. This age is inconsistent with that estimated on the hypothesis that the

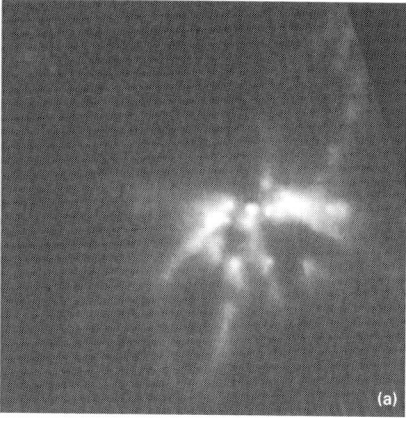 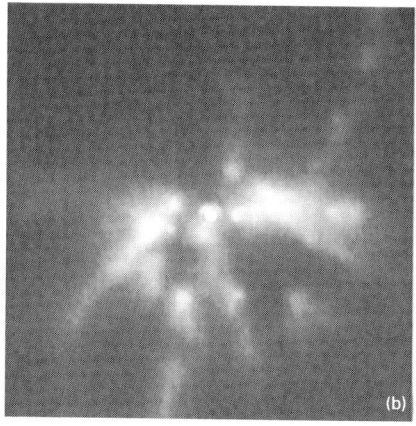

Fig. 7.10. (a) Deconvolved *HST* post-COSTAR image of LMC N66 in the [OIII] line, showing the faint nebulosities far from the main body. (b) A close-up of the main structure, showing details of the many lobes, loops and knots within the object. (Images supplied by J.C. Blades.)

central star left the AGB as a hydrogen-burning star; it suggests that it may be a helium burner.

Among the Cloud PNe, LMC N66, with a core mass of ~ 1.1 M$_\odot$, is only surpassed by the nucleus of the type I PN SMC N67. This has been detected through thermal emission of the central star at X-ray wavelengths, placing it at $\log(L/L_\odot) = 4.6 \pm 0.7$; $\log(T_{eff} \sim 5.5$ (Wang 1991a), outside the area shown in Fig. 7.9.

Using *HST* imagery in [OIII] λ 5007 and UV spectrophotometry in combination with previously secured measurements, Dopita *et al.* (1994) were able to construct a fully self-consistent nebular model of <u>LMC SMP85</u>. It is a dense, young carbon-rich PN which started to be ionized about 500–1000 yr ago. It contains a substantial inner reservoir of atomic or molecular gas, probably as small cloudlets. These cloudlets have been ejected, just before the termination of the AGB phase of evolution, at a velocity ≤ 6 km s^{-1}, an important clue to the mass loss during the late AGB evolution. The central star is directly reached through its UV continuum emission. It has a $T_{eff} = 46\,000 \pm 2000$ K, a luminosity $L = 7300 \pm 700$ L$_\odot$ and a core mass of 0.63–0.67 M$_\odot$. There is a severe depletion of the nebular gases onto dust grains, most likely of the Ca Mg silicate variety; a surprising result in view of the carbon-rich nature of the ionized nebula.

There are a few PNe in the Magellanic Clouds with properties which set them apart from the general PNe population. The nebula <u>SMC SMP28</u> has been found to be remarkably deficient in carbon (Meatheringham *et al.* 1990); the carbon abundance is less than 1/450th solar. The electron temperature is very high (25 200 K). This may be the consequence of the lack of efficient cooling by carbon. The central star has an effective temperature, $T_{eff} = 1.8 \times 10^5$ K and a radius of 0.09 R$_\odot$. A nebular mass of 0.71 M$_\odot$ and a stellar mass of about 0.65–0.71 M$_\odot$ are inferred. It appears likely that it has evolved from a massive progenitor, with a main-sequence mass greater than at least 5 M$_\odot$, which underwent both second and third nuclear dredge-up, as well as very efficient hot-bottom burning. These processes raised the surface abundances of

7.8 Novae

Table 7.3. *HST data for LMC N66*

Structure	Dimensions		Ref.
	arcsec	10^{17} cm	
Ring, inner radius	0.69 ± 0.08	4.8	Dopita *et al.* 1993a
Ring, outer radius	1.09 ± 0.08	7.6	
Filaments	1.86	13.0	
Component	Distance to central star arcsec	Relative intensity in [OIII]	
Lobe A	0.50	37	Blades *et al.* 1992
Lobe B	0.66	33	
Knot 1	0.67	5	
Knot 2	0.12	3	
Knot 3	0.64	3	
Knot 4	0.94	1	
Knot 5	0.37	2	
Loop 1	2.05	6	
Loop 2	2.27	3	
Loop 3	1.22	5	
Star	0	2	

He and N while depleting O and reducing C drastically. It is thus an extreme type I PN.

The PN LMC SMP64 has been shown to have an extremely high electron density, a strong radial density gradient, and a central star with a very low effective temperature, $T_{eff} = 31\,500$ K (Dopita and Meatheringham 1991b). The stellar luminosity is 6400 $L_\odot$, implying a core mass of 0.62 $M_\odot$, typical of the LMC population of PNe. The star may only just have reached a temperature high enough to ionize a portion of the material ejected during its AGB evolution. The age of the PN shell would be as low as ~ 310 yr if the expansion rate were 7 km s^{-1} and the mass loss $\sim 1.3 \times 10^{-6}$ $M_\odot$ yr^{-1}. The authors find this unlikely and suggest that a reservoir of dense gas exists close to the star and that the ionized material is being replenished by ionization of this gas. LMC SMP64 may perhaps qualify as a true transition object between an AGB star and a PN nucleus.

7.8 Novae

The use of novae for determinations of the distances of the Clouds has been discussed in Sect. 2.6.3, where the super-bright novae were also considered. Here, a few other remarkable objects will be described.

Nova LMC 1988 No. 2 was a very fast nova ($t_2 = 5^d$) of large amplitude $\Delta m > 11.2$ mag). The visual maximum was $m_V = 10.3$ mag October 13.75, 1988 UT. A spectrum taken 1.34 days after the maximum showed strong Balmer, He I, and Fe II lines with P Cyg profiles. Two principal absorption systems at -1700 and -2500 km s^{-1} and higher velocity diffuse enhanced absorption systems extending to

-5000 km s^{-1} were seen (Sekiguchi et al. 1989). 50 days after maximum the nova had developed into the early nebular phase and showed very strong [NeIII] emission. Its characteristics resembled those of V1500 Cyg and V1370 Aql.

The first extragalactic recurrent nova observed is Nova LMC 1990 No. 2. Its optical outburst was discovered on February 14.1, 1990, by Liller (1990) ($m_V = 11.2$), who suggested a correspondence in position between this nova and the one reported in the LMC in 1968 (Sievers 1970), which reached a maximum of ~ 10.9 mag. Shore et al. (1991) confirmed that the two novae coincide in position to within $2''.4$ in right ascension and $6''$ in declination and concluded that it is a single object, a recurrent nova. Because its distance is well known it is possible to determine its total outburst energy and thus understand the outbursts of these objects better. The very short recurrence time scale requires a high-mass white dwarf and a high mass-accretion rate. The secondary is therefore assumed to be an evolved star rather than a main-sequence object. At the position of Nova LMC 1990 No. 2 no remnant brighter than $B = 17$ mag can be seen. Any red component in the system must have $M_V > -1.5$ mag. Its nature will have to be determined by direct observations.

IUE observations by Shore et al. (1991) show that the luminosity between 1200 and 3300 Å during the first two days of the outburst was 3.1×10^{38} erg s^{-1}, or 7.8×10^4 L$_\odot$. The bolometric luminosity was thus greater than the Eddington luminosity, 3.1×10^4 L$_\odot$, for a 1.0 M$_\odot$ white dwarf, assuming a solar mixture of the elements and electron scattering as the principal source of opacity. If helium and nitrogen were enhanced over the solar mixture by large factors, the nova's luminosity would become less 'super-Eddington'.

8

The interstellar medium

The interaction in a galaxy between the stars and the interstellar medium (ISM) through mass loss, accretion, and the formation of new objects is exceedingly important for its evolution. The composition of its ISM tells us about its past history and, correctly understood, also about its future. The ISM represents the youngest generation and has therefore, and as a consequence of experience from our Galaxy, been considered to be basic in studies of the kinematics of external galaxies. It has played, and plays, a fundamental role in all discussions of the Magellanic Clouds. Historically, the Lick Expedition radial velocities of a few emission nebulae in the LMC and one in the SMC (see Chap. 1) served chiefly to prove that the Clouds were outside the main body of the Galaxy. An indication of a gradient of velocities in the LMC pointed towards rotation or translation. The existence of this gradient was put beyond doubt in the 1950s, when 21 cm HI line observations began. Since then, other components of the ISM have been observed, making important contributions regarding the composition and kinematics of the Clouds.

8.1 The neutral hydrogen

The continuous radio emission from the Magellanic Clouds was first detected by Mills in 1953. Radio isophotes at 3.5 m were obtained in 1954 (Mills 1954). The line radiation from neutral hydrogen in the Clouds was detected in 1953 and a preliminary survey of its distribution was presented (Kerr *et al.* 1954). At 3.5 m both Clouds appeared to be about the same size, but the surface brightness of the LMC was about three times as large as that of the SMC. The derived HI isophotes showed a general similarity to the optical equidensity contours and the total HI mass was found to be 6×10^8 and 4×10^8 $M_\odot$ for the LMC and the SMC, respectively. Since then a multitude of radio surveys have been carried out of the Clouds.

8.1.1 The Large Magellanic Cloud

The HI structure in the LMC is complicated. In the fundamental work by McGee and Milton (1966) 52 large HI complexes were identified, embedded in a fairly low-level background. The complexes were shown to be well correlated with HII regions in positions as well as in velocities. It was found impossible to fit a unique rotation curve and inclination to all the radial velocity data, and a structure with spiral arms in different planes was suggested.

More recent observations with a better signal-to-noise (S/N) ratio and velocity resolution have shown that the total HI in the line of sight has a much smoother distribution (Rohlfs et al. 1984). Among the large complexes seen in McGee and Milton's Fig. 1 only two appear in the new map; one around and south of 30 Doradus and one forming an arc through the Shapley Constellation I. The two voids first noted by Westerlund (1964) are seen in both surveys. One, centred at 5^h32^m, $-66°40'$ is now referred to as SGS LMC 4 (Meaburn 1980). The other has its centre at 5^h27^m, $-68°52'$, on Constellation II. A SGS with that centre would have the just-mentioned arc as part of its envelope.

A certain cloudiness of the neutral hydrogen is, however, also seen in the Rohlfs et al. maps (1984, Fig. 4) for a series of velocity ranges as closed contours. The more complete smoothing in their total hydrogen map may be partly due to the inclusion of more of the hydrogen outside the plane of the LMC, partly to the slight undersampling mentioned by the authors.

Evidence for extensive, partly overlapping HI sheets, 1–3 kpc across and with different radial velocities, was found by Meaburn et al. (1987) in profiles of the 21 cm HI line obtained with high S/N ratios along tracks intersecting several of the SGSs. The velocities of the HI sheet were interpreted as showing motions away from the plane of the LMC. The energy needed for this, estimated as $\approx 4.5 \times 10^{54}$ erg, could be supplied by supernovae and stellar winds over 2×10^7 yr. These HI sheets may result from processes similar to those forming the supergiant shells. To account for this, simultaneous bursts of star formation over the whole central region of the LMC would be needed. The bursts would have begun at about the time of the latest close encounter between the LMC and the Galaxy, i.e. about 50 Myr ago.

In their 21 cm line survey of the LMC (a field of $8.4° \times 8.4°$) Luks and Rohlfs (1992) identified two separate features: a gas disk extending over all of the LMC with 72% of the HI gas and a low-velocity (L) component of 19%. The kinematics of the two components will be discussed in Sect. 12.1. The L-component consists of two large complexes, each extending over more than 2.5 kpc. The southern lobe contains, apparently, part of the large complex south of 30 Doradus but extends much further south than SGS LMC 2, and covers an area where star formation has not yet begun. Towards the west it becomes indistinguishable from the disk. The other lobe is north of 30 Doradus, extending west and then north. It may contain SGS LMC 3, but does not appear to contain any of the superassociations (Fig. 8.1). The lobes stay clear of the Bar. They are situated 0–500 pc above the disk; the gas has a radial velocity 20–60 km s^{-1} lower than that of the disk.

The strong UV radiation region (1500 Å), of which 30 Doradus is a part, links the two lobes. A long filamentary structure of UV radiation extends along the southern edge of the northern lobe. If this is indicative of star formation connected with the lobe, then the northern HI lobe shows more activity than the southern where much less UV radiation is seen..

The time needed to separate the L-component from the disk is estimated to be ~ 20 Myr. This is a factor of 10 less less than the time proposed by Fujimoto and Noguchi (1990) to move the gas cloud formed by the SMC-LMC collision 0.2 Gyr ago from the collision point, at 5^h40^m, $-68°$, to its present position. The latter model has the advantage that it explains the extension of the observed HI cloud towards the south and southwest as tracing the later part of the orbit of the giant cloud. The northern lobe may have a similar history of formation and evolution with possibly more than one cloud being involved.

8.1 The neutral hydrogen

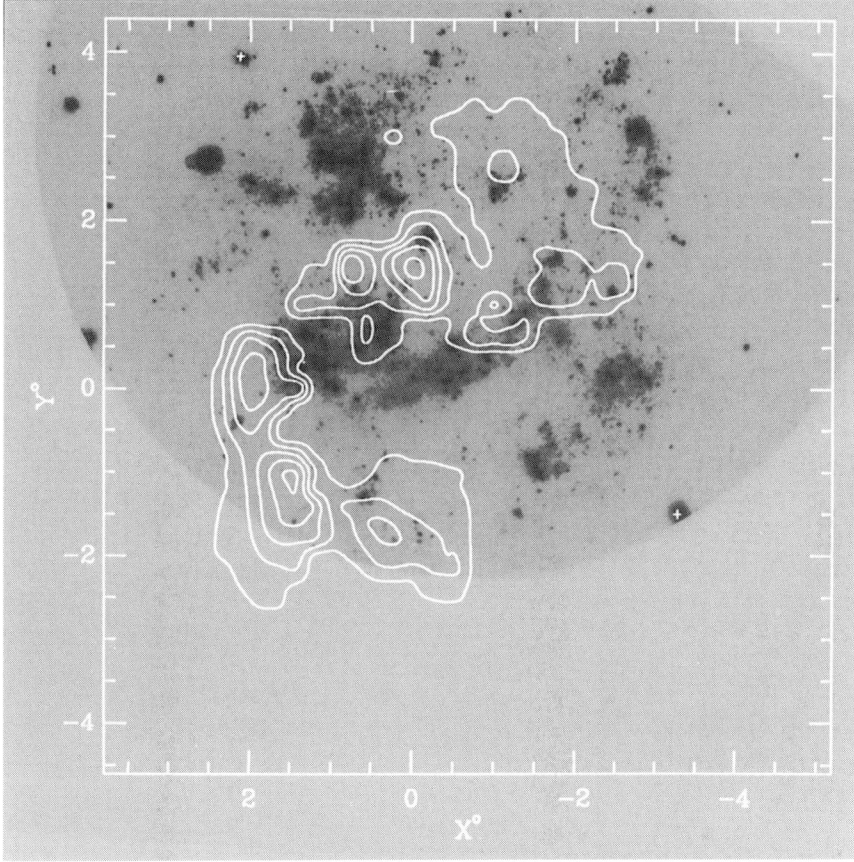

Fig. 8.1. HI contours of the L-component superposed on an UV-map at 1500 Å (Luks and Rohlfs 1992). 30 Doradus is between the two loops.

The gas structure in the 30 Doradus region is more complex than revealed by the HI emission. In a $30' \times 30'$ region around SN 1987A, including the core of 30 Doradus, high-resolution NaI and CaII spectra show absorption profiles in several lines of sight with at least five interstellar components spanning the range 260 km s^{-1} $< RV_{hel} <$ 300 km s^{-1} (Vladilo et al. 1993). The most intense absorptions are seen at 280–290 km s^{-1}. This gas may be considered as the LMC galactic disk disrupted into several components. The most intense HI emission in the field is at $\sim$ 270 km s^{-1}. There is a good coincidence in velocity and sky distribution with the NaI and CaII gas but the latter absorptions are weak, indicating either that the 270 km s^{-1} gas lies in large part beyond the observed stars or that it has a high degree of Na ionization.

8.1.2 The Small Magellanic Cloud

The structure of the SMC as displayed by the neutral hydrogen is apparently smoother than that of the LMC as far as the surface distribution is concerned (Hindman 1967). There are no real complexes of the kind seen in the LMC and in which HII regions and stellar associations are embedded. The OB stars are located in zones of

146 *The interstellar medium*

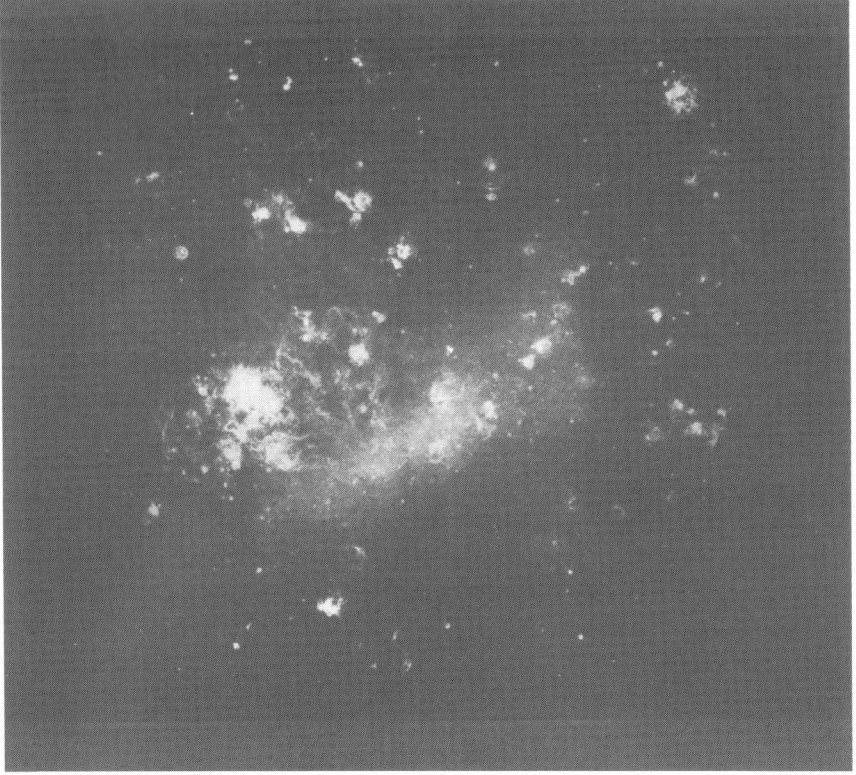

Fig. 8.2. Deep Hα + [NII] image of the LMC through a 100 Å mosaic filter. North is up, east to the left. Below 30 Doradus, the brightest nebula in the field, are seen N158, N160 and N159 (see Fig. 8.7). SGS LMC 2 and SGS LMC 3 are the filamentary structures to the SE and NW of 30 Doradus, respectively. The scattered nebulae in the upper left belong to SGS LMC4. (Reprinted from Davies *et al.* (1976) by permission of Blackwell Science Ltd.)

high gradient rather than in zones of high gaseous density (Azzopardi and Vigneau 1977). The HI emission covers a larger area than the optical isophotes. It displays the Bar and the Wing as the most prominent features.

The complexity of the SMC structure becomes apparent when the HI line profiles are studied with their double or multiple peaks over much of the Cloud. The interpretation of these peaks has a direct bearing on the extent of the SMC in the line of sight and will be dealt with in Chap. 12.

8.2 The ionized hydrogen

The large-scale structure of the ionized gas in the LMC and the SMC (Figs. 8.2 and 8.3) has mainly been investigated with the aid of Hα photographs, first those by Henize (1956), and later on those by Davies *et al.* (1976). Quantitative information on individual HII regions has been provided by Caplan and Deharveng (1985), who carried out Hα and Hβ photometry of several HII regions in the LMC. The average Hα/Hβ ratio was 3.5 ± 0.4, about 15 % higher than that determined spectroscopically.

8.2 The ionized hydrogen

Fig. 8.3. Deep Hα + [NII] image of the SMC through a 100 Å mosaic filter. SGS SMC 1 is the sequence of nebulosities in the lower left, marking the end of the classical Wing. (Reprinted from Davies *et al.* (1976) by permission of Blackwell Science Ltd.)

Among early observations of the radial velocities of HII regions in the Clouds may be mentioned those by Feast (1964, 1970) and Smith and Weedman (1971, 1973).

Kennicutt and Hodge (1986) measured the integrated Hα fluxes for several hundred HII regions in the Magellanic Clouds. The HII regions span a range of over 10^4 in luminosity, from compact HII regions to the 30 Doradus giant nebula. The faintest HII regions may be ionized by single stars, the majority by stellar associations and clusters. The extended networks of filaments in the LMC probably contribute 15–25 % of the total Hα luminosity. This is comparable to some of the largest HII regions (see Table 8.1). 30 Doradus accounts for ≈ 25 % of the total emission of the LMC and is twice as luminous as the entire SMC. The integrated Hα luminosities and equivalent widths of the LMC and the SMC are close to the average for Magellanic Cloud irregulars. This suggests that their current rate of star formation is higher than the average past rate. The Lyman photon production rates derived from the Hα emission are significantly lower than those predicted by radio continuum measurements. It appears likely that this is mainly due to an underestimate of the interstellar extinction. The total population of massive stars predicted by the Hα emission is $\approx$ 2–4 times higher than the catalogued number of stars (in 1986). This difference may be due to

Table 8.1. *Hα filaments in the Magellanic Clouds*

	DEM No.	Dimension arcmin	Flux 10^{-12} erg cm^{-2} s^{-1}
LMC	177	30 × 20	370 ± 100
LM	203	25 × 20	530 ± 150
LMC	224	25 × 20	300 ± 100
LM	232	30 × 15	1300 ± 400
LMC	310	60 × 40	2200 ± 800
SMC	167	35 × 30	400 ± 200

The listed DEM objects in the LMC are all in either
SGS LMC 2 or SGS LMC 3; the one in the SMC is in SGS SMC 1.

incomplete surveys but it could also be explained if the initial mass function in the 30–100 M$_\odot$ range were considerably shallower than the Salpeter function, or if the stellar mass limit in the Clouds were considerably higher than 100 M$_\odot$.

The large-scale Hα filamentary structures, 'the froth', in Magellanic Cloud irregular galaxies have been studied by Hunter and Gallagher (1990). Their main conclusions are as follows. (i) The level of ionization of the froth is lower than in HII regions in the same galaxies with [SII], [NII], and [OII] relative to Hα being enhanced while [OIII]/Hβ is reduced in the froth. (ii) Frothy features have large line velocity widths that often exceed those found in classical giant HII regions in the same galaxies.
(iii) The froth has Hα surface brightnesses of ∼ 100 rayleighs, 10–100 times brighter than typical diffuse galactic emission. (iv) The dimensions are large, a typical filament is 700 × 50 pc.

Shock models for the ionization are ruled out because of the absence of strong [OI] emission. Pure photoionization seems implausible because of the large distance to the nearest OB associations, 0.2–0.8 kpc. A combination of the two mechanisms may best explain the structures.

The SMC is, relative to the LMC, deficient in large nebulae of high surface brightness as well as in filamentary structures and in shells.

The equivalent widths, $W_{H\beta}$, the [OIII]/Hβ ratio, and the Hβ line flux have been measured for 30 HII regions in the SMC by Copetti and Dottori (1989). Using the Kennicutt and Hodge Hα fluxes they derived a ratio Hα/Hβ = 4.0 ± 0.3, which is about 20 % higher than the average of the values derived spectroscopically. This may, at least partly, be due to a higher interstellar extinction in the SMC than is commonly assumed (it may partly result from the differences in the two observing sequences), assuming that the spectroscopic observations refer to the brightest parts (least absorbed) of the HII regions. Larger interstellar extinctions towards the HII regions in the Clouds are also suggested by comparisons between optical and radioastronomical measurements (Israel 1980, Kennicutt and Hodge 1986).

Torres and Carranza (1987) have shown, by interferometric Hα observations, that diffuse emission covers most of the SMC and that it is separated into four velocity groups partly overlapping in the sky. The overall radial velocity gradient is in excellent agreement with that of neutral hydrogen. These results were confirmed by Le Coarer *et al.*(1993) in a kinematic Hα survey of the SMC with scanning Fabry–

8.2 The ionized hydrogen

Table 8.2. *Integrated properties of the ionized gas in the Clouds*

Parameter	LMC	SMC	Unit
$F(H\alpha)$	9.0	1.0	10^{-8} erg cm^{-2} s^{-1}
$F_{HII}(H\alpha)$	5.8	0.6	10^{-8} erg cm^{-2} s^{-1}
Diffuse fraction	0.35	0.41	
$L(H\alpha)$	2.7	0.48	10^{40} erg s^{-1}
SFR	0.26	0.046	M$_\odot$ yr^{-1}
N(O5 V)	540	95	equivalent
Area	40.7	5.4	deg^2

The contribution of the 30 Doradus nebula to $F_{HII}(H\alpha)$ is $\sim 33\%$.
The diffuse flux from the nine SGSs in the LMC amounts to about 40% of the total. It contributes on the average about 50–60% of the total flux of the SGSs.

Perot interferometers. They also found that the distribution of the radial velocities of the individual HII regions shows the four components (VL, L, H, VH) found by Martin *et al.*(1989). Their data indicate a velocity gradient along the SMC Bar; the velocity at the northern end is about 50 km s^{-1} higher than in the southern end. This gradient is, however, not real. It is caused by the three velocity components VL, L, and H partly overlapping (as seen in Martin *et al.*, Fig. 2J).

Kennicutt *et al.*(1995a) have used calibrated CCD images, obtained with the 'Parking Lot Camera', to investigate the large-scale structure of the ionized gas in the LMC and the SMC. In both galaxies the bright HII regions are embedded in diffuse nebulosity. Around 30 Doradus this cloud has a diameter of ~ 2 kpc. Diffuse emission can be detected over most of the LMC. The filamentary structure and diffuse emission is less prominent in the SMC, but extended emission can be seen in the Bar, in the SE region and in the shells in the Wing, confirming the results obtained by Torres and Carranza.

The integrated properties of the ionized gas in the Clouds are summarized in Table 8.2.

The estimates of the fractions of Hα fluxes produced by the diffuse emission may be overestimated if the interstellar extinction in the HII regions is significantly larger than for the diffuse emission. The mean extinction, $A_{H\alpha}$, in the HII regions is ~ 0.3 mag for the SMC and ~ 0.4 mag for the LMC (Caplan and Deharveng 1985, Caplan *et al.*1996). The 30 Doradus nebula has a higher than average extinction, $A_{H\alpha} \sim 1$ mag. The values in Table 8.2 are derived assuming that the extinction in the diffuse emission is the same as in the HII regions; if it is zero in the diffuse gas, the values fall to 25% for the LMC and 34% for the SMC.

Much of the diffuse emission comes from filamentary and shell-like structures. In the LMC the nine supergiant shells (see Sect. 6.1) may contribute at most $\sim 20\%$ of the total Hα luminosity. This leaves a significant amount of Hα luminosity and ionized gas mass in the truly diffuse medium. The mechanism for ionization may be either photon leakage from the large HII regions, which would require a certain porosity of the ISM, or, perhaps more likely, some kind of shock ionization.

The star formation rates (SFR) given in Table 8.2 are derived from the Hα luminosities using Kennicutt's updated model (Kennicutt *et al.*1995b). It is assumed that all Hα emission results from photoionization by massive stars and that no ionizing radiation escapes.The derived SFRs are modest compared with that for our Galaxy ($\sim$ a few $M_\odot$ yr^{-1}). They reflect the low masses of the Clouds and may also indicate that their current SFRs are below their averages. This conclusion is the opposite to the one drawn by Kennicutt and Hodge from a previous Hα survey (see above).

By using the Rohlfs *et al.*(1984) and Hindman (1967) HI maps, Kennicutt *et al.*(1995a) found that most of the discrete HII regions appear in regions where the HI column density exceeds 7 and 25×10^{20} H cm^{-2} for the LMC and the SMC, respectively, in agreement with previous results (Davies *et al.* 1976).

8.3 Some well-studied emission nebulae in the LMC

The nebulae in the Magellanic Clouds appear in all kinds of shapes and sizes, from the giant 30 Doradus complex to small, virtually stellar objects. Many are ring-shaped or bubble-like. If the supergiant shells are included, their dimensions range from diameters of about 1 pc to over 1 kpc. They may have formed through the effects of SSPSF, resulted from SN explosions, or been blown up by stellar winds. The initial conditions must be expected to differ from one type to another. In many cases the ionizing stars have been found to be small clusters. Examples of that kind have been given in Sect. 5.6. A few interesting nebulae are discussed here.

8.3.1 The N11 nebula

Less famous than the 30 Doradus nebula but prominent in many ways is the giant HII region N11, located in the NW part of the LMC. It has the second highest Hα luminosity of the LMC HII regions and is also the second strongest radio continuum source (see maps by e.g. Haynes *et al.* 1991 and Xu *et al.* 1992). It has a significant IR radiation (Schwering 1989a,b, Xu *et al.* 1992) and a CO complex of 21 clouds (Israel and de Graauw 1991, Booth 1993) in a region of about 500×500 pc; their diameters are typically around 20–30 pc. The CO clouds follow the *IRAS* maps closely, especially those at 25 and 60 μm. In the SW the CO clouds form a ring around a rich cluster of luminous stars. The hole at the cluster position indicates that the CO has been destroyed or driven out.

It has been suggested that (the optically defined) N11 has a distinct, dual structural morphology remarkably analogous to that emerging from current IR imaging of 30 Doradus (Walborn and Parker 1992). A central relatively emission-free cavity contains the rich OB association LH 9 (see Chap. 5). In the faint emission that covers the cavity, five separate velocity components have been observed with radial velocities between $210 \leq V_{hel} \leq 370$ km s^{-1} (Meaburn *et al.* 1989). This is indicative of expansions more complex than of a single, radially expanding spherical shell.

The surrounding filamentary structure contains bright HII regions in which the associations LH10, in N11B, and LH13, in N11C, are embedded. Somewhat more detached, to the NE, is LH14, in N11E. As pointed out in Chap. 5, the stellar content of the nebular-embedded associations is significantly younger than that of LH9 and the indication of sequential star formation is very strong.

The 30 Doradus and N11 complexes may be compared with the associations in the Scorpius and Carina regions in the Galaxy. Thus, LH9 in N11 represents the Sco OB1 stage with an age of $\sim$ 5 Myr, LH10 and the 30 Doradus central cluster correspond to the Carina Nebula phase, age 2–3 Myr, though LH10 may be slightly younger as WN objects have not yet appeared. Finally, the youngest 30 Doradus generation, surrounding its central cluster, is in the Orion nebula phase, containing IR protostars much less than 1 Myr old. In both regions an initial central starburst triggered a second starburst around its periphery about 2 Myr later. In N11 the whole process is further advanced than in 30 Doradus by 2 Myr. Its dominating object, HD 32228 ($\equiv$ R64) may be considered an evolved R136. The similarities between the two objects have been further emphasized by Walborn et al. (1995), who, with the aid of the *HST* PC, have resolved HD 32228 into 15 components with $12 \leq V \leq 15$ mag and 72 with $V < 18$ mag (Fig. 8.4). Similar results have been achieved by Schertl et al. (1995), who in speckle masking observations of HD 32228 detected 25 stellar components with $12.5 < V < 17.1$ in a $6''.4 \times 6''.4$ field. R136 is, however, far more massive than HD 32228; the former contains a few times 10^4 $M_\odot$, the latter less than 10^3 $M_\odot$.

Walborn and Parker further speculate that the positions of 30 Doradus and N11, diametrically opposite in the LMC and north of the Bar, may be significant. They refer to a study by Elmegreen and Elmegreen (1980), who found that approximately half of all barred Magellanic irregular and spiral galaxies (larger than 4′) have their largest HII regions near the ends of their bars. Star formation in 30 Doradus may have been triggered by the large-scale compression of gas moving around the Bar in the LMC. However, there is no evidence for any gas motion connected with the Bar and, furthermore, 30 Doradus is most likely above the LMC main plane. N11 is not part of any supergiant shell, though it is a significant member of a superassociation (see Sect. 6.1).

8.3.2 The N44 complex

The emission nebula N44 ($\equiv$ NGC 1929, 34, 35, 36, 37, IC 2126, 27, 28 $\equiv$ Shapley's Constellation I $\equiv$ DEM150, 151, 152) is rich in bright emission arcs, of which the dominating one forms a shell around the OB association LH47 ($5^h22^m.1$, $-67°59'$), containing ~ 60 OB stars (Lucke and Hodge 1970, Lucke 1974). Also the associations LH48 and LH49 are embedded in N 44, to the north and to the south of LH47, respectively. Bok (1964) determined the mass of the ionized hydrogen in N44 to 6×10^4 $M_\odot$ from optical and radio data. It is surrounded by an HI cloud with a diameter of ~ 350 pc and a total mass of $\sim 5 \times 10^6$ $M_\odot$. The age of the stellar group is estimated as 2 Myr, the stellar mass as $\sim 2 \times 10^4$ $M_\odot$. The stellar association may be expanding.

The motions in the N44 nebulosity have been analysed by Meaburn and Laspias (1991). The near side of the 70×50 pc primary shell expands towards us with a velocity of 46 km s^{-1} and the far side away at 36 km s^{-1} (relative to a heliocentric radial (perimeter) velocity of 304 km s^{-1}). The ionized mass involved in the expansion is $M_{HII} = 10^4 \alpha^{1/2}$ $M_\odot$, where α is a filling factor which can be 0.1 or as low as 0.01.

In a secondary shell, to the west of the main body, only a receding velocity with a maximum of 16 km s^{-1} has been detected. NaI absorption line profiles in the spectra of the two brightest stars in LH47 indicate expansions of -17 and $+4$ km s^{-1}. The profile of the HI 21 cm emission line from the whole region (beamwidth 14′.5) presents three components with expansion velocities of -57, -31 and 0 ± 2 km s^{-1}. The CaII interstellar absorption line profile for a star, Sk86-67, at the northern edge of the primary shell covers the same velocity range as the HI profile.

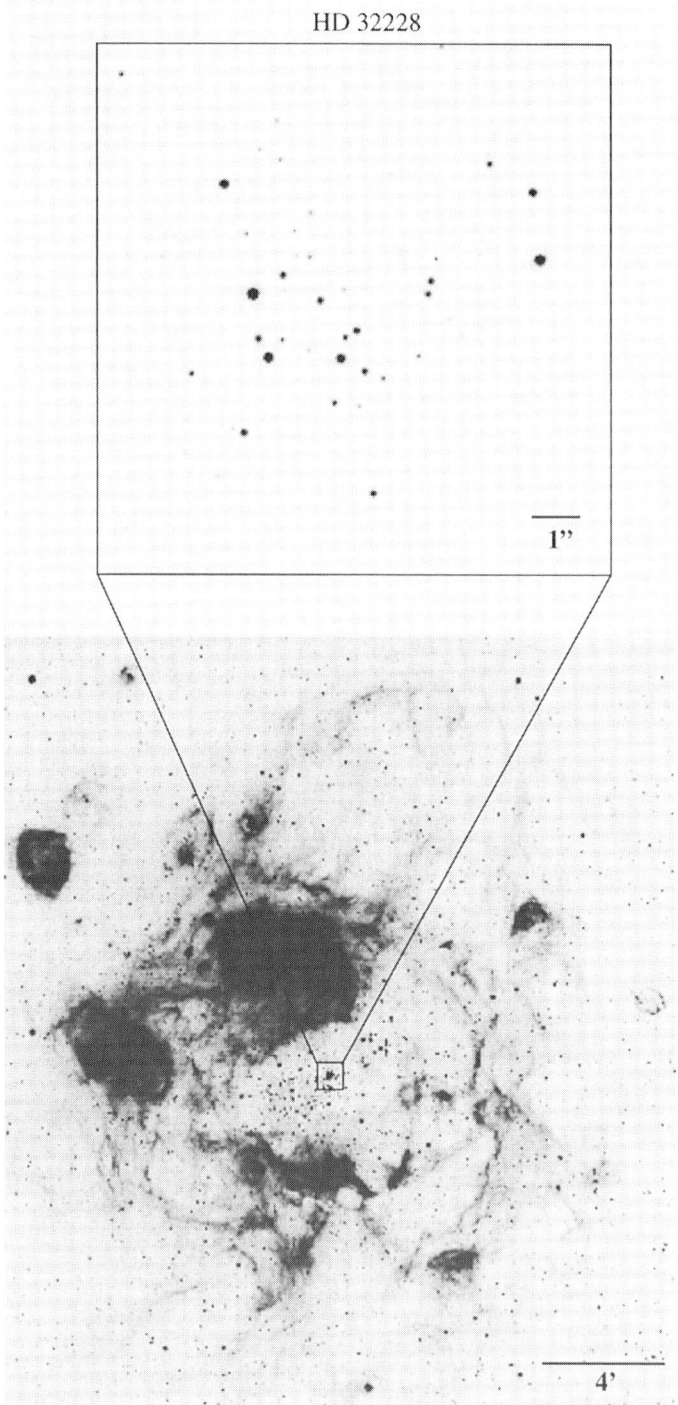

Fig. 8.4. N11 in Hα light (from a CTIO 4m prime focus photograph) with the massive central object HD 32228 identified and, in the upper V image, resolved in an *HST* PC2 image. LH10 ionizes the large HII region to the north of HD 32228. The latter belongs to the association LH9. (Figure supplied by N. Walborn.)

8.3 Some well-studied emission nebulae in the LMC

Table 8.3. *The brightest stars in N120*

Star	Sp.	V	E_{B-V}	Location	Remarks
Sk107-69	O9.5Ib	11.24	0.12	In C1	multiple, R89
Sk106-69	WC 6	12.00		In C1	double?, R90
BI141	(B3 Iab)	12.24	0.23	In N120A	see text
Sk98-69	B0	12.29	0.04	In spur (HIJ)	
Sk103-69	B0	12.48	0.08	20" from SNR 120	
Sk105-69	B0 comp	12.79	(0.08)	In centre of ring	

IRAS maps of the N44 region give fluxes in the far-infrared (FIR) bands at 12, 25, 60 and 100 μm of 51, 179, 990 and 1815 Jy, respectively. This FIR emission is energized by the LH 47, 48 and 49 OB association stars.

For a flux of Lyman continuum photons of 6×10^{50} s^{-1}, ≡ 10 O5 stars, the stellar mass of LH47 is ≈ 3 10^4 M$_\odot$ with a Salpeter distribution (Meaburn and Laspias 1991). The age has been estimated as 3.5 Myr (Braunsfurth and Feitzinger 1983). Using the same arguments as for N11, Meaburn and Laspias find that around 280 ± 140 type II supernova explosions may have occurred in LH47 during its existence. The soft X-ray emission found by Chu and Mac Low (1990) and ascribed to SNRs in collision with the perimeter of its primary shell around LH47 strengthens this interpretation.

It is possible to explain the motions in N44 even with a much smaller SN rate as those of a pressure-driven, energy-conserving bubble if the stellar winds and SN explosions combine to generate a shocked superheated gas inside the primary shell during the 3.5 Myr. Many such bubbles are known in the LMC around individual WR stars (Chu 1983).

8.3.3 The nebular complex N120

The HII region N120, ≡ DEM134, at 5^h19^m, $-69°43'$, is one in a sequence of nebulae (N113, N114, N119, N120) seen projected on the southern part of the LMC Bar. It is composed of several bubbles of different types arranged in an arc (Fig. 8.5) with the dimensions 129 × 100 pc (distance 50 kpc). No stellar association can be seen on the cavity side. Embedded in the nebula is the OB association LH42, ≡ NGC 1918. The brightest stars are listed in Table 8.3.

Henize (1956) identified four concentrations, N120 A, B, C, D. A more detailed description has been given by Laval *et al.* (1992a), who split the 'C' section into six parts, C1–C6, and also identified a 'spur' HIJ in the NW part (Fig. 8.5). The complex consists of an ionization front associated with the interior stars, a SNR, two wind-blown bubbles, N120A and N120B, and several 'normal' HII regions. The kinematical data show that these nebulae are associated, forming an incomplete ring. The ambient medium is clumpy, of high density, and is probably related to the molecular cloud seen in its vicinity. Violent motions exist inside N120A, and, to a lesser degree, inside N120B.

The ionizing flux of the stars in LH42 has been estimated as 1.53×10^{50} photons s^{-1} (Smith *et al.* 1987). Of this, the two evolved stars Sk107-69 and Sk106-69 (Table 8.3)

154 *The interstellar medium*

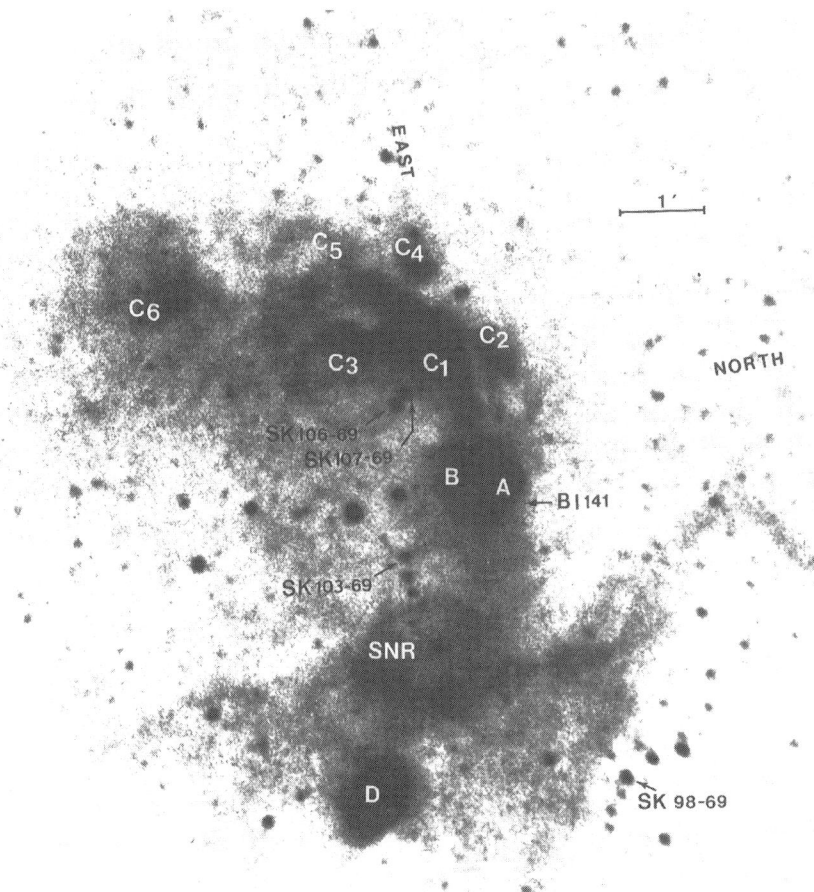

Fig. 8.5. Hα monochromatic photograph of the nebular arc containing N120A, B, C, D. The beginning of the spur HIJ is seen above SK98-69 (Laval *et al.* 1992a). (Photograph made by Observatoire de Marseille.)

supply only $\sim 12\%$. The remaining flux must come from undetected O stars. Two extended non-thermal sources and one thermal source have been located in N120 (Milne *et al.* 1980). One of the former is SNR 0519-696; the other, SNR 0518-696, is controversial. The thermal source is associated with the nebula C1 (Fig. 8.5). There is only one velocity component of HI in the direction of N120, $V_{hel} = 256$ km s^{-1} (McGee and Milton 1966).

There is a large CO cloud in the vicinity of N120 (Cohen *et al.* 1988). The CO (1-0) emission at 115 GHz around the star BI141 has been mapped by Laval *et al.* (1994). It showed the highest intensity immediately north of the star, which is probably at the southern edge of the CO cloud. The velocity of the cloud, $V_{LSR} = 238$ km s^{-1}, differs from that of the Hα emission of the bubble, $V_{LSR} = 224$ km s^{-1}.

The interstellar absorption towards N120A, derived from the Hα/Hβ ratio, is $A_V \approx$ 0.20–0.30 mag (Caplan and Deharveng 1986), whereas $E_{B-V} = 0.36$ from photometry of the brightest star in LH42, Sk106-69, O9.5Ib. Walraven photometry of BI141

8.4 The molecular content

has given $A_V \approx 0.7$ mag (Laval et al. 1994). Using a mean galactic colour excess in this direction of 0.03 mag, the reddening due to the LMC is $E_{B-V} = 0.17$ mag. The differences between the reddening determinations may be due to peculiarities of BI141 (see Sect. 5.6.1).

8.4 The molecular content

The CO observations of a galaxy are expected to measure its molecular content rather accurately. Other parameters of the interstellar gas are relatively undefined because CO is excited at low density, 100–1000 cm^{-3}, and the lines of its main isotopic form are often saturated. Temperatures are most accurately estimated from multi-level observations of individual molecules with unsaturated lines (see e.g. Johansson et al. 1984). Regions of various densities are traced by observations of molecular transitions requiring different excitation. A significant number of different molecular species need to be observed for a determination of the chemical composition.

8.4.1 The CO content

The first detection of molecular emission in the Magellanic Clouds was that of CO (1–0) by Huggins et al. (1975). A more systematic survey for CO, in the (2–1) line at 230 GHz, was carried out by Israel et al. for a small fraction of the Clouds (Israel 1984). In 1982 ^{12}CO (2–1) observations were carried out with the ESO 3.6 m telescope at La Silla, Chile, at 22 positions in the LMC and 16 in the SMC; emission was detected at 11 positions in the LMC and 6 in the SMC (Israel et al. 1986), mainly at the positions of dark clouds or HII regions. Most detections of CO emission in the LMC were made in the region south of 30 Doradus and in the SMC in the SW part of the Bar, notably associated with the HII regions N12, N27 and N88. In the SMC the *IRAS* sources appear to be the only reliable criteria for molecular emission.

The CO emission from Cloud objects tends to be weaker than that from galactic objects. The principal cause for this is relatively low C and O abundances, relatively low dust-to-gas ratios, and relatively strong mean UV radiation fields in the Clouds. The use of the galactic CO to H$_2$ ratio would lead to a significant underestimate of the H$_2$ content in the Magellanic Clouds.

8.4.1.1 Low-resolution surveys

A coarse resolution (12′) CO (1-0) map of a 6° × 6° central area of the LMC was derived by Cohen et al. (1988). 40 molecular clouds were listed of which a dominating complex extended south from 30 Doradus for nearly 2.4 kpc (Fig. 8.6). Generally, there is a very close correspondence between the molecular clouds and other Population I features, in particular dark nebulae. The total mass of molecular clouds in the LMC is estimated to be 1.4×10^8 M$_\odot$; the complex south of 30 Doradus contains nearly half of it, 0.6×10^8 M$_\odot$. This may be compared with the distribution of neutral hydrogen where 0.6×10^8 M$_\odot$ is in that region out of a total of 5×10^8 M$_\odot$ (Kerr 1989). A remarkable feature is a clumpy ring of molecular gas expanding from the 30 Doradus region (see Sect. 10.2).

The agreement between the molecular clouds and the HII regions, the SNRs, and other Population I objects was found to be unusually poor in SGS LMC 4. The molecular clouds there may have been dispersed during the star-forming events. There is, however, still some CO left close to the SNRs in this region also, as shown by

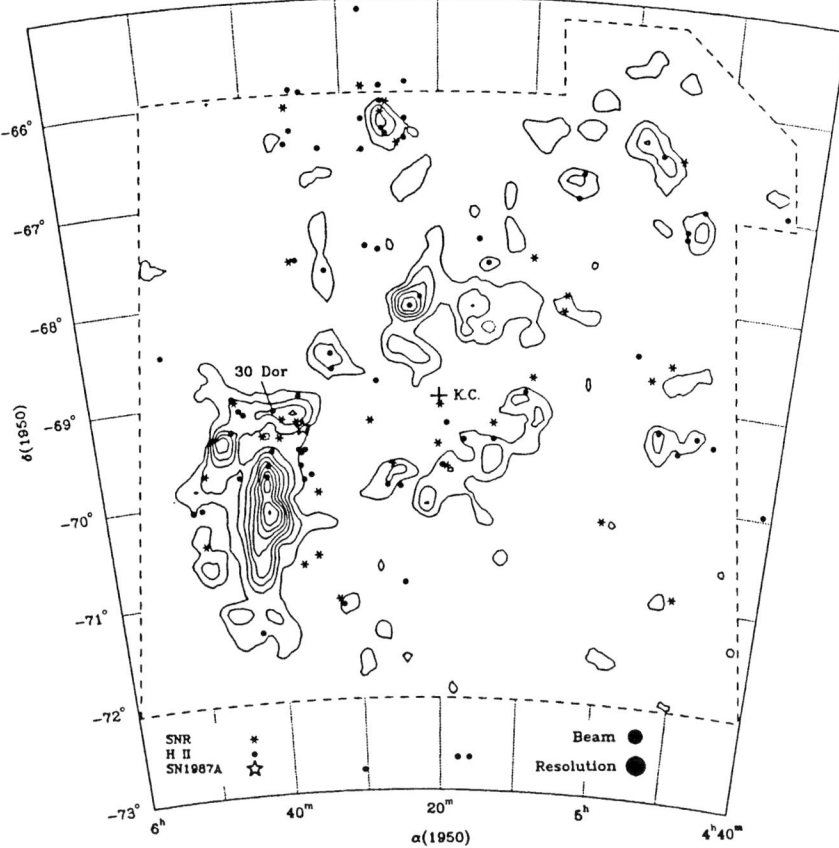

Fig. 8.6. A low-resolution CO map of the LMC (Cohen *et al.* 1988). SNRs, HII regions, SN 1987A, and the radio centre of rotation (K.C.) are identified.

Hughes *et al.* (1989). Using the high spatial resolution of the *SEST* they found a CO cloud centred on the brightest part of the X-ray and optical emission of the SNR N 49 and with the same velocity as the SNR, $V_{LSR} = 286$ km s^{-1}.

The star forming regions south of 30 Doradus, on the edge of SGS LMC 2, are still embedded in molecular clouds and are presumably more recent formations than the envelope of SGS LMC 4.

The central $2° \times 2°$ of the SMC has been fully surveyed in the ^{12}CO ($J = 1$–0) line at an angular resolution of $8'.8$ by Rubio *et al.* (1991). Two large complexes produce most of the emission, one in the SW and one in the NE region of the Bar, with two (SW-1 and SW-2) and three individual components, respectively. They are usually projected *in* the direction of HI clouds with the largest column densities ($N(HI) > 10^{22}$ cm^{-2}). They are also projected *near* bright, compact HII regions (optical or radio continuum sources). The CO radial velocities are about the same as the HI radial velocities of the associated atomic gas clouds. In the SW region the CO velocities agree with that of the low (L); in the NE with that of the high (H) velocity component of the SMC (see Chap. 12).

8.4 The molecular content

It appears likely that the molecular clouds in the SMC are the dense cores of large and massive regions of HI gas and are shielded from the UV radiation. They are mostly associated with regions of recent star formation. Only one cloud, but the most luminous in CO, appears to be a massive but quiescent molecular cloud with few signs of star formation.

The ^{12}CO antenna temperatures of the SMC molecular clouds are very low, ~ 0.035 K. (A typical galactic large cloud at the distance of the SMC would produce an antenna temperature of ~ 2 K.) The reason for the low ^{12}CO antenna temperatures is most likely that the characteristic size of the CO clumps of the molecular clouds in the SMC is smaller (by a factor of 8) than in our Galaxy.

8.4.1.2 High-resolution observations

In the ESO-*SEST* Key Programme on CO in the Magellanic Clouds (Israel *et al.* 1993; Rubio *et al.* 1993a,b; Lequeux *et al.* 1994) observations of a number of selected regions have so far been carried out.

^{12}CO ($J = 1$–0) has been observed towards 92 positions in the LMC and 42 in the SMC. The detection rates are 87 and 57%, respectively. In the SMC emission was searched for in HII regions, dark clouds and *IRAS* FIR sources. As this survey showed that CO is found mainly towards *IRAS* sources, the LMC survey was limited to catalogued *IRAS* sources over a significant range of FIR intensities. The most direct result is the weakness of the CO emission from SMC and LMC objects as compared to that from galactic clouds. (It is likely that the survey is biased towards the brightest CO clouds in the SMC and the LMC.) In the LMC significant numbers of CO clouds are detected only for $T_{mb} < 3$ K and in the SMC for $T_{mb} < 1$ K or for $I_{CO} < 20$ km s^{-1} and $I_{CO} < 2$ km s^{-1}, respectively.* In the LMC the $^{12}I_{CO}$ (as well as the $^{13}I_{CO}$) values are lower than in the Galaxy by at least a factor of 3; in the SMC they are lower by at least a factor of 10. The CO weakness is of the same order of magnitude as the underabundance of oxygen, carbon and dust in the Clouds.

The brightest CO clouds in the LMC are associated with the HII regions N44 and N159. In the SMC the brightest sources are associated with the intense but small HII regions N12 and N27. In a few cases CO clouds are found without any *IRAS* source counterpart. Strong CO emission anti-correlates with the presence of *IRAS* 12 μm excess emission ('cirrus'). CO sources in the SMC have a smaller linewidth than those in the LMC.

^{13}CO ($J = 1$–0) was observed towards the brighter ^{12}CO sources in the LMC (37) and the SMC (9). The mean velocity-integrated isotopic intensity ratios I_{12}/I_{13} are 12.5 for the LMC and ~ 15 for the SMC. In the Galaxy this ratio is 3–8.

Two regions, one near 30 Doradus and one covering the N11 complex, have been mapped (Booth 1993). The distribution of CO in two fields in the former area is shown in Fig. 8.7. 23 clouds are identified in the 30 Doradus field; their diameters range from 5 to 30 pc. Three small, fairly luminous CO clouds are within a few minutes of the 30 Doradus core. It is difficult to say to what extent the expanding ring feature, mentioned above, is confirmed or not. It is, anyhow, obvious that the CO content is relatively low in that field (for further discussion, see Sect. 10.2). In the second field, centred on N159, weak CO is associated with N158C and N160A.

---

* T_{mb} = observed main beam brightness temperature, I_{CO} = velocity-integrated ^{12}CO intensity.

158 *The interstellar medium*

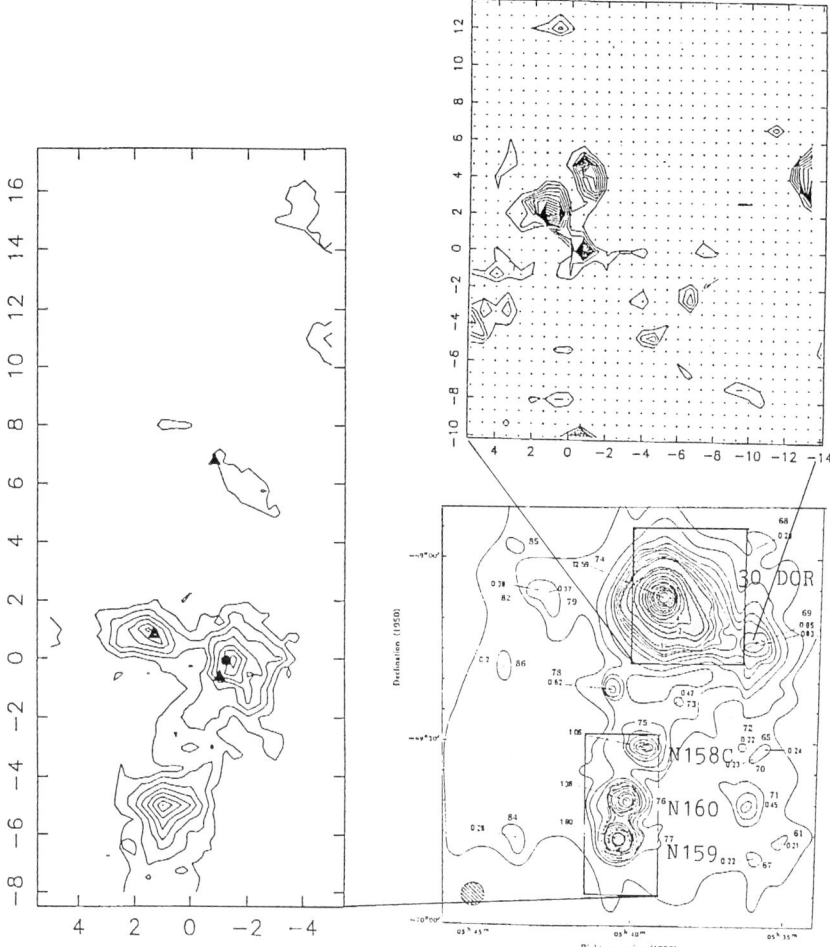

Fig. 8.7. Maps of CO in the parts of the 30 Doradus region indicated on the 5 GHz continuum map of McGee *et al.* (1972). The filled triangles identify the strongest *IRAS* point sources in this region. (Reprinted from Booth (1993) by permission of Springer-Verlag.)

A few luminous CO clouds are associated with N159 and a luminous cloud is seen just south of it where there are no optically visible signs of star formation.

Two fields in the SW Bar region in the SMC have been fully sampled in the ^{12}CO ($J = 1$–0) line at 45″ resolution with the *SEST* (Rubio *et al.* 1993a,b). The fields, SMC-B1 and SMC-B2, fall inside the SW-1 and SW-2 clouds, respectively (see above). The main CO features correspond to optical dark clouds; some are associated with HII regions and FIR sources. The spatial and velocity distribution of CO is complex with structures on all scales and large-scale velocity gradients. For two resolved CO clouds associated with HII regions comparable to the Orion nebula, four CO lines have been observed at a similar angular resolution. The ^{12}CO (1–0)/^{13}CO (1–0) line intensity ratios are between 10 and 16; the ^{12}CO (2–1)/^{12}CO (1–0) are slightly larger than 1 and the ^{13}CO (2–1)/^{13}CO (1–0) are even larger.

8.4 The molecular content

In SMC-B1 a molecular complex 70 pc in diameter surrounds the bright HII region DEM16 with an ionizing flux comparable to that of the Orion nebula (Kennicutt and Hodge 1986). It coincides in size and position with the dark cloud H1 (Hodge 1974a). Van den Bergh (1974) considered this cloud as composed of three absorption patches. They coincide roughly with the main CO concentrations around DEM16. The mean LSR radial velocity of the molecular complex is $V_{LSR} = 120$ km s^{-1}; it agrees reasonably well with that of DEM16, 122.9 km s^{-1} (Maurice et al. 1989).

In SMC-B2 the molecular clouds in the southern part surround but avoid the main extended HII region DEM37 (also comparable in flux to the Orion nebula). Their mean $V_{LSR} = 120$ km s^{-1} is slightly smaller than V_{LSR} (DEM 37), 126.7 km s^{-1}. The main CO cloud coincides well with the dark cloud H 14. Further north, the CO complex with $V_{LSR} = 121$ km s^{-1} appears associated with the compact HII region DEM38 with $V_{LSR} = 123.6$ km s^{-1}. Some of the CO clouds are projected towards the dark nebulae H 12 and H 13, W and N of DEM38, respectively. A faint CO emission component in this region has $V_{LSR} = 139$ km s^{-1}. In the northern part of the field there are no strong HII regions.

Lequeux et al. (1994) have shown that it is possible to reproduce the CO line ratios observed in SMC molecular clouds by simple plane-parallel, single-density models. Real clouds must, of course, be expected to contain a broad range of densities. Low-density gas ($n_H = 100$ cm^{-3}) is photodissociated and does not contain CO, in extreme cases not even H$_2$; the latter may not yet have formed in this gas. In a given cloud, the density at which the CO lines start to be emitted increases efficiently with increasing UV radiation field. The medium close to this critical density (the 'clumps') dominates the ^{12}CO line emission. The high ^{12}CO (1–0)/^{13}CO (1–0) line intensity ratios (see above) imply a high total absorption, perhaps an A_V of a few magnitudes. The ^{13}CO (2–1)/^{13}CO (1–0) line intensity ratios imply high excitation temperatures for the emitting parts of the clumps.

8.4.1.3 The CO intensity in the SMC, the LMC and the Galaxy

From a comparison of the CO luminosity versus line-width relation for molecular clouds in the SMC, the LMC and the Galaxy, Rubio et al. (1991) concluded that for the SMC the conversion factor to derive molecular masses from the velocity-integrated emission is 6×10^{21} cm^{-2} K^{-1} km^{-1}, i.e. ~ 20 times larger than the value adopted for our Galaxy. In the LMC the factor is ~ 6 times larger than the galactic value. The total mass of molecular gas in the SMC is then $\sim 3 \times 10^7$ M$_\odot$, and the ratio of molecular to atomic gas is about 7 %, 15 times smaller than observed in our Galaxy.

Preliminary results from *SEST* observations confirmed the Cohen et al. (1988) estimates that for the same amount of H$_2$, CO in the LMC is five times weaker than in the Galaxy (Johansson and Booth 1989). It may come closer to the galactic ratio in the largest and most massive clouds in the 30 Doradus region. The isotopic ratio ^{12}CO/^{13}CO/C^{18}O has been determined in N159 to 500: 70: 1, which is unusually weak for C^{18}O.

8.4.2 Molecular hydrogen

Weak emission of molecular hydrogen at 2 μm was detected by Koornneef and Israel (1985) in N159 in the LMC and in N81 in the SMC. In 1988 H$_2$ had been observed in seven compact HII regions in the two Clouds (Israel and Koornneef 1988). The observed H_2 must be either shock-excited by stars embedded in the molecular clouds or radiatively excited by the UV emission from the stars exciting the nebulae. The H_2 clouds are then within 0.15 pc or 2 pc, respectively, of the stars. It is expected that molecular hydrogen clouds should be quite common in both Magellanic Clouds. Information about the amount of H_2 there, will, as mentioned above, eventually come from the amount of observed CO.

Lequeux (1994) has proposed that large amounts of molecular gas, H_2, exist in the SMC. In this way the extinction observed in its outer layers, from galaxy counts, may be explained. This extinction is much larger than predicted from the column density of atomic hydrogen.

8.4.3 Ionized carbon

The Magellanic Clouds are rich in gas and young, hot stars. They are underabundant in heavy elements relative to the solar value (see Chap. 11) and have a relatively low dust content. This creates an environment in which neutral gas clouds have little shielding against UV radiation above the Lyman limit. The lifetimes of the neutral clumps are shorter than those of their galactic counterparts. The UV radiation will ionize or dissociate molecules and atoms with ionization and dissociation potentials less than the hydrogen ionization potential of 13.6 eV. Carbon, with its ionization potential of 11.3 eV, is the most abundant atom that can be ionized in this way. Because of the low dust content the UV photons will penetrate into the constituent clumps of the molecular clouds and dissociate the CO molecules and form C^+ regions. Coincident CO and [CII] radiation is therefore to be expected in the youngest star forming regions such as N159 in the LMC. As the beam filling factor of the molecular clumps in regions with strong UV radiation is small, small CO line intensities are observed. Most of the low-J CO emission will come from cold molecular gas far from the sites of star formation which produce the dominant [CII] radiation.

A comparison of the situation in the LMC and in the Galaxy shows that the gas-to-dust ratio in the LMC is about a factor of four higher (Koornneef 1984) and carbon is depleted by about the same factor (Cohen et al. 1988). The mean UV energy density is about five times higher in the LMC; the dust there is consequently hotter than in the Galaxy (Lequeux 1989). Molecular clouds appear to be larger in the LMC but relatively scarce. A consequence of this is that the column density of C^+ in a photodissociation region should be similar in the LMC and the Galaxy. The larger sizes of the LMC complexes mean that they contain more ionized carbon than the galactic molecular clouds.

The typical [CII] emitting clouds in the LMC are thus immersed in a rather strong UV field. The thickness of the C^+ 'shells' depends upon the local protection by dust grains against the UV radiation. The abundant C^+ ions will radiate the [CII] 158 μm fine-structure ($^2P_{3/2} \rightarrow {}^2P_{1/2}$) line.

Observations of the [CII] 158 μm line have been restricted to small areas near bright HII regions where the UV radiation is strong. Boreiko and Betz (1991) obtained spectra towards 17 locations in the LMC, chosen to coincide with FIR emission peaks or relatively strong CO ($J = 2-1$) detections. All but two of these locations were within the 30 Doradus–N159 region. CII emission was detected in over

8.4 The molecular content

half of the locations observed. A good spatial correlation between FIR continuum and [CII] emission was found, as would be expected if the dust emission comes from a relatively dense region near early-type stars. The correlation between the locations of strong C^+ and CO ($J = 2$–1) emission is, on the other hand, poor, in contrast to the situation in our Galaxy. No [CII] was detected at the positions of N11 and N160A, where the CO ($J = 2$–1) line has been observed (Israel et al. 1986) or in N159, where the CO ($J = 1$–0) line has been mapped (Booth 1993). Conversely, no CO ($J = 2$–1) emission was detected from the FIR peak of N160A or in 30 Doradus at the [CII] peak. The CO ($J = 1$–0) line is, however, seen throughout the region and is particularly strong in the ridge south from 30 Doradus (see Sect. 8.4.1.2). The absence of significant [CII] emission about $5'.5$ south of N159 may be explained by star formation propagating from SGS LMC 2 and now having reached N159. Until the activity has extended further south, the CO in the region will remain cold, and, until OB stars have formed, the C^+ will not be abundant.

Israel and Maloney (1993) have detected and mapped [CII] in three LMC and six SMC HII region/CO cloud complexes. The sources were selected as having high FIR intensities, preferably also ^{12}CO ($J = 1$–0) observations (30 Doradus was not considered as being observed earlier; see below). Of the LMC sources, N11 was mapped only around the brightest 100 μm peak. The peak [CII] intensities were somewhat lower than those observed in galactic sources such as W3. In N160 a bright peak was found surrounded by a large region of lower surface brightness. In N159 three major [CII] peaks were identified, surrounded by a weak emission. The westernmost peak coincides with the bright nebula N159A. The [CII] peak in N160 coincides with an unresolved 60 μm source. It is displaced ~ 15 pc NE of the CO maximum. The thermal radio emission peak, the HII region, is another 15 pc further NE. The eastern and western [CII] peaks in N159 have similar displacements from the CO maxima; the central peak has no associated CO emission at all. The former coincide with 60 μm peaks, the latter with a weak extended FIR source.

Of the SMC sources N27 and N66 ($\equiv$ NGC 346) are of special interest. N 27 is the strongest CO detection so far and N66 the strongest *IRAS* source as well as the dominating HII region in the northern part of the Bar. The [CII] map of N27 shows a featureless, resolved source with low surface brightness extensions to the south and the west. Its peak coincides with an unresolved *IRAS* source. The CO source peaks about 10 pc west of the [CII] peak; it has a diameter of about 25 pc (Rubio et al. 1992). A radio continuum source (Mills and Turtle 1984) is close to the CO peak and has a weak extension to the east, coinciding with the [CII] peak. The situation is consistent with a molecular cloud complex in the process of being eroded by radiation from the exciting stars in N27.

The N66 map shows an extended [CII] source with several peaks, all of low intensity. A strong extended FIR source covers about $3' \times 3'$ and peaks close to the [CII] peaks. There is a small and weak CO source to the NE of the [CII] complex; it is much smaller than the [CII] emitting region. Evidently, the strong UV radiation of the rich ionizing N66 cluster has destroyed most of its parent molecular cloud complex leaving the C^+ cloud as a remnant. Its distribution is similar to that of the ionized hydrogen (Ye et al. 1991). A strong local HI maximum is also associated with N66 (Bajaja and Loiseau 1982). There are some indications that ionized hydrogen and C^+ may exist in the same volume in N66.

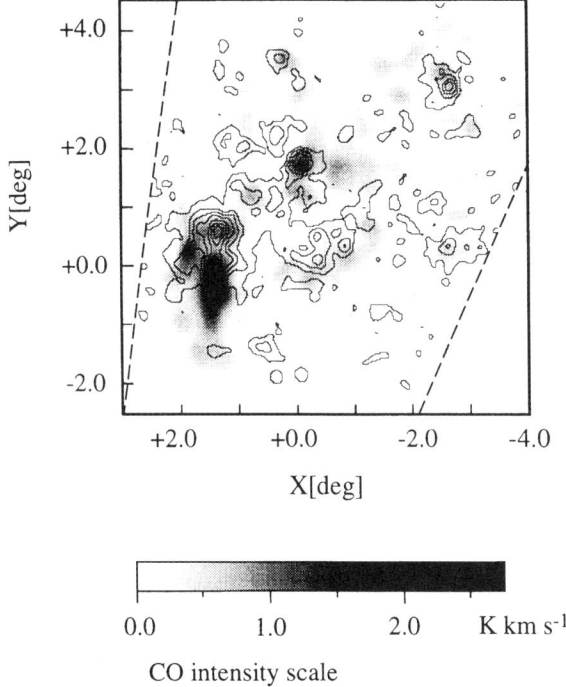

Fig. 8.8. Velocity-integrated [CII] intensity map (contours) of the LMC (Mochizuki et al. 1994) superposed on a CO intensity map (gray scales; Cohen et al. 1988). The contour levels are 1, 2, 4, 6, 8, 10 and 12×10^{-5} erg s^{-1} cm^{-2} sr^{-1}. The angular resolution is 15'.0. The dashed lines show the boundaries of the [CII] observations.

The results confirm that C$^+$ has a distribution in the Clouds well correlated with that of the far-IR emitting dust complexes but not with that of the molecular clouds, traced by CO emission. The frequent coincidences between [CII] and the far-IR peaks indicate a close coupling between ionized carbon and warm dust.

Mochizuki et al. (1994) have carried out a large-scale mapping of the LMC in the [CII] 158 μm fine-structure line with the *Balloon-borne Infrared Carbon Explorer (BICE)* system. [CII] line emission was detected over most of the LMC (Fig. 8.8). The mean [CII]/CO($J = 1$–0) line intensity ratio is 23 000, 18 times larger than the typical value observed in the galactic plane (1300). This result implies that each clump of the molecular clouds in the LMC has a larger C$^+$ envelope relative to its CO core than those in our Galaxy. It confirms the conclusions above that the lower dust abundance, due to its lower metallicity, allows UV photons, which convert CO molecules into C$^+$ ions, to penetrate deeper into the clumps in the LMC than in our Galaxy.

The [CII]/CO($J = 1$–0) emission ratio towards the 30 Doradus nebula is a factor of 5 higher than the best fit to starburst galaxies and galactic OB star forming regions, and another factor of 2–5 higher than that of less active galaxies and galactic molecular clouds (Stacey et al. 1991). There is, consequently, in 30 Doradus a high fraction of photodissociated gas relative to neutral gas, a consequence of the lower metallicity in the LMC. The situation is not quite as extreme in N159. There, the

8.4 The molecular content

[CII]/CO emission ratio is a factor of 2 in excess of that in the active galaxies. The same trend can be seen in the HCO^+/CO ratio which is a factor of 10 higher towards 30 Doradus than in the less active region south of N159.

8.4.4 Other molecules

H_2CO absorption was observed by Whiteoak and Gardner (1976a) towards the LMC HII region N159; they also found OH in absorption, at 1.557 GHz, in the same region (Whiteoak and Gardner 1976b). The velocities were 254 km s^{-1} and 253 km s^{-1}, respectively, close to the velocity known for CO in that direction.

Several OH and H_2O masers have been observed in the Clouds. The hydroxyl radical is sufficiently abundant in molecular clouds to be used for estimating their physical properties. The maser emission from OH can be used as an even more precise probe. The Zeeman doublets are easiest recognized in 6.035 GHz masers, providing a useful tool for measuring magnetic fields. In conjunction with the 1.665 GHz measurements and specific pumping schemes they can indicate the physical conditions in the masing region.

H_2O masers were detected in N159 and in N105A in the LMC and in two HII regions in the SMC by Scalise and Braz (1981, 1982). A total of 32 positions in the two Clouds had been searched. The strengths of the detected sources are comparable with galactic sources with luminosities $L \leq 10^{-2}$ $L_\odot$.

An OH maser, strongly polarized in the right-hand circular sense, was detected in the N105 region by Haynes and Caswell (1981).

OH maser emission at 1.665 GHz (ground-state) was observed in the N160A (first given as N159) region by Caswell and Haynes (1981). It had two narrow velocity features, separated by $\sim$ 4 km s^{-1} and displaying opposite senses of circular polarization. At this position, 50″ SW of the peak of the HII region N160A, an OH maser was detected by Caswell (1995) at the 6.035 GHz transition. It is at the location of a 3 mJy continuum source, probably a weak HII region. Three right- and two left-hand circular polarization features were seen in the spectrum. Some intensity variations were noted between two observations separated by seven weeks. A very accurate position was obtained: $05^h 39^m 38^s.92$, $-69°39'10''.9$ (2000). The positional errors are estimated to be $< 0''.5$.

The N160A OH maser is similar to several galactic objects. Two of its characteristics, excited OH stronger than the ground-state maser and the absence of a 6.6 GHz methanol maser, are both rare but not unprecedented in our Galaxy.

H_2O maser emission was searched for by Whiteoak et al. (1983) at a set of radio continuum sources in the LMC. Detections were made near three HII regions, N105A, N157A, and N160A. The Scalise and Bras result for N105A was thus confirmed, but the suggested maser in N159 could not be seen. The luminosities were comparable to typical H_2O masers in galactic HII regions. A further search for H_2O maser emission toward 80 HII regions in the LMC (Whiteoak and Gardner 1986), with a limiting luminosity of about $10^{29.5}$ erg s^{-1} added only one source, 0513-694B, in N113.

OH maser emission was searched for by Gardner and Whiteoak (1985) at five HII regions in the LMC, which were associated with H_2O emission. Circularly polarized maser emission was detected at two sources, 0510-689 and 0540-696B, but not at the one in N159 (0540-697A). Weak OH absorption was seen at 0539-691, and 0540-697A and possibly also at 0540-696. The optical depths of these sources were very low,

Table 8.4. *Molecules identified in the Magellanic Clouds*

Molecule	Frequency GHz	Region	Ref.
H_2	IR	LMC, SMC	Koorneef, Israel 1985
OH	1.7	LMC	Whiteoak, Gardner 1976b
OH Maser	1.7	LMC	Caswell,Haynes 1981
CH	3.3	LMC	Whiteoak *et al.* 1980
CO	115	LMC	Huggins *et al.* 1975
CO	230	LMC, SMC	Israel 1984(review)
HCO^+	89	LMC	Batchelor *et al.* 1981
H_2CO	4.8	LMC	Whiteoak, Gardner 1976a
H_2O Maser	22	LMC, SMC	Scalise, Braz 1981,82

Molecular lines detected with *SEST* up to 1991

CS(2–1)	98	LMC, SMC	
CS(5-4)	245	LMC	
CN(1–0)	113	LMC, SMC	
CN(2–1)	227	LMC	
$SO(3_2-2_1)$	99	LMC	
HCN(1–0)	89	LMC, SMC	
HNC(1–0)	91	LMC	
$H_2CO(3_{1,2}-2_{1,1})$	226	LMC	
$C_3H_2(2_{1,2}-1_{0,1})$	85	LMC, SMC	
$C_2H(1-0)$	87	LMC	

~ 0.01, showing that the molecular densities are significantly lower in the LMC than in the Galaxy.

Molecules detected in the Magellanic Clouds until 1983 are listed by Israel (1984). In 1991, when the *SEST* had been in operation for a few years, eight molecules were added to the seven previously known (Booth and de Graauw 1991; Table 8.4). Since then, a few molecules have been added (see below).

Two methanol, CH_3OH masers are known in the LMC (Sinclair *et al.* 1992; Ellingsen *et al.* 1994). One is centred near the OH emission source on the SE boundary of N105A, the other near the continuum emission peak of the HII region N11. They are the only ones found in a search covering 35 HII regions in the LMC and 13 in the SMC and, on a 3' grid, an area $0°.3$ square, south of the 30 Doradus nebula. The Clouds appear to be deficient in this emission compared to our Galaxy, possibly resulting from a general deficiency of complex molecules.

Of the three masers in the LMC with OH or CH_3OH emission the latter is five times stronger than the former (upper limit) in N11; in N105A the ratio is the opposite, and in N159 the OH emission is stronger than the CH_3OH emission (upper

8.5 The thermal and non-thermal components

limit) by a factor > 1.5. These ratios are inside the range found for galactic star forming regions. The two CH_3OH emission masers show no signs of variability.

The central part of the LMC HII region N159 has been searched for molecules by Johansson et al.(1994) on mm waves. Isotopomers of CO, CS, SO, CN, HCN, HCO^+ and H_2CO were detected, some of them in two transitions allowing excitation temperatures to be estimated. For ^{13}CO and ^{12}CO the temperatures were close to or in excess of 10 K. $C^{18}O$, CS, and HCO^+ showed temperatures between 5 and 10 K. All observed isotopomer intensity ratios suggested small optical depths provided that galactic isotopic abundance ratios do apply. From the ^{12}C-and ^{13}C-isotopomer data of CS and HCO^+ combined with the $C^{34}S$ line emission the gas phase $^{12}C/^{13}C$ abundance ratio was estimated to be 50 $^{+25}_{-20}$ in N159. The $^{18}O/^{17}O$ abundance ratio is a factor of ~ 2 lower than in the Galaxy and another factor of 2 lower than in starburst galaxies.

The detected lines in the LMC are weaker by a factor of 1.5–3 relative to those in the galactic S138 molecular cloud except HCO^+. The normalized luminosities of the S138 molecular cloud are similar to those observed in a sample of strong CO galaxies. Fractional abundances of the detected molecules have been estimated using the conversion factor from CO emission to H_2 column density. While the derived abundances of S138 agree to within a factor of 3 with those valid for Orion KL and TMC-1, all molecular concentrations observed in the LMC are typically a factor of 10 lower.

Optical and IR observations of N159 indicate that its molecular cloud is associated with the formation of massive stars, consistent with the relatively high HCN/HNC ratio observed. The excitation temperatures derived here indicate kinetic temperatures typical of cold dark cloud cores (~ 10 K) if the observed species are assumed not to be subthermally excited. The higher temperature derived from the ^{12}CO emission could result from the higher UV field and the lower amount of cooling agents in the LMC. The HCN/HNC ratio of about 3 in N159, on the other hand, indicates an active star-forming region (dark clouds in the Galaxy have a ratio ~ 1; see e.g. Goldsmith et al. 1981).

8.5 The thermal and non-thermal components

The 1.4 GHz continuum maps of the Clouds presented by Haynes et al.(1986) show in the LMC a disk, probably non-thermal, covering about 7 kpc (NS) $\times$ 5.5 kpc (EW). Superposed on the disk are a number of sources and source complexes. The bulk of the sources have flat spectra. They are associated with bright HII regions so that the morphology mimics the well-known asymmetric pattern of the young stellar constituent (Fig. 8.9).

Also in the SMC the agreement is good between the optical and the radio emission at 1.4 GHz (Fig. 8.10). Here, as well as in the LMC, the radio contours extend beyond the optical sites of obvious star formation, and the disk is probably non-thermal. The Wing is clearly seen in the SE but divided into two complexes. The inner one, with two strong sources, corresponds to the NGC 456-460-465 HII/OB associations. The outer one is at the position of SGS SMC 1.

The Haynes et al. data were used by Klein et al.(1989) together with recent observations at 0.408, 1.4 and 2.3 GHz for a discussion of the overall radio morphology of

166   *The interstellar medium*

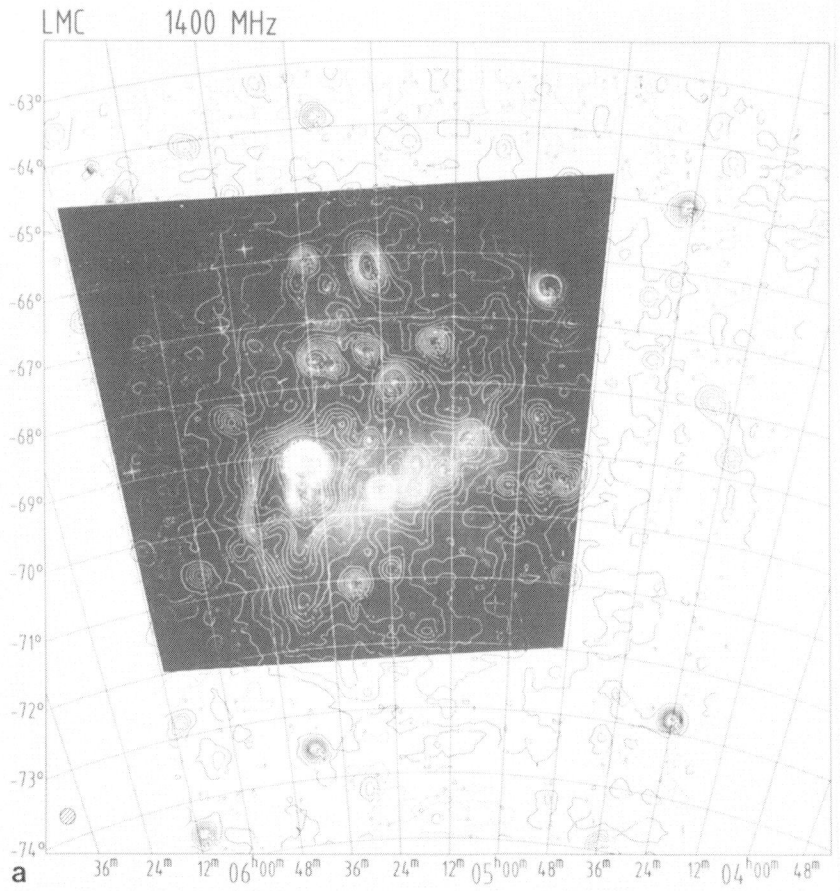

Fig. 8.9. Contour map of the LMC at 1.4 GHz (Haynes *et al.* 1986). The contours are: 0.0 (dashed), 0.10, 0.25,...1.0, 1.2,...2.0, 2.5,...4.0, 4.5,...10, 12,...20, 22, 30, 32 Jy/beam area. The beam size is indicated by the hatched circle in the lower left.

the LMC. An east–west asymmetry is noticeable. In the west, south-west faint radio emission extends in the direction of the SMC, in the east there is a strong gradient in the radio intensity. The effect is most likely due to ram pressure, caused by the motion of the LMC through the halo of our Galaxy. It compresses the magnetic field in the east, the direction of the translational motion of the LMC, and drags it out in the south-west. A similar effect is also seen in the HI distribution (Hindman *et al.* 1963).

The mean radio spectral index in the LMC is significantly flatter, -0.56 ± 0.05, than that of normal spirals, indicating a larger amount of thermal emission. The spectral index distribution reveals a good spatial correlation with the Hα emission, giving a flat radio spectrum in regions of intense Balmer emission. A close correlation is also found between the radio continuum maps and the FIR *IRAS* maps. The ratio of radio/FIR luminosity for the LMC HII regions is significantly higher than for

8.5 The thermal and non-thermal components

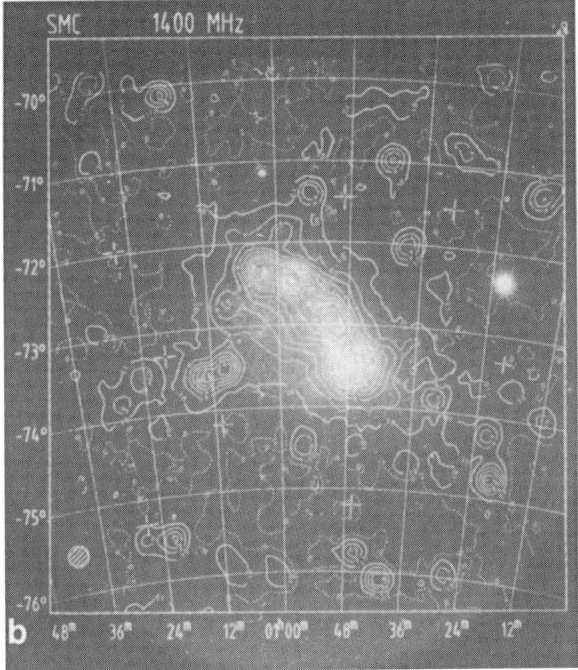

Fig. 8.10. Contour map of the SMC at 1.4 GHz (Haynes et al. 1986). The contours are the same for the LMC. The beam size is indicated by the hatched circle in the lower left.

galactic HII regions, in agreement with the low CO content, the low metallicity and the high gas/dust ratio in the LMC.

The Haynes et al. and Klein et al. results have been largely confirmed in the new continuum and polarization studies by Haynes et al. (1991) and Xu et al. (1992). The new frequencies observed are 2.45, 4.75, and 8.55 GHz. A non-thermal spectral index of -0.70 ± 0.06 is derived for the LMC; for the SMC it is steeper, -0.85 ± 0.10. The non-thermal emission is relatively smoothly distributed and resembles that derived at low frequencies.

A low-resolution map of the LMC at 45 MHz shows that the synchrotron radiation has a remarkably symmetric structure (Alvarez et al. 1987). The isophotes are concentric ellipses (Fig. 8.11) with an axial ratio of 0.71 which gives $i = 45°$ for the LMC plane (see Table 3.5), if it is a flat circular disk. The centre of the distribution is at $05^h 28^m, -69.°4$ (see Table 3.7). The line of nodes has a PA of 180° within 3.°6 from the centre. Beyond this distance, there are abrupt changes at equidistant points towards the north as well as towards the south, which could be caused by warps.

The total extent of the thermal component in the SMC is seen when the continuum maps are compared with the Hα interferometric observations by Torres and Carranza (1987). The latter show strong HII regions against the diffuse Hα background emission, and the thermal sources coincide with the most prominent HII regions in the Bar as well as in the Wing. There is an extended radio continuum component of predominently non-thermal origin (see Loiseau et al. 1987) covering the whole main body.

168 *The interstellar medium*

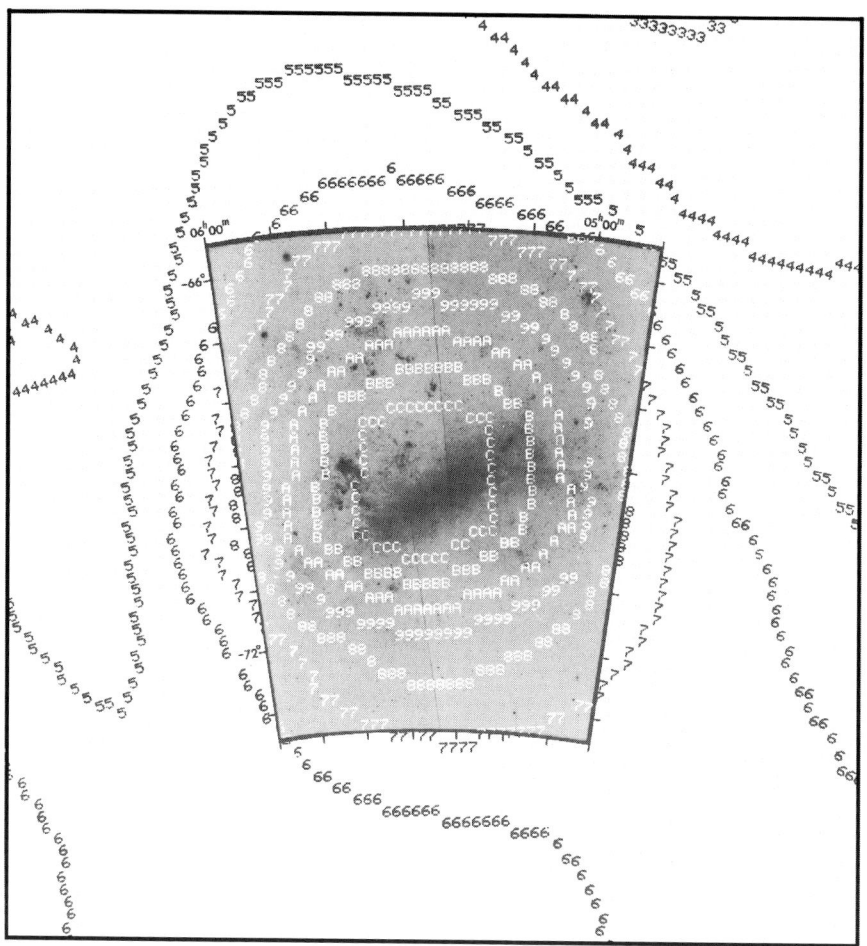

Fig. 8.11. The LMC at 45 MHz; antenna temperature distribution according to Alvarez et al. (1987). The contours are in steps of 300 K over an arbitrary base of 3000 K.

The synchrotron radiation in the SMC (Alvarez et al. 1989) has symmetrical elliptical contours (Fig. 8.12) with a centre at $0^h47^m \pm 5^m$, $-72°.7 \pm 0°.2$ (Table 3.9), an axial ratio of 0.48 and a PA of 38°. The spectral index between 45 and 2300 MHz is -0.58, the same as for the LMC.

As for the LMC, the distributions of the radio continuum and the FIR emission in the SMC resemble each others closely. The brightest FIR sources coincide with the thermal HII regions. Possibly the SMC HII regions are underluminous in the FIR.

Polarization data are available from optical (Wayte 1990) and radio (at 2.3, 4.75 and 8.4 GHz, Wielebinski 1989, Haynes et al. 1991) measurements. The magnetic field vectors determined optically (smoothed) and those determined from radio continuum emission match well, implying that the Davies–Greenstein mechanism probes the same magnetic field (in the LMC) as does the synchrotron radiation

Polarization is strong and complicated in the 30 Doradus region with no unambiguous association of velocity features with polarization directions. A particularly

8.5 The thermal and non-thermal components 169

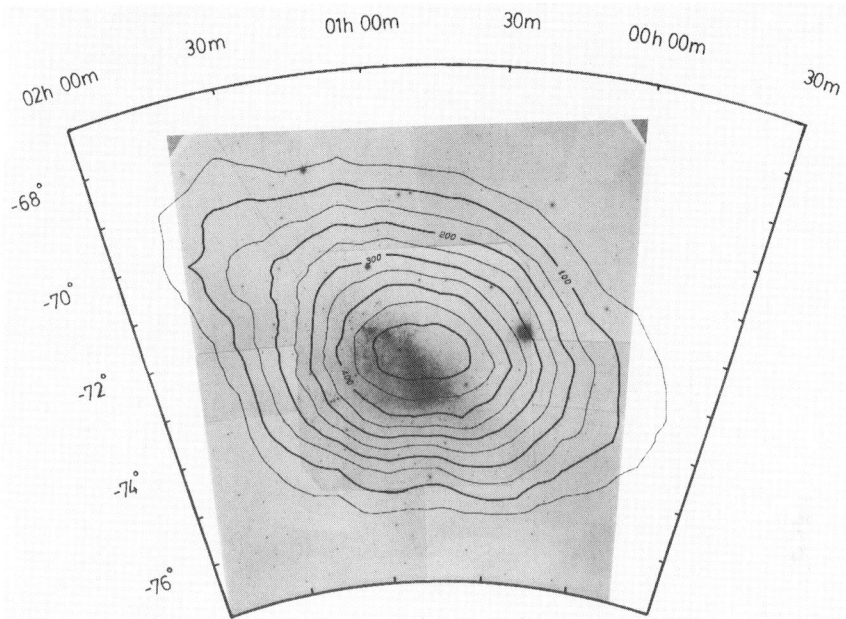

Fig. 8.12. The SMC at 45 MHz according to Alvarez *et al.*(1989). Observed antenna temperature contours with the Milky Way foreground and one extragalactic source removed.

strong linear radio polarization extends south, south-west of 30 Doradus with two features (at 2.5 GHz), the western being ~ 2.2 kpc long. It begins at a point where the strong radio continuum 'protuberance' seen at 4.75 MHz suddenly weakens and where there is strong optical obscuration. The magnetic field here has a predominantly south-westerly direction, pointing more or less towards the SMC.

An explanation of the magnetic field structure south of 30 Doradus has been suggested by Klein *et al.*(1993): the magnetic field bends out of the LMC plane in the eastern filament and into the LMC in the western one. The two features commence in the region of highest atomic (HI) and molecular gas density (CO) where also a sharp dust lane starts running N–S. Here, the diffuse X-ray emission is also brightest. The unusual magnetic field morphology may be connected with the HI L-lobe (see Sect. 8.1.1). If, however, the huge gas cloud in this region has arrived there following the orbit suggested by Fujimoto and Noguchi (1990), then the magnetic field may mark part of this orbit.

The optical data give some indication of a magnetic field parallel to the Bar. In the west a pan-Magellanic magnetic field is clearly evident. The magnetic field appears to follow the HI features, indicating a field trapped in the gas. Wayte considers this as a spiral field centred on 30 Doradus. However, disregarding 30 Doradus itself with a highly confused field, and the well organized pan-Magellanic structure, the remaining features may equally well outline supergiant shells.

The radio polarization of the SMC is very weak. Within the errors they are consistent with the results from optical polarimetry. The low level is likely to be a consequence of the large line-of-sight depth of the SMC. Optically, strong pan-

Magellanic polarization is seen in the centre of the SMC Bar. It continues in the near Wing along the (radio) continuum and HI ridge. In the south-west, in its most luminous part, the magnetic field goes virtually north–south, perpendicular to the pan-Magellanic field. Wayte suggests that this part is the 'original' SMC and that the extension towards the north-east as well as the Wing is the result of the collision with the LMC a few 10^8 yr ago.

8.6   The dust content

68 dark nebulae in the LMC were identified by Hodge (1972) on plates in blue (B) and visual (V) light, taken with the ADH and the Curtis Schmidt telescopes. The sizes vary from $1' \times 1'$ to $20' \times 8'$, most of them are elongated parallel to the LMC Bar. All of them are seen, by necessity, in the star- and HII-rich regions. In a similar survey of the SMC, 45 dust clouds were identified, with two exceptions all in the Bar (Hodge 1974a). They are generally smaller than those found in the LMC; their sizes range from $1'$ to $6'$ in the major dimension. Nearly 75% of the dust clouds are in the southern half of the Bar. This correlates well with the identified CO sources, which are also mainly seen in that region. The main results of the Hodge (1972, 1974a) surveys were confirmed by van den Bergh (1974), who, however, did not confirm the elongation of the LMC dust clouds parallel to the Bar.

Using the CTIO 4 m telescope Hodge (1988a) extended the survey for dark nebulae in the LMC to objects a factor of 2 smaller than those in the previous survey. The 146 dust clouds listed have a mean dimension of 27 pc ($m-M = 18.5$ mag). A comparison with the Cohen et al. (1988) CO data shows the overall good agreement between CO and dust clouds but also obvious cases of non-detection in one or the other survey. A higher-resolution CO mapping and more quantitative optical measures of the dust content are needed for an explanation of those cases.

From galaxy counts in an 85 deg^2 area centred on the SMC Hodge (1974b) concluded that the 'field dust', i.e. the dust outside the small dark clouds, has a distribution similar to HI. The dust appears, however, to be distributed over a larger area. In the direction of the Bar it can be traced from about $-68°$ to $-74°$. The mass of the field dust was estimated to $\sim 5 \times 10^5$ M$_\odot$ and the ratio HI/dust to ~ 300. The maximum extinction amounted to $A_V = 1.2$ mag.

The dust, heated by stellar radiation, is seen as FIR emission. It has been observed in the Magellanic Clouds with *IRAS* at 100 μm, 60 μm, 25 μm and 12 μm. The 100 μm radiation is considered as thermal emission by normal grains at low temperature, 20–30 K. The 12 μm radiation is emitted by small grains at higher temperatures, 300–1000 K. Maps have been produced and discussed by Schwering (1988, 1989a,b) and Schwering and Israel (1989). There is less diffuse emission present in the 25 μm and 12 μm maps in both Clouds. The contrast between small discrete sources and the background is therefore higher in these than in the 100 μm and 60 μm maps. (See Fig. 8.13: contours in the upper map are 5, 10, 20, 40, 100, 200, 500, 1000, 2000 $\times 10^{-8}$ Watt m^{-2} sr^{-1}, and in the lower 2, 6, 10, 20, 40, 100, 200, 500, 1000, 2000 $\times 10^{-8}$ Watt m^{-2} sr^{-1} and Fig. 8.14: contours in the upper map are 5, 10, 15, 20, 50 $\times 10^{-8}$ Watt m^{-2} sr^{-1}, and in the lower 0.5, 1, 2, 4, 6, 8, 10, 15, 20, 40, 80 $\times 10^{-8}$ Watt m^{-2} sr^{-1}.)

For the LMC 1823 infrared sources are listed, for the SMC 219. Many are foreground stars. In the LMC a few stars, PNe and SNRs are identified. A good

8.6 The dust content 171

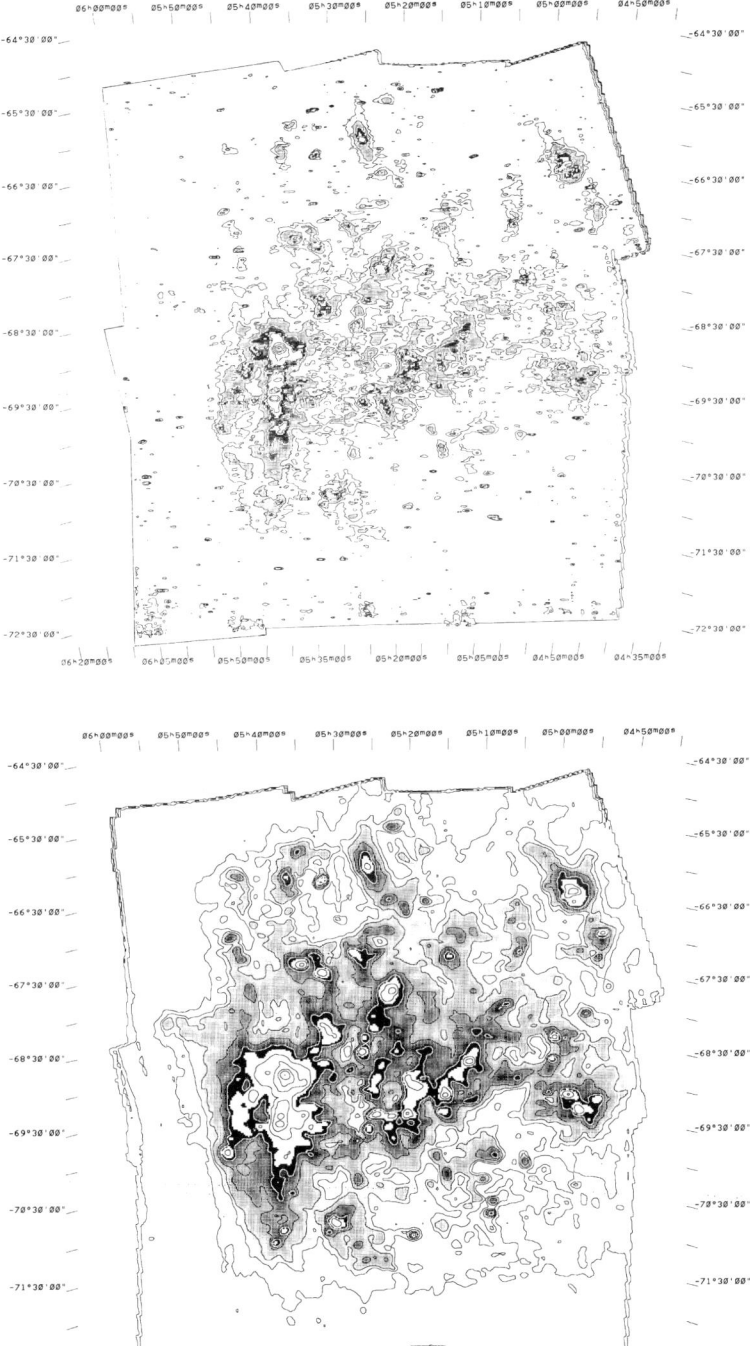

Fig. 8.13. FIR maps of the LMC (Schwering 1989). The upper figure shows the 12 μm map, the lower the 100 μm map, both from NS scans.

172   *The interstellar medium*

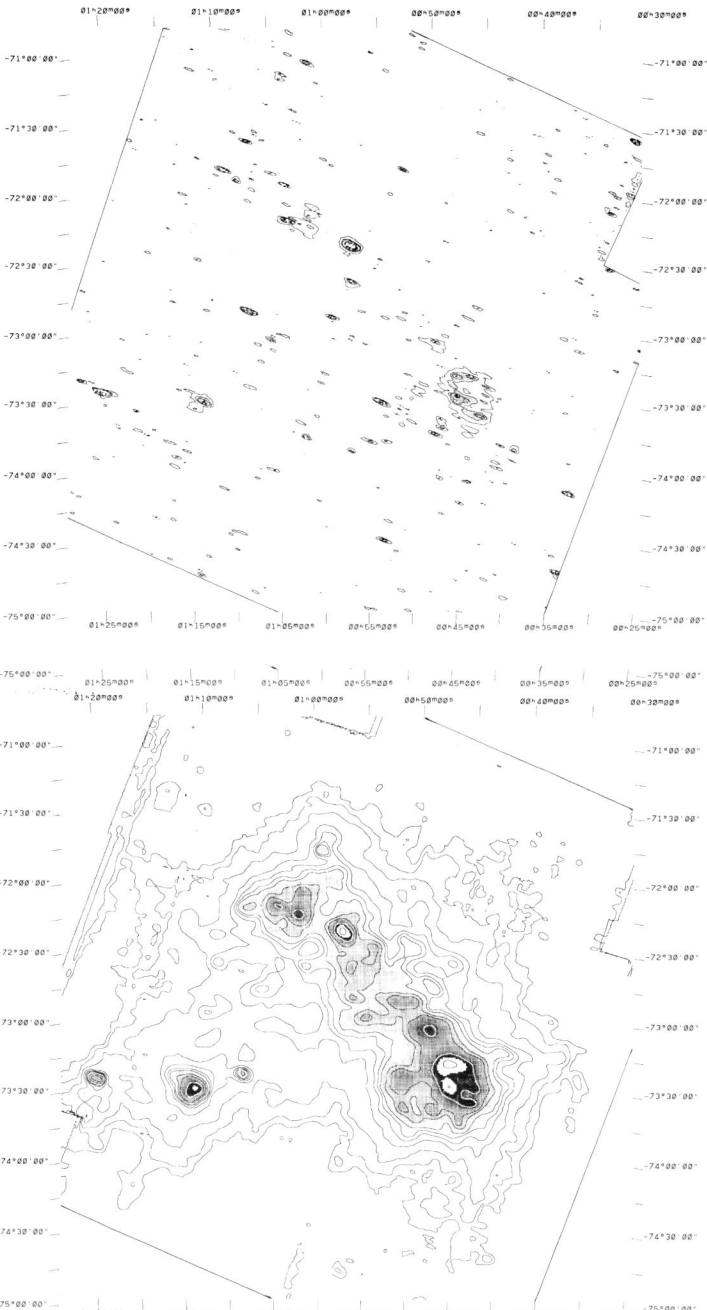

Fig. 8.14. FIR maps of the SMC (Schwering and Israel 1989). The upper figure shows the 12 μm map, the lower the 100 μm map, both from NS scans. The resolution is higher in the scan direction.

8.6 The dust content

correlation with dust clouds and HII regions is found. The extended 30 Doradus region emits about 40 % of all infrared radiation.

In the SMC two blue globular clusters and seven PNe have been identified with FIR sources but no individual SMC stars nor any SNRs. A good correlation is found between the sources and HII regions and dust clouds.

From the analyses by Schwering (1988) and Lequeux (1989) of the *IRAS* data it may be concluded that (i) the average Magellanic normal dust is hotter than the galactic dust due to a high interstellar radiation field. The 12 μm radiation is very low, indicating that the small grains may have been destroyed by the abundant UV radiation. (ii) There are remarkable differences between the colour–colour diagrams, 60μm/100μm vs 12μm/25μm, of the Magellanic Clouds and the Galaxy, indicating appreciable differences in the properties of the dust.

The conditions in the SMC are similar to those in the LMC but more pronounced (Lequeux 1989). Particularly revealing is the colour–colour diagram, which shows that, as for the LMC, there are no cold molecular clouds. In addition the SMC lacks points in the diagram indicating the existence of HI and H_2 clouds heated by the far-UV radiation of hot stars. The lack of such regions may be due to either different dust properties or to a more even distribution of the hot stars, or both. There are in fact few stellar associations in the SMC of the kind that is abundant in the LMC; the only rich ones are NGC 346 in the Bar, NGC 456, NGC 460, NGC 465 in the Wing, and the SGS SMC 1 region still further out.

The Milky Way, the LMC and the SMC are the only galaxies with observed interstellar extinction curves. Their mean curves are similar in the infrared and visual bands but differ substantially in the far-UV: the far-UV extinction is smallest in the Milky Way, largest in the SMC. The most conspicuous differences are the sequential changes in the strength of the 2175 Å feature and the level of the far-UV extinction from the Milky Way to the LMC to the SMC: the former decreases while the latter increases. Variations in the extinction curves are caused by variations in the size distribution and/or chemical composition of the dust grains. Graphite is the carrier of the 2175 Å feature and silicates are responsible for the 9.7 μm and 18 μm features. Most of the carbon may be produced in carbon stars, most of the silicates are from stars with normal abundances (C < O).

The 2200 Å bump in the LMC appears to have been first observed by Nandy and Morgan (1978) in the spectrum of the reddened star Sk69-108, $E_{B-V} \sim 0.45$, when compared with that of Sk67-108, $E_{B-V} \sim 0.05$. In the Milky Way extinction curve the bump has a width (FWHM) of nearly 500 Å. It is present in the LMC extinction curve but has only half the strength. In the typical SMC extinction curve it is almost missing; its strength is at most 1/8 of that in the Milky Way curve. The SMC extinction curve can be fitted by silicate grains alone. The abundance of graphite to silicates diminishes from the Milky Way to the LMC to the SMC.

There are, however, some remarkable exceptions. The star Sk143 is a relatively heavily reddened star in the SMC. The UV extinction curve has been derived from IUE observations by Lequeux *et al.*(1982). It is surprisingly similar to the galactic one, with a large bump at λ2200, and a smaller rise shortward of λ2000 as compared with the normal SMC extinction curve. Also the $N(H)/E_{B-V}$ ratio is 'galactic', 6 × 10^{21} cm^{-2} mag^{-1}, about ten times smaller than the SMC gas-to-dust ratio (see below).

Also in the LMC at least one star, Sk69-108, is known with an extinction curve similar to the galactic one (Nandy et al. 1980), with a much smaller far-UV extinction than is normal for LMC stars.

The low carbon content in the (HII regions in the) Clouds is surprising. Both Clouds contain large numbers of carbon stars so that a fair amount of graphite should be produced, unless they only produce CO which, being poorly shielded, is then partly dissociated by the intense radiation fields. As noted above, [CII] has a distribution in the Clouds that is well correlated with that of the FIR emitting dust complexes but not with that of the molecular clouds, traced by CO emission.

The dust to gas (HI) ratios are (Pei 1992), for the Galaxy : the LMC : the SMC 1 : 1/5 : 1/10. The heavy element abundances are 1 : 1/3 : 1/8 (Lequeux et al. 1979). Other ratios are (in the same order) the 2175 Å feature, 1 : 1/2 : 1/8; the 9.7 μm feature , 1 : 2 : 5; the 18 μm feature, 1 : 3 : 6; the FIR extinction 1 : 2 : 3 and the far-UV extinction 1 : 2 : 3. The carbon to silicon ratios are 0.95 for the Galaxy and 0.22 for the LMC.

8.6.1 Dust in the emission nebulae

The optical properties of the dust, and its spatial distribution with regard to other components of the ISM and the stars, are essential for our understanding of many astrophysical phenomena such as the FIR emission and the UV extinction of stellar light, as well as the physics of HII regions. We would like to know how the dust is distributed in an emission nebula: how much of it is inside the ionized volume and how much is outside. Is the dust clumped or more evenly distributed? What are the properties of the dust grains? Some of these questions may be answered by studies of nebulae in the Magellanic Clouds.

It is necessary to determine the interstellar extinction in all parts of the Magellanic Clouds so that eventually no (poorly) estimated mean values have to be used. It must be realized that the extinction varies much more over the Clouds than is so far accepted or assumed. Caplan and Deharveng (1985, 1986) have used absolute Hα and Hβ fluxes for 51 HII regions in the LMC to describe the extinction in these directions and to discuss the location of the dust in these regions. Thermal radio-continuum fluxes of the HII regions and HI column densities as well as colour excesses of nearby stars are used to support the investigation. They conclude that some of the observed extinction is due to clumped dust well outside the emission zone. The rest is caused by dust located just outside the boundary of the ionized gas; some of the dust may be inside the nebula. In some of the nebulae the observed Balmer decrements deviate significantly from the theoretical value predicted from case B recombination-line model calculations. Stars in 13 of these nebulae have been observed by Greve et al. (1988, 1990) using the Walraven photometric system. The stars are mainly OB main-sequence stars with ages of the order 3–10 Myr. The average photometric extinction for each nebula agrees with that derived by Caplan and Deharveng from the Balmer decrement and seems to indicate a rather uniform distribution of dust in and/or around the nebulae. Individual stellar extinction values indicate some clumpiness. The average photometric extinction derived is $A_V \geq 0.4$ mag for HII regions embedded in isolated CO clouds; in the 30 Doradus complex it is $A_V \sim 1.1$–1.5 mag. The latter region will be further discussed in Chap. 10.

8.7 Interstellar absorption lines

Spectrophotometry at 8–13 μm of the HII regions LMC N44A and SMC N88A showed a prominent silicate emission feature in the former but no evidence of any silicate dust in the latter (Roche *et al.* 1987).

8.7 Interstellar absorption lines

Interstellar absorption lines in the Magellanic Clouds were first detected by Feast (1960) in the spectra of several supergiants. Since then many investigations have been carried out aimed at the structure of the ISM and at the structure and kinematics of the Clouds (see e.g. Molaro *et al.* 1993). Here (see also Chaps. 3 and 12) some recent studies will be discussed.

The structure of the ISM in the LMC and the SMC has been discussed by Wayte (1990). He combined observations of CaII absorption spectra of 25 early-type supergiants in the Clouds with HI observations and optical and radio polarization measurements. In the direction of the LMC there are, at intermediate velocities, galactic ($\sim$ 60 and 130 km s^{-1}) as well as Magellanic (170 km s^{-1}) contributions. These components had been previously seen in the UV by Savage and de Boer (1979, 1981) and considered to be galactic. However, the low-velocity components have been ascribed to the Magellanic Clouds by Songaila *et al.* (1986). From comparisons of their Ca II K absorption velocities with HI measurements they concluded that there is lower column density material, similar to the 60 and 130 km s^{-1} features, within the Clouds. There is then no compelling reason to assign the 60 and 130 km s^{-1} weak absorption features to the halo of our Galaxy. Wayte showed that these components exist at positions up to 20° from the Clouds and are thus unlikely to be Magellanic gas. The 130 km s^{-1} component could possibly be material tidally torn from the LMC. De Boer *et al.* (1990) consider that their HI 21 cm spectra in the direction of the LMC, covering 14° along declination $-67°$, unambiguously prove that the intermediate-velocity absorption lines, at 70 and 140 km s^{-1}, originate in gas *not* associated with the Magellanic Clouds.

Among the intermediate components, $25 < V < 235$ km s^{-1} and W(NaI D$_2$)/W(CII K) < 1,[*] those at velocities $\sim$ 73 and 125 km s^{-1} are also seen in HI 21 cm data (Vidal-Madjar *et al.* 1987). The ratios derived, N(CaII)/N(HI) $\simeq 2.3 \times 10^{-7}$ and N(NaI)/N(HI) $\sim 14 \times 10^{-9}$, show that the sodium to hydrogen ratio is virtually galactic, whereas the calcium to hydrogen ratio is about 100 times larger than the galactic one. This is indicative of a total absence of depletion. It can therefore at present not be decided if the intermediate velocity gas components are produced in the galactic halo or in the LMC (halo).

An indication that the intermediate velocity clouds are non-Magellanic is found in an analysis of the interstellar NaI and CaII lines in 13 lines of sight in a $30' \times 30'$ field centred on SN 1987A and including 30 Doradus (Molaro *et al.* 1993, Vladilo *et al.* 1993). A number of intermediate velocity clouds are detected in almost all the Ca II spectra but only seldom in NaI. Clouds at $\sim$ 64, $\sim$ 90, and $\sim$ 170 km s^{-1} are seen along more than one line of sight and at different intensities. The cloud morphology is clumpy on an arcmin scale. Some of the clouds are definitely in front of the LMC. In particular, the $\sim$ 64 km s^{-1} cloud shows a remarkably constant radial velocity over 20 arcmin of the sky, ± 0.3 km s^{-1}, proving that it must be close to our Galaxy.

---

[*] W = equivalent width in mÅ.

Thus, in all likelihood, the intermediate velocity clouds are not part of the main body of the LMC.

In high-resolution spectra of SN 1987A Vidal-Madjar et al. (1987) identified 24 components of interstellar CaII, 13 of NaI, 3 of KI, 4 of CaI and 3 of LiI (see Sect. 10.3). Of the components, those with velocities $235 \leq V_{hel} \leq 280$ km s^{-1} and $1 \leq W(\text{NaI D}_2)/W(\text{CaII K}) \leq 2$ were considered to have formed in LMC gas. The seven CaII components in this range represent the resolved components of features seen toward stars in and around the 30 Doradus nebula. SN 1987A must then be located in or behind the LMC main plane, and at least at the distance of 30 Doradus.

The structure of the interstellar gas in the LMC is extremely more complex than expected from the 21 cm line data which show only two components in the $30' \times 30'$ field centred on SN 1987A and including 30 Doradus (Molaro et al. 1993, Vladilo et al. 1993). Here, in several lines of sight the absorption profiles show more than five components spanning a velocity range of $260 \leq V_{hel} \leq 300$ km s^{-1}. The most intense components are seen between 280 and 290 km s^{-1}. This gas may be looked upon as representing several components in the main HI disk of the LMC (note that a very small portion of the LMC is considered!). The HI emission at 271 km s^{-1}, spread over a large fraction of the LMC, is identified in absorption over several hundred parsec but in modest intensity. This suggests that this gas is either beyond the main layer at 280–290 km s^{-1} or it has a higher degree of Na I ionization.

In their interpretations of the absorption line and HI velocities Songaila et al. (1986) suggested tentatively that both the LMC and the SMC are extended and fragmented along the line of sight. The fragments may be spatially and kinematically quite distinct. For the SMC, this is generally accepted, whereas the LMC is generally considered as a rather flat system (see Chaps. 3 and 12).

To the south and south-west of the LMC, absorption occurs mainly around 220 km s^{-1} (Wayte 1990). Absorption at these velocities is also seen in the Bar and near 30 Doradus, and it is strong toward SN 1987A. Its distribution is very similar to that of the 170 km s^{-1} component. In the south-west a 270 km s^{-1} velocity component is seen in HI but not in absorption; it must be behind the low-velocity components.

To the north of 30 Doradus strong absorption is seen at a velocity ~ 25 km s^{-1} lower than that of the disk component visible also in HI. Several stars appear to be in front of the HI disk. Further towards the north-east, the disk component is seen in front of a higher-velocity component.

South of 30 Doradus the picture is confused; low-velocity material may be behind the higher-velocity disk component. It appears as if gaseous complexes may be colliding in that part of the LMC (SGS LMC 2, see Chap. 6). The double-velocity profile can be traced in HI along an arc towards the west, and Wayte suggests that the SGS LMC 2 region may be phenomenologically connected with the Bridge between the LMC and the SMC. This is to some extent supported by polarization measurements.

The forbidden line [Fe] $\lambda 6374.51$ in absorption was detected in the spectrum of SN 1987A by D'Odorico et al. (1987). It arises in coronal gas, $T = 1 \times 10^6$ K, distributed between the heliocentric velocities 205 and ~ 370 km s^{-1}, with an equivalent width $EW = 16.4 \pm 0.6$ mÅ. The column density of Fe^{9+} implied is of the order of 10^{17} cm^{-2}. This highly ionized gas may be widely distributed within the 30 Doradus complex and may have its origin in the combined effects of previous supernovae in the region.

8.8 Diffuse interstellar bands

Alternatively, it may be due to the circumstellar environment of SN 1987A itself and a result of the highly ionizing UV flash at the shock breakout.

In the SMC, Cohen (1984) has observed the interstellar Ca II line at 3933 Å in 31 reddened early-type stars. The SMC components are stronger than would be seen in galactic stars of the same reddening. The most reasonable explanation is that less Ca is bound to grains in the SMC than in a typical disk region in the Galaxy. The radial velocities of the observed stars correspond as a rule to that of the higher-velocity HI gas, those of the CaII lines to that of the lower-velocity HI gas (see Sect. 12.2).

HI emission and CaII absorption in the spectrum of Sk35 on the western side of the SMC Bar cover the velocity range from 200 to 90 km s^{-1} continuously. In the Wing, five stars in the region of NGC 456–465 have a main velocity component at 145 km s^{-1} (Wayte 1990). Additional components are seen at 130 and 160 km s^{-1} in Sk151 and Sk155, and in the latter also at 220 km s^{-1}. These stars may be behind the others. Suggested distances are 69 kpc for Sk155, 62 kpc for Sk151, and 54–46 kpc for the remaining three. Of the two main peaks in HI, at 135 and 170 km s^{-1}, the former is strongest in the south-west, the other in the north-east, being equal in between. In the south-west the 135 km s^{-1} component clearly dominates in the optical absorption lines, whereas in all other regions the absorption spectrum is more confused and does not match the HI peaks so well. All facts support the conclusion that the main lower-velocity component is closer than the main high-velocity component.

Songaila et al. (1986) have suggested that the three velocity components in the SMC correspond to material ordered so that the 135 km s^{-1} component is the closest, the 170 km s^{-1} component the most distant and the 220 km s^{-1} component in between the former two. Wayte considers that this applies only in the south-west; in all other parts the highest-velocity components are the most distant. In the south-west, the velocity distribution is enigmatic.

8.8 Diffuse interstellar bands

The diffuse interstellar band (DIB) at λ4430 band was first observed in the spectra of reddened LMC and SMC stars by Hutchings (1966). It was found to be surprisingly strong in both Clouds. Dachs (1970) found that the dust content of neutral interstellar material in the SMC was too low for such strong bands and argued that the λ4430 band originated in galactic material. Further observations by Blades and Madore (1979), Hutchings (1980) and Houziaux et al. (1980, 1985) showed, however, that weak λ4430 bands were formed in the Clouds. The first measured strengths were in error due to interfering stellar lines. Hutchings also reported the probable presence of the DIBs at λ5780 and λ6284 in some spectra. Their equivalent widths were generally $\leq$ 0.6 Å and the correlation with E_{B-V} was poor.

The carrier(s) of the DIBs is (are) still unknown. Houziaux et al. (1985) found that five reddened SMC stars observed by them all showed λ4430, whereas only one of the stars had a measured λ2200 feature. This eliminates graphite as a carrier for the former, assuming, as above, that the λ2200 feature is caused by graphite grains. Similarly, the LMC star Sk69-108 (see Sect. 8.6), with a much smaller far-UV extinction than normal for the LMC but with a λ4430 band similar to that in other LMC stars with normal LMC-type extinction curves, indicated that this band is not associated with the small particles assumed to cause the far-UV extinction.

The interstellar medium

Table 8.5. *Galactic and LMC DIBs in SN 1987A spectra*

DIB	Galactic			LMC		
	λ	EW mÅ	Depth %	λ	EW mÅ	Depth %
5780	5780.87	39.0	1.6	5785.87	33.0	1.4
5797	5797.48	10.5	0.7	5802.57	6.7	0.7
6269		<1.4	<0.5	6275.62	8.7	0.6
6284	6284.30	55.0	1.3	6289.39	45.0	1.6
6376	6376.56	4.9	0.3	—	< 2.0	< 0.7
6379	6379.49	4.6	0.6	—	< 1.3	< 0.7

Note: All data are from Vladilo *et al.*(1987).

The DIBs at $\lambda\lambda$5778, 5780, 5797, 6269, and 6284 have been detected in spectra of SN 1987A by Vladilo *et al.*(1987). The DIB at λ 6613 was previously observed by Vidal-Madjar *et al.*(1987). A comparison of the measured equivalent widths (EW) of the LMC and galactic features in the SN 1987A spectra is given in Table 8.5.

With the exception of λ6269, the galactic DIBs are stronger than the LMC DIBs (Table 8.5). If the carriers of the DIBs could be located to some of the clouds defined by the radial velocities of the interstellar lines in this direction, some clues to the formation of the bands could possibly be obtained. It is also necessary to separate the portions of the reddening in this direction, $E_{B-V} = 0.17$ mag (see below), into a LMC and a galactic contribution. The latter may be between 0.03 and 0.11 mag. There is reason to believe that the correlation between the EWs of the DIBs and the colour excess is rather good, provided they can be the located to the same absorbing cloud. The DIBs may be divided into families. λ5780 and λ6284 are likely to belong to the same, whereas λ5797 is of another family (Westerlund and Krelowski 1989). The ratios DIB(LMC)/DIB(Gal) = 0.85, 0.82, and 0.64, for the three DIBs, respectively, support this division and indicate that the bands may be formed in different clouds in the LMC as well as in the Galaxy. Dividing $E_{B-V} = 0.17$ mag correspondingly between the clouds producing the λ5780 and λ6284 DIBs gives 0.09 mag as reddening due to galactic dust and 0.08 mag due to LMC dust. In the case of the λ5797 band, the corresponding values are 0.10 and 0.07 mag, respectively. The divided-up colour excesses and the corresponding band strengths of λ6284 match galactic low-reddening ($E_{B-V} \leq 0.08$ mag) data well (see e.g. Jenniskens *et al.*1994).

9

X-ray emission and supernova remnants

X-ray emission from the Magellanic Clouds was first observed in a five-minute rocket flight from Johnston Atoll in the South Pacific on October 29, 1968. The LMC was detected as an $\sim 4\,\sigma$ excess in two adjacent 5° bins, the flux was $\sim 1.5 \times 10^{-9}$ erg cm^{-2} s^{-1} and the spectrum was slightly softer than that of the diffuse background (Mark et al. 1969). Two years later two source regions in the LMC were identified by Price et al. (1971) and emission from the SMC was recorded. The same year Leong et al. (1971) showed that the LMC emission could be resolved by the collimated detector system of the *Uhuru* satellite into three steady and one possible highly variable source; they were designated LMC X-1, X-2, X-3, and X-4. There was also a possible diffuse emission extending over much of the Cloud. The SMC emission was located to a single, highly variable source, called SMC X-1, in the Wing region. It was the first stellar X-ray source to be confirmed in an external galaxy.

Confirmation of the existence of LMC X-4 was presented in the second *Uhuru* catalogue (Giaconi et al. 1972).

The LMC sources X-1, X-2, and X-3 were confirmed by *Copernicus* satellite observations (Rapley and Tuohy 1974) and given more accurate positions. Also Markert and Clark (1975), using the *OSO 7* satellite confirmed these three sources, defined an upper limit for LMC X-4, and introduced a fifth source, LMC X-5. LMC X-1, X-2 and X-3 are close to luminous stars with which they may be connected; LMC X-2 and X-3 are in relatively remote regions of the Cloud.

Over the next few years it became clear that the five sources LMC X-1, X-2, X-3, X-4, and SMC X-1 are persistent bright X-ray binaries, consisting of a normal star and a collapsed companion. Non-periodic variability on various time scales was established through observations with the collimated detector systems of the *SAS 3*, *Copernicus*, *Ariel V*, and *HEAO* satellites. The transient sources LMC X-5, X-6, A0538-66, X0544−665, and SMC X-2, X-3 were detected.

The existence of soft (0.15–1.5 keV) X-ray emission from the LMC Bar was first evidenced in a rocket survey by Rappaport et al. (1975). A number of soft X-ray (0.5–2 keV) sources were found in the Bar by McKee et al. (1980); their luminosities are between 10^{37} and 10^{38} erg s^{-1}. One of the sources was tentatively identified with the SNR N132D. Emission in the 0.15–0.28 keV band was detected from LMC X-3 and possibly also from the Bar region.

The pointed surveys with the IPC and HRI detectors in the X-ray telescope on board the *Einstein* satellite had enough sensitivity to detect supernova remnants, extremely soft sources and diffuse emission in the Magellanic Clouds. Seward and

180 *X-ray emission and supernova remnants*

Mitchell (1981) reported on 26 SMC field sources in ~ 40 deg^2 to a limiting luminosity of 10^{36} erg s^{-1}. They did not detect the two very luminous transient sources SMC X-2 and X-3.

Long et al. (1981) carried out an X-ray survey of the LMC with the IPC of the *Einstein* Observatory over a 37 deg^2 region, finding 97 point sources ('CAL') to $L_x \sim 3 \times 10^{35}$ erg s^{-1} in the energy range 0.15–4 keV. They noted the soft X-ray spectra of CAL83 and 87.

In 1984 the X-ray population in the field of the Magellanic Clouds amounted to ~ 150. Of these about half were interlopers, foreground galactic stars or background active galactic nuclei or galaxy clusters. Helfand (1984) noted the large number of foreground stars in the direction of the LMC; otherwise the number of interlopers appeared normal. Of the LMC and SMC member candidates 24 and 9, respectively, were considered supernova remnants, six LMC sources and one SMC object (SMC X-1) are confirmed binaries. Two candidate binaries were recorded in the LMC.

9.1 X-ray observations of the LMC

Wang et al. (1991) have reanalysed the *Einstein* IPC data and increased the number of discrete sources in the LMC field to 105. Of the sources in their list 33 are new. Of the 97 sources reported by Long et al. (1981) 26 failed to appear because of the higher threshold source acceptance and more strictly selected data base in the Wang et al. survey. LMC X-3 falls outside the latter survey. Of the remaining 25 sources, CAL39, identified with SNR DEM204, and CAL85, coincident with a bright foreground star, are real sources. A few of the remaining missing sources may be real variable X-ray emitters, falling below the threshold. Most of them are, however, likely to be instrumental artefacts.

The survey was expanded from point sources (angular sizes $\leq 1'$) to include larger-diameter sources (≤ 60 pc). Nine such sources were found, two of which are inside SGS LMC 2 and three coincide with LH associations.

About 10% of the Wang et al. sources were identified with galactic stars and three with background Active Galactic Nuclei (AGN). 28 SNRs and nine X-ray binaries in the LMC have been confirmed through optical and radio observations. Nearly 20 X-ray sources are associated with OB associations and HII regions. 35 unidentified sources remain; they are shown to be likely background sources.

The distribution of the 54 discrete X-ray sources identified with LMC objects shows a strong correlation with the Hα emission, which traces the regions of active star formation. The X-rays from the LMC obviously arise primarily from the (extreme) Population I component, in sharp contrast to the situation in our Galaxy and in M 31 where the integrated X-ray luminosity is dominated by Population II low-mass X-ray binaries.

9.1.1 Stellar X-ray sources in the LMC

In 1991 only eight stellar X-ray sources were confirmed SMC or LMC members: the famous massive X-ray binaries (MXRB) SMC X-1, LMC X-1, LMC X-3, LMC X-4, and 0538-66, and three low-mass X-ray binaries (LMXRB) LMC X-2, CAL83 and CAL87. The LMXRBs are engaged in mass transfer from an evolved main-sequence star to a compact stellar source. They are also called supersoft X-ray sources (SUSOs $\equiv$ SSS and (related) to EUS) because they emit predominantly below the 0.5 keV

9.1 X-ray observations of the LMC

Table 9.1. *Parameters for LMC X-1*

Parameter	BBXRT	Ginga
$M_{BH}(M_\odot)$	$4.7\sqrt{\cos i}$	$\sim 5.0\sqrt{\cos i}$
L (erg s^{-1}) (2–10 keV)	$1.0 \times 10^{38}/\cos i$	$\sim 1.22 \times 10^{38}/\cos i$
$\dot{M}$ ($M_\odot$ yr^{-1})	$2.9 \times 10^{-8}/\cos i$	$\sim 1.9 \times 10^{-8}/\cos i$

energy range of soft X-ray telescopes. The SUSOs are unique in their very high X-ray luminosity. Their X-ray emission fits a black-body curve with 1.5–5×10^5 K and is not due to thermal bremsstrahlung or to any non-thermal mechanism producing X-rays. They are therefore among the hottest stellar objects observed in an X-ray band.

CAL83 and CAL87 appear to belong to the new class of X-ray binaries, the SSS sources, first identified as such in the LMC in the *ROSAT* All-Sky Survey (Trümper et al. 1991). A third source of this kind, RX J0513.9−6951 was identified by Pakull et al. (1993). The *ROSAT* observations* have led to the discovery of ten more SSS in the Magellanic Clouds, five in the LMC and five in the SMC (Kahabka et al. 1994).

LMC X-1, 0540.1−6946, was difficult to identify optically because of two excellent candidate stars within $\sim 3''$. Both have variable velocity and lie in a nebula with λ 4686 emission. The fainter star, an O-type binary with a $\sim 4^d$ period and antiphased NIII λ 4640 emission, appears to be the correct identification (Hutchings et al. 1983). The nebula centred on the star is photoionized by the X-rays, making it a tool for studying the EUV flux of the source (Pakull and Angebault 1986). All orbital data indicate a mass of 2.5–8 $M_\odot$, placing it in the LMXRB class. It is also a black-hole candidate (BHC) similar to Cyg X-1. The X-ray spectrum cannot be fitted by a single-component model. It requires at least two components: an ultrasoft black body with $kT \sim 0.8$ keV and a hard power law with a photon index of ~ 2.5. A quasi-periodic oscillation (QPO) of 0.075 Hz was discovered in *Ginga* data (Ebisawa et al. 1989), apparently associated with the hard component. The ultrasoft component appears to be a characteristic unique to BHCs (White and Marshall 1984).

In 1990 observations of LMC X-1 were made using the Broad Band X-Ray Telescope (*BBXRT*) flown in the shuttle Columbia in the *Astro-1* mission. The spectral coverage was broad, 0.3–12 keV, the energy resolution moderate, ~ 90 eV at 1 keV and ~ 160 eV at 6 keV. The spectrum obtained confirmed previous results: a soft disk black-body component and a power-law tail are needed to fit the spectrum (Schlegel et al. 1994).

LMC X-2; 0521.3−7201, has been identified with a faint blue star ($\sim 18^m$) with He II emission (Pakull and Swings 1979). Its velocity is somewhat low for the LMC; it could possibly be a galactic halo object. It has a long period, $12^d.5$ (Crampton et al. 1990), and presumably contains an evolved subgiant secondary. Weak absorption

---

* The first *ROSAT* results were summarized at the Heidelberg meeting on Magellanic Cloud Research (Pietsch and Kahabka 1993, Kahabka and Pietsch 1993, and Dennerl et al. 1993). A few hundred sources were reported, including SSS sources. Among the extended sources were candidates for new supernova remnants.

lines of the Ca II infrared triplet at $\sim \lambda\, 8600$ are present. The velocity behaviour of the He II $\lambda\, 4686$ emission line suggests that its origin is near the X-ray star and that the masses are typical of LMXBs.

LMC X-3, 0538.8-6406, has been identified with a B-type star, in a binary orbit of 1.7 day period with a very large (235 km s^{-1}) orbital velocity. The X-ray star in this system has a mass of 7–13 M$_\odot$ (Cowley et al. 1983) and is perhaps the strongest case so far for a stellar black hole. It belongs to the LMXRB class. There is evidence of an accretion disk in the system (weak absorption lines, large variations).

Optical photometry (van Paradijs et al. 1987, Kuiper et al. 1988) revealed a double-peaked ellipsoidal light-curve supporting the high mass of the collapsed star suggested by Cowley et al.. The light curve changes systematically, with longer-term changes in the mean light level. The interpretation of the light curve is, however, subject to a number of free parameters describing the disk shadowing and irradiation.

Simultaneous optical and UV observations were used by Treves et al. (1988) to deconvolve the energy distribution of the LMC X-3 accretion disk. The bulk of the disk emission is due to X-rays reprocessed in its outer parts. Its luminosity is variable. An orbital inclination of 50°–60° is suggested.

X-ray monitoring with the Ginga in combination with HEAO-1 archival data suggest a $\sim 99^d$ period (Hutchings and Cowley 1991). The mean optical light level varies, possibly with the X-ray intensity. This may be caused by the disk precession.

LMC X-4, 0532−66, exhibits intensity variations over a wide range of time scales. A high intensity state of ~ 40 minutes duration has been observed (Epstein et al. 1977) during which time short time scale flares (~ 20 seconds) occurred. Other observers have reported flaring behaviour on time scales of minutes to hours as well as apparently random variability with a time scale of days (see e.g. Skinner et al. 1980). A proposed optical counterpart (Sanduleak and Philip 1977a) was shown to be in a close binary system with a $1^d.408$ orbital period (Chevalier and Ilovaisky 1977). X-ray eclipses with ~ 5 hour duration occurred every $1^d.408$ (Li et al. 1978, White 1978). A $30^d.48$ periodic variability in the ~ 13 to ~ 80 keV flux was discovered by Lang et al. (1981). The source intensity was modulated by at least a factor of 5 through the cycle, being low for $\sim 40\%$ of the time. Low-resolution IUE spectra indicated the existence of a disk around the compact star (van der Klis et al. 1982). The UV light curves obtained at two differed epochs, separated by ~ 8 months, differed appreciably and obviously depend on the phase in the $30^d.5$ period. Whereas the primary is heated by a considerable X-ray flux all the time, we are, at least now and then, shielded from the X-rays by a precessing disk. Also the intensities of the UV resonance lines vary orbitally. This is understood in terms of ionization of the companion stellar wind by the X-ray flux (Cf. SMC X-1 below).

CAL83, 0543.8−6823, was identified optically by Pakull et al. (1985). Its orbital period is 1.0436 days, with a sinusoidal modulation of 0.22 mag around a mean $M_V \sim -1.5$ (Smale et al. 1988). The source might be a black-hole candidate similar to LMC X-3. The spectrum shows HeII and Balmer emission and weak CIII, CIV, and NV in the far UV. Radial velocity variations indicate typical LMXB masses. The HeII emission lines show an asymmetry which may have a long period (69^d?), possibly indicating a disk precession. Based on the HeII $\lambda\, 4686$ line strength, the source-intrinsic X-ray luminosity may be very high. It is probably shielded from us by the accretion disk (Hutchings and Cowley 1991).

9.1 X-ray observations of the LMC

CAL87, 0547.5−7110, was first identified optically by Pakull *et al.* (1987). Optical data showed it to be an eclipsing system (Naylor *et al.* 1989). Cowley *et al.* (1990) measured a variation of 1.2 mag in V with a period of $10^h.6$; the M_V(max) was ∼ 0.5 mag. They suggested a mass ratio of 10 between the compact source and its companion and considered it to be an eclipsing black-hole binary. It belongs to the LMXRB class. In the near infrared a late-type spectrum is marginally detected (Hutchings and Cowley 1991) but it is due to a nearby, ∼ 0.5 arcsec, F–G star. A partial X-ray eclipse has been detected by *ROSAT* (Schmidtke *et al.* 1993). It is interpreted as due to a grazing eclipse of an accretion-disk corona surrounding the central X-ray source, which is hidden from direct view by the disk itself. However, the much softer X-ray spectrum of CAL87 argues against the existence of an accretion disk corona (Kahabka *et al.* 1994). Instead, a model is suggested with a hot region occulted by the optical companion star

A0538−668 (Pakull *et al.* 1985) also belongs to the LMXRB class. It is a Be transient source with a period of $16^d.5$. During extended high states its spectrum shows strong emission lines during and after X-ray maximum. They are assumed to be due to periastron effects in an eccentric orbit. The radial velocity curve of the primary star has been determined (Corbet *et al.* 1985, Hutchings *et al.* 1985) with different results due to the crucial spectra near the periastron passage. This may be due to weak variable line emission (Smale and Charles 1989). It appears, anyhow, obvious that the orbit is highly eccentric ($e > 0.8$) and that the component masses are normal for a B star and a neutron star.

IUE UV spectra of A0538−668 over a long interval confirm the Be nature of the primary and show some of its changes with X-ray state (Howarth *et al.* 1984). The optical outbursts can be powered solely by reprocessed X-rays. The mass-loss rate of the primary via its stellar wind is estimated as ∼ 4×10^{-9} $M_\odot$ yr^{-1}. There is an excess in the infrared H and K passbands which is too large to be explained by free–free emission from this wind alone.

The archival *Einstein* Observatory data on the three ultrasoft X-ray emitting sources CAL83 and CAL87 in the LMC and the one possibly connected with the PN N67 in the SMC (see below) have been analysed by Brown *et al.* (1994). CAL83 and CAL87 have X-ray light curves which exhibit random fluctuations in intensity, while the SMC source, 1E 0056.8−7154, appears to exhibit fluctuations which may be non-random. All three sources can be modelled with extremely low black-body temperatures: the best fit values of kT are 13 or 9 eV for CAL83, 63 eV for CAL87 and 38 eV for the SMC source.

Recent *ROSAT* observations have led to the discovery of new ultrasoft X-ray sources in the LMC: two of the sources, RX J0537.7−7034 and RX J0534.6-7056, have spectra that indicate black-body temperatures in the range from 20 to 40 eV (Orio and Ogelman 1993). A third source, RX J0527.8−6954, showed a count rate level a factor of 4 lower than in an earlier discovery (Trümper *et al.* 1991). RX J0513.9−6951 was discovered by Schaeidt *et al.* (1993) during a search for time variability in the ecliptic pole regions using the *ROSAT* All Sky Survey Data (see Table 9.3). The source brightened by a factor of ∼ 20 during a ten day period and showed an X-ray spectrum very similar to that of CAL83 and RX J0527.8−6954. RX J0439.8−6809 was reobserved about one year after the survey (Greiner *et al.* 1994) and found to belong to the SSS class (see Table 9.2).

Table 9.2. *Ultrasoft X-ray sources in the LMC and the SMC*

Source	N_H atoms cm^{-2} $\times 10^{20}$	kT_{bb} eV	L_{X-ray} erg s^{-1} $\times 10^{37}$	Remarks and Ref.
CAL83	7.1	9–13	19–110	random fluct.; 1
CAL87	7.3	63	20	random fluct.; 1
SMC(N67)	4.7	38	0.23	nonrandom fluct. ;1
RX J0059.2−7138	8.8	35	3.5	pulsar;2
RX J0439.8−6809	4.3	18	0.05	3

References: 1, Brown *et al.*(1994); 2, Hughes (1994a); 3, Greiner *et al.*(1994).

Table 9.3. *Ultrasoft X-ray sources in the LMC and the SMC*

Source	N_H atoms cm^{-2} $\times 10^{20}$	kT_{bb} eV	L_{bol} erg s^{-1} $\times 10^{38}$	Remark	Ref.
CAL 87	150	34	3.2×10^4	LMC, eclipse	1
1E0035.4−7230	5.0	41	0.24	SMC	1
1E0056.8−7154	7.7	28	1.9	SMC N 67 ?	1
RX J0048.4−7332	58	8.3	770	nova SMC 3	1
RX J0058.6−7146	2.8	60	0.0035	SMC?	1
RX J0122.9−7521	4.5	15	210	SMC?	1
RX J0513.9−6951	9.0	39.3	2.33	variable, LMC	2

References: 1, Kahabka *et al.*(1994); 2, Schaeidt *et al.*(1993).

Einstein LMC X-ray point sources have been observed using *ROSAT*'s High-Resolution Imager by Schmidtke *et al.*(1994) and accurate positions have been obtained for a search for optical counterparts. X-ray positions and fluxes, information about variability, optical finding charts for each source, a list of identified counterparts were presented as well as information about candidates which have been observed spectroscopically in each of the fields. Sixteen point sources were measured at a level greater than 3σ; fifteen other sources were either extended or less significant detections. About 50% of the sources had not been seen in previous surveys. More than half of the sources are variable. Of the sixteen sources optically identified or confirmed, six are foreground cool stars, four are Seyfert galaxies, two are SNRs in the LMC, and four are peculiar hot LMC stars, probably binaries.

9.1.2 Diffuse X-ray emission in the LMC

The LMC provides an excellent laboratory for studying the X-ray properties of HII regions. A classical HII region of 10^4 K ionized gas should not emit X-rays. It may, however, contain luminous early-type stars – single, in clusters or in associations – with strong stellar winds and ionizing fluxes. The winds may interact with the ambient medium to produce $\sim 10^6$ K X-ray emitting plasma. Moreover, the most

9.1 X-ray observations of the LMC

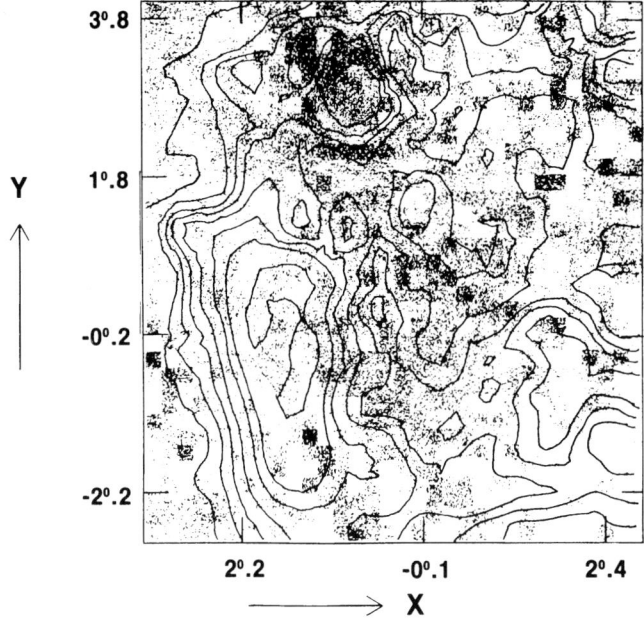

Fig. 9.1. Grayscale map of the LMC in the X-ray band 0.18–0.28 keV (Singh et al. 1987) superposed on the HI contours (Rohlfs et al. 1984). The shading becomes darker with intensity increasing from 4 to 12 counts s^{-1}.

massive stars may become supernovae and heat the associated HII region. Thus, X-ray emission may well be detected in HII regions.

Singh et al. (1987) have presented a 1/4 keV X-ray map of the LMC from scanning observations with the *HEAO 1* low-energy detectors. There is a complete lack of modulation of the 1/4 keV X-rays due to absorption by the neutral matter in the LMC. There is also no evidence for X-ray emission from either the disk or a halo of the LMC. Two large regions were identified, Shapley III (or more correctly SGS LMC 4) and the Bar with X-ray luminosities of 3.4×10^{37} erg s^{-1} and 7.2×10^{37} erg s^{-1} (0.1–1.0 keV), respectively (see Fig. 9.1). The former may be considered as an X-ray super(super)bubble. About one-third of the X-ray emission in SGS LMC 4 may come from the two SNRs N63A and N49. The stellar contribution is estimated to about 5×10^{35} erg s^{-1}, assuming that 470 OB stars contribute. This falls short of the observed emission by a factor of 70. The X-rays have to be diffuse in origin.

The emission identified with the Bar is probably due to the numerous point sources and SNRs in the Bar, sources such as N132D, N157B, 0540−69.3 and N103B. It is also possible that the X-ray emission here, too, is caused by a superbubble. The Bar has superposed on it a sequence of HII nebulae and young star forming regions which may well be an arc in a supergiant shell (see Chap. 6).

It is, however, also conceivable that a large population of discrete sources with $L_x < 10^{35}$ erg s^{-1} remain to be discovered in the LMC. Such a population could well produce all the X-ray enhancement described here. The Wang et al. (1991) discrete source survey seems, however, to make a certain amount of diffuse emission likely (see below).

Table 9.4. *X-ray emitting OB/HII systems in the LMC*

Nebula no.	OB association LH no.	X-ray luminosity 10^{35} erg s^{-1}	Stellar contribution 10^{32} erg s^{-1}	Expansion velocity km s^{-1}
N44	47, 48	14	< 88	30
N51D	51, 54	3.3	< 6.9	30–35
N57A	76	0.74	< 7.7	complex
N70	114	1.8	< 5.0	30–60
N154	81, 87	8.6	< 86	
N157	100	70	400	see note
N158	101, 104	11	< 62	40–45

N57A is a filamentary shell with complex internal motion, complicated by nearby HI sheets; up to 7 velocity components have been observed. A bright X-ray source to the north of N57A is SNR 0532−67.5 in LH75 (Fig. 9.2).
N70 shows an X-ray emission that suggests that the structure consists of a SNR within a wind-blown bubble.
N157 is the giant HII region 30 Doradus. At least 4 X-ray sources are catalogued by Long *et al.* (1981), only one is a confirmed SNR, 30 DorB (see Chap.10).
30 DorB contributes 3×10^{36} erg s^{-1}. In N157 there is a general correspondence between the optical Hα morphology and the X-ray contours. The stellar contribution is counted for 15 O3 stars, only.

Chu and Mac Low (1990) inspected the 75 of the 122 LH OB associations for which *Einstein* data were available. Diffuse X-rays were found around 10–15 ssociations, several of which were in the same HII region complex. Seven 'OB/HII systems'emitting diffuse X-rays were not associated with SNRs: N44, N51D, N57A, N70, N154, N157 (30 Doradus), and N158 (Table 9.4). All of these systems except 30 Doradus have a structure in the HII region that can best be described as a giant shell – a shell of gas swept up by the stellar winds and SNRs from the OB associations. The shell diameters fall in the range 75–150 pc. The HII regions have mostly thermal radio emission and low [SII]/Hα ratios. There is no spatial correlation between the X-ray emission and the nebular Hα emission. The X-ray emission is, rather, correlated with the central cavities of the shells. The OB/HII systems contain a large number of early-type stars which are known to emit X-rays.They are, however, not likely to be responsible for the diffuse X-ray emission: the surface brightness of the X-ray emission does not follow the distribution of the OB stars. The stellar contribution could, anyhow, not produce more than $\sim 1\%$ of the observed X-ray emission (see Table 9.4, col. 4). The observed X-rays must therefore be emitted by interstellar gas heated to $\sim 10^6$ K. The energy may be supplied by SNRs or stellar winds.

If modelled as superbubbles, the observed X-ray luminosities of the OB/HII systems, ranging from 7×10^{34} erg s^{-1} to 7×10^{36} erg s^{-1} (in 30 Doradus) (0.2–3.5 keV), are about an order of magnitude above that expected. Off-centre SNRs hitting the ionized shell could explain the observed emission.

Some OB associations have strong X-ray emission that can be explained by known SNRs or SNR candidates. Known SNRs exist in LH42, LH53, LH83, LH88, LH90, LH96 and LH99 (and, as mentioned above, probably also in LH111, ≡ N70). The identification of SNR 0538−69.3 in LH96 is considered doubtful; the X-ray emission

9.1 X-ray observations of the LMC

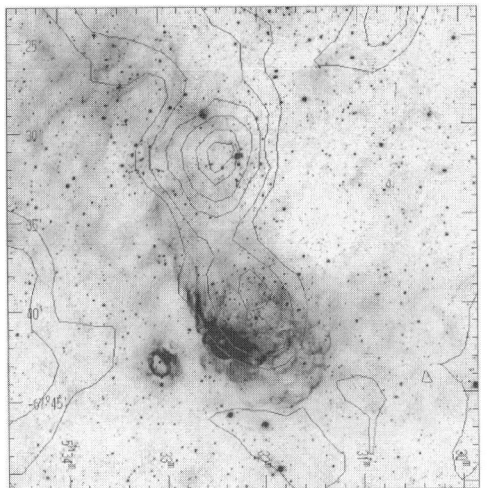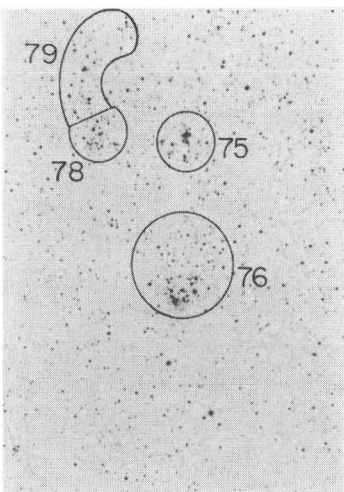

Fig. 9.2. Hα image of N57A–LH76 with overlaid X-ray contour map and a red continuum image to show the stellar associations. X-ray contour levels are 10, 25, 50, 75, and 90 % of the peak value of the brightest source in the field, the SNR candidate 0532−67.5 in LH75 (Chu and Low 1990). (Hα image supplied by Y.-H. Chu. *ROSAT* PSPC data are used for the plot, reduced by S. Snowden.)

is dominated by a variable foreground point source. The X-ray emission near LH80 is dominated by SNR 0534-69.9. Its connection with LH80 is uncertain and LH 80 is a questionable OB association. LH39, LH65 and LH75 contain discrete X-ray sources that could be SNRs.

The Wang *et al.*(1991) discrete source survey (see above) is complete to $\sim 3 \times 10^{35}$ erg s^{-1}. It appears to be highly unlikely that a combination of discrete X-ray emitters can account for the diffuse emission remaining when all known discrete sources are subtracted. (Such undetected emitters could be old SNRs, single stars in various stages of evolution, X-ray binaries, low-mass Population II binaries. Reasonable assumptions show, however, that they would be insufficient to explain the observed energy of $\sim 2 \times 0^{38}$ erg s^{-1}.) The spectrum of the remaining emission is well fitted by an optically thin plasma with temperatures ranging from $\sim 2\times 10^6$ K in the western part of the LMC to $\sim 10^7$ K in the vicinity of the active star forming regions near 30 Doradus. An anticorrelation between HI and diffuse X-rays on the scale of $\leq$ 1kpc is consistent with a picture in which superbubbles of hot gas are created in the neutral ISM by the combined action of stellar winds and SN in massive stellar associations. Such superbubbles are the two LMC large regions, Shapley III (or SGS LMC 4) and the sequence of HII and young star forming regions (part of the supergiant shell connected with the northern HI lobe) superposed on the Bar (Singh *et al.*1987, see also Chap. 6).

Over a substantial part of the LMC, the hot ISM has an emissivity and a pressure significantly in excess of that in the medium in the solar neighbourhood. Estimates of the cooling and heating rates for the hot gas suggest that a substantial fraction of the energy input from stellar winds and SNRs either radiates in other wavelengths or leaves the LMC in the form of a wind.

The highly structured diffuse X-ray emission of the LMC has been imaged in detail by *ROSAT*. The brightest regions are east of LMC X-1 and in the 30 Doradus region. In the latter it is strikingly similar to the optical picture. There is a strong correlation between diffuse features in the X-ray image and ESO colour images of the LMC in visible light. Bright knots in the X-ray map correspond to HII emission in the optical.

The low temperature of the X-ray emitting gas implies that its total cooling rate could be sufficient to balancing the supernova heating in the galaxy.

9.2 X-ray observations of the SMC

Following the early detections of the bright X-ray sources SMC X-1, X-2, and X-3 an extension to fainter sources was first done in the Seward and Mitchell (1981) survey mentioned above. It is approximately complete to $L_x = 10^{36}$ erg s^{-1}. Of the 26 sources seen, 5 were identified with objects not associated with the SMC. SMC X-1, the brightest X-ray source in the SMC, was the only previous source detected. The second brightest source, 0102.2−7219, had $L_x = 2 \times 10^{37}$ erg s^{-1} and is a previously unknown SNR in the central part of the SMC. Four other weak sources ($L_x \approx 10^{36}$ erg s^{-1}) are probably also SNRs. The remaining 15 sources were not identified; some of them are far from the centre of the Cloud and may not be SMC members.

70 discrete X-ray sources were identified in a 32 deg^2 field in the vicinity of the SMC by Wang and Wu (1992). 24 objects were classified as SMC objects, 13 as galactic stars, AGNs and clusters of galaxies. These discrete emitters account for less than half of the excess emission associated with the SMC. Of the discrete sources in the SMC 12 are identified as SNRs, 5 are stellar binaries and 5 are OB associations or HII filaments.

One source was found to vary in X-ray luminosity by a factor larger than 10 over a period of about one year and showed a relatively hard X-ray spectrum which changed with the X-ray luminosity. It is associated with a B1 star and is most likely to be a high-mass X-ray binary in the SMC.

Two pointlike X-ray sources (1E 0035.4−7230 and 1E 0056.8−7154) have extremely soft spectral characteristics; one, 1E 0035.4−7230, is probably a member of the LMXRB class.

In the SMC the *ROSAT* survey has so far led to the detection of > 40 sources (limit 4×10^{35} erg s^{-1}). Among these is a new supersoft source, RX J0048.4−7332, which could be identified with the symbiotic star SMC 3 in the cluster NGC 269. Two other SSSs, RX J0058.6−7146 and RX J0122.9−7521, are not yet confirmed as SMC members (Kahabka *et al.* 1994). The high luminosities deduced from the X-ray observations favour white dwarfs accreting at high rates ($\geq 10^{-7}$ M$_\odot$ yr^{-1}) and stable burning of hydrogen or helium. If accretion occurs over system evolutionary time scales, neutron stars may form by induced collapse. The observation of an X-ray turn-on from RX J0058.6-7146 with a duration of about one day is an important signature for short-time variability of the SSSs. Differences between the behaviour of the individual SSSs argue against a unique class of objects.

9.2.1 *Stellar X-ray sources in the SMC*

SMC X-1 is a well-studied and exceedingly luminous pulsating binary. The orbit is very circular; nevertheless, the optical radial velocities are best fit with an appreciable eccentricity. Some of these can be modelled with X-ray heating but not all. The

9.2 X-ray observations of the SMC

determination of orbital elements from optical data must then be considered a first approximation, only. The HeII λ 4686 emission originates near but not at the X-ray star. The entire Roche lobe of the X-ray star may be filled by an accretion disk. The existence of a disk was evidenced by analyses of the optical (van Paradijs and Zuiderwijk 1977) and UV (van der Klis et al. 1982) light curves. In the UV, the resonance-line doublets of NV, CIV and SiIV were particularly weak when the X-ray star was in front of the primary. The changes in line strengths may be due to an anisotropic ionization structure in the expanding atmosphere of the primary caused by the presence of the X-ray source companion.

SMC X-2 and X-3 have hard spectra and their luminosities in the energy range 2–11 keV are 1.0 and 0.7×10^{38} erg cm^{-2} s^{-1}, respectively. They are both highly variable in the X-ray band. Observations with the *SAS 3* X-ray observatory failed to detect the two sources; the limit of the survey was 1.0×10^{37} erg s^{-1} (Clark et al. 1978, 1979). This appears to show that the maximum luminosities of early-type X-ray stars in the SMC congregate near 10^{38} erg s^{-1} which is about five times the maximum luminosity of similar sources in the Galaxy. Also the Seward and Mitchell (1981) X-ray survey, complete to $\sim 10^{26}$ erg s^{-1}, failed to detect SMC X-2 and X-3, indicating that they are highly transient.

Optical identifications have been suggested by Sanduleak and Philip (1977 b,c) and further investigated by Crampton et al. (1978). Both optical counterparts, if correctly identified, are ~ 1.5 mag less luminous than the optical counterpart of SMC X-1. It appears possible that at least the SMC X-2 optical counterpart is slightly variable.

1E 0035.4−7230 has been optically identified with a variable star with $B = 19.9$–20.2 mag (Orio et al. 1994). The star shows strong UV excess, a hot, blue continuum and weak lines of high ionization. The lines are redshifted by 3–4 Å, verifying SMC membership.

1E 0056.8−7154 corresponds in position to the planetary nebula N67, $\equiv$ SMP 22. Its central star is likely to be responsible for the X-ray emission. With an effective temperature of $T_{eff} \sim 3 \times 10^5$ K it is the hottest ever observed in an X-ray band.

The identification of 1E 0056.8−7154 with the central star of the PN N67 has been doubted by Inoue et al. (1983), who consider it a foreground X-ray source. Nevertheless, so far the PN appears to be the most obvious optical counterpart. The two other objects in the HRI error circle are optically non-variable stars of types A and B and are therefore not likely to be highly luminous soft X-ray sources. N67 is thus the only unusual object within the circle. It is a hot, optically thick nebula, excited by a central star of 210 000 K and a luminosity of 2740 $L_\odot$ (Dopita and Meatheringham 1991a). This is not too exceptional; Kaler and Jacoby (1989) list several galactic PN in the same class as N67. However, if it is the X-ray source, its X-ray luminosity, 3.0×10^{37} erg s^{-1}, is exceptional for a PN, and an evolutionary theory for this class of PN is required (Brown et al. 1994).

RX J0059.2−7148 has been reported by Hughes (1994a) as a luminous (3.5×10^{37} erg s^{-1} over the 0.2–2 keV band) transient X-ray pulsar with an extremely soft component in its X-ray spectrum. It is the first time such a spectrum has been observed in this class of objects. The pulse period is $2^s.7632$, and the pulse modulation appears to vary with energy from nearly unpulsed in the low-energy band of the *ROSAT* PSPC (0.07–0.4 keV) to about 50 % in the high-energy band (1.0–2.4 keV). The source also shows flickering variability in its X-ray emission on time scales of 50–100 s. The pulse-averaged PSPC spectrum can be described by a two-component source model seen through an absorbing column density of $\sim 10^{21}$ atoms cm^{-2}. One spectrum component is a power law with photon index 2.4.The other is significantly softer and can be described either by a steeply falling power law or a black body with $kT_{BB} \sim$ 35 eV. This component is transient and unpulsed and requires, for blackbody model fits, a large bolometric luminosity (near or greater than the Eddington luminosity for a 1.4 M$_\odot$ object). When these characteristics are considered, RX J0059.2−7138 is similar to other X-ray sources in the Magellanic Clouds which show only ultrasoft or extreme ultrasoft X-ray spectra (EUS). The characterisitics of RX J0059.2−7138 show that EUS spectra may arise from accretion-powered neutron-star X-ray sources. The presence of a pulsar in this source, as well as in the galactic X-ray binary pulsar Her X-1, argues convincingly against the accreting white dwarf model as the explanation for the EUS emission.

9.2.2 Diffuse X-ray emission in the SMC

The extensive diffuse X-ray emission from the SMC constitutes about 60 % of the total observed X-ray flux in a 0.16–3.5 keV band (Wang 1991b). It extends beyond the active main body of the SMC, suggesting appreciable emission from its halo. The emission spectrum is substantially softer than that of the discrete sources, consistent with the presence of X-ray emitting coronal gas associated with the SMC (Wang and Wu 1992). It may be fitted to an optically thin thermal plasma with a temperature $\sim 10^6$ K, which is significantly cooler than the hot gas detected from recent star formation regions in the eastern part of the LMC. This suggests that most of the detected diffuse X-rays from the SMC have an origin in the gaseous halo around the SMC. It becomes soft away from the main body; this is consistent with a picture in which hot gas, created in recent star formation regions, cools off on the way out into the halo. The low temperature of the X-ray emitting gas implies that its total cooling rate could be sufficient to balance the supernova heating in the galaxy.

However, there are contradictory conclusions regarding the diffuse X-ray emission from the SMC. The extended emission left in the SMC after removal of the *detected* discrete sources may, at least partially, be accounted for by point sources with a luminosity below 4×10^{35} erg s^{-1} seen in the deep pointed observations (99 detected sources) (Kahabka and Pietsch 1993, Kahabka et al. 1994).

9.3 Supernova remnants in the Clouds

The supernova remnants (SNRs) in the Magellanic Clouds are important not only because they provide information about the mechanisms of supernova explosions but also about the ISMs of those galaxies. The exploding stars are important sources of heavy elements and energy.

The first SNR identified in the LMC was N49 (Mathewson et al. 1963). Since then the number of detected and optically identified SNRs has increased appreciably. Of a total of 32 SNR candidates in the LMC, 28 are well identified (Fig. 9.3). In the SMC, 11 candidates exist (Rosado et al. 1993b), of which at least 6 appear to be well established, based on observations at optical, radio, and X-ray wavelengths (Mathewson et al. 1983, 1984, 1985a,b). The 6 optically identified SNRs in the SMC are all inside the Bar, in two well-separated groups, one in the southern and one in the northern end (Fig. 9.4).

9.3 Supernova remnants in the Clouds

A catalogue containg 38 confirmed radio supernova remnants in both Clouds was presented by Mills *et al.* (1984) with maps of the LMC remnants at 843 MHz and a resolution of $43'' \times 45''$.

Criteria for the identification of SNRs in the Clouds are:

1. the source must have optical emission with
(a) $[SII]/(H\alpha + [NII]) \geq 0.6$;
(b) a looped or filamentary structure in $H\alpha$;
(c) an $H\alpha$ flux different from the one expected from the radio flux for a thermal source;
2. the source must be a non-thermal radio source;
3. the source should be an extended soft X-ray source.

Examples of the application of these criteria are the recent detections of two extended X-ray sources in the LMC HII regions N4A and N9 during *ROSAT* PSPC observations (Smith *et al.* 1994b). The X-ray characteristics suggest that they may both be SNRs. The optical nebulae show high $[SII]/H\alpha$ ratios, confirming the presence of high-velocity shocks. Radio continuum observations at 8.55 and 4.75 GHz show a non-thermal spectrum. The detection of these sources demonstrates that many SNRs in or near HII regions may have been overlooked in previous surveys. Also overlooked (in X-ray surveys) may be small-diameter SNRs with weak X-ray emission. Very large diameter SNRs may have escaped detection because their non-thermal radio emission and X-ray emission are below the sensitivity limits of available detectors.

A search for hidden SNRs in the LMC was undertaken by Chu *et al.* (1994) using the *IUE* archives of high-dispersion spectra of stars in X-ray bright superbubbles. They attempted to separate out the effects of SNR shocks from those of local stellar winds and of a global hot halo around the LMC by using suitable control objects. It was found that almost all interstellar absorption properties could be explained by the immediate environment of the objects. The low-ionization lines (SII, SiII, CII*) had velocities that agreed with those of surrounding HII regions; the high-ionization lines were all connected with either supergiant shells, superbubbles or HII regions around OB associations. There was no evidence for a uniform hot-gas halo around the LMC. Hidden SNR shocks were revealed by large velocity shifts between the high-ionization (CIV and SiIV) and low-ionization species; four objects within X-ray bright superbubbles were observed.

Initially, 20 X-ray sources in the LMC were identified with optical SNRs (Mathewson *et al.* 1983a, Cowley *et al.* 1984). Recently, Hughes *et al.* (1995) have noted that 27 of the LMC SNRs emit X-rays at a detectable level ($F_x \geq 5 \times 10^{-13}$ erg s^{-1} in the 0.15–4.5 keV band). The remnants range in size from diameters of ~ 2 pc to ~ 100 pc. They cover a range of ages from young ejecta-dominated remnants to old remnants in the radiative phase.

Three basic morpholgial types can be distinguished among the X-ray surface brightness distributions of the SNRs: a limb-brightened shell (examples N63A and N132D), a more or less uniform image (N49B), and a highly centrally peaked distribution (N157B). The latter is an extragalactic example of Crablike remnants.

The three types of X-ray morphology are mirrored in the optical appearances though discrepancies exist. Several, e.g. N63A, N49B and N132D, have X-ray images that are significantly larger than their optical extents.

192 *X-ray emission and supernova remnants*

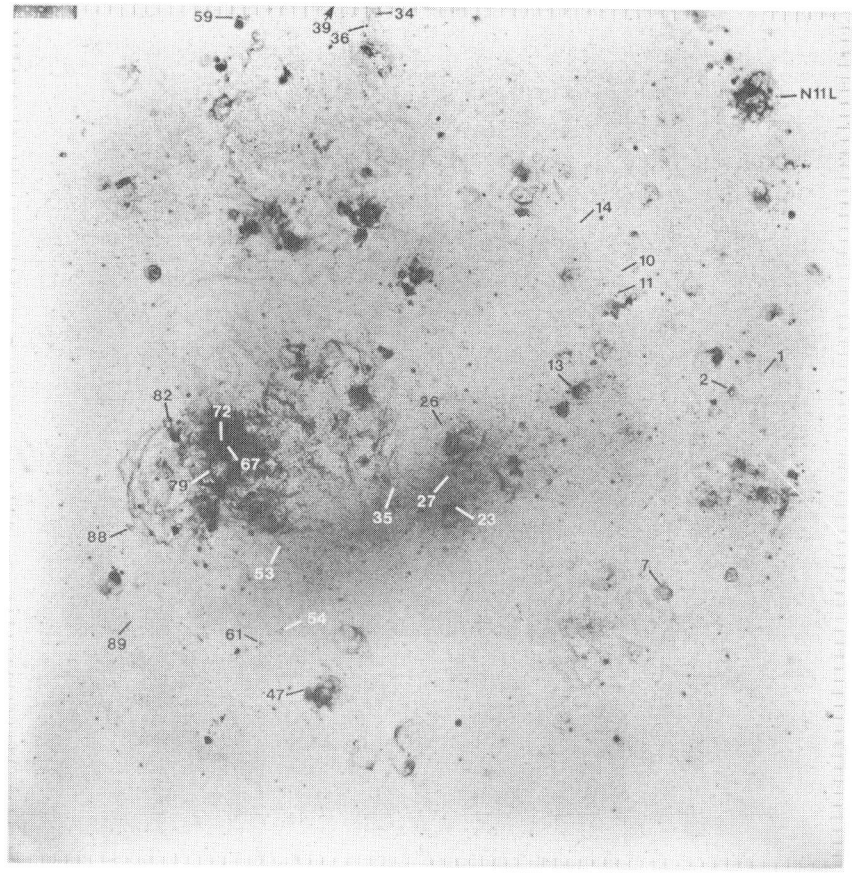

Fig. 9.3. Supernova remnants in the LMC. The SNRs are identified by their X-ray source numbers. (Mathewson *et al.* 1983a.)

Only four type I SNRs* have been identified in the LMC. The number of type I SNRs is about half that of type II SNRs for diameters < 30 pc. If the velocity of expansion is the same for the two types, then the frequency of occurrence of type I SNRs is about half that of type II SNRs in the LMC.

The four type I SNRs are scattered in the field with no apparent connection to young features except for 0505−67.9, which coincides with DEM71. High-resolution spectra of the four type I SNRs have been discussed by Smith *et al.*(1994a) with the particular aim of studying the non-radiative filaments, i.e. filaments behind the shock front. The attempts to find the broadening mechanism for the Hα emission lines from these filaments led to a precursor associated with cosmic-ray acceleration as being the most likely.

The distribution of the other SNRs in the LMC shows (i) a clumping of objects in the 30 Doradus region, (ii) several remnants within the Bar of the LMC, and (iii) the

---

* Type I SN have no hydrogen lines in their spectra whereas type II SN have. Type I SN are thought to result from explosions of compact white dwarf stars, whereas type II SN form in the explosions of massive stars, generally red supergiants, at the end of their lifetimes. The progenitor of SN1987A may have been a B3 I star, a post-red supergiant (Walborn *et al.* 1987).

9.3 Supernova remnants in the Clouds

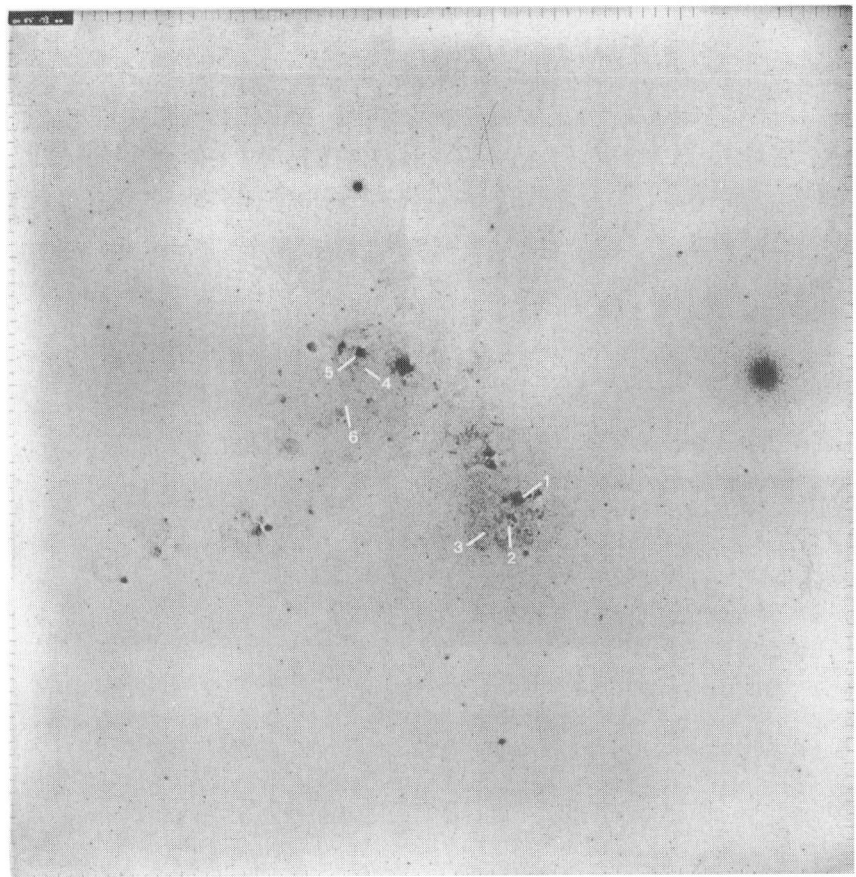

Fig. 9.4. Supernova remnants in the SMC. The SNRs are identified by map numbers assigned by Mathewson *et al.* (1983a).

remainder of the remnants in identified superassociations. A clustering of remnants in the north-western part of SGS LMC 4 is particularly noticeable. The progenitors of the SNRs in (i) and (iii) must have been young massive stars. On the basis of the content of Cepheids in the Bar, van den Bergh (1988) suggested that some of the Bar objects, (ii), may have progenitors aged over 4×10^7 yr. However, most SNRs in the Bar are in regions where many young clusters are located, clusters as young as $\leq$ 10 Myr (see Bica *et al.* 1992), so that the progenitors of these SNRs may also have been young massive stars.

9.3.1 Individual supernova remnants

The SNR 30 DorB is located in N157B ($\equiv$ NGC 2060), an HII region ionized by the OB association LH99. N157B appears to be separated from the main body of the 30 Doradus nebula, N157A, by a dust lane. Similarities in velocity indicate that they originate from the same parent cloud. The SNR 30 DorB is sufficiently isolated to permit its interaction with its environment to be studied. It is, however, in too complex a region for an accurate determination of its basic properties. Several dust clouds within N157B contribute to the confusion.

SNR 30 DorB was initially identified by its non-thermal radio emission and its high radio/Hβ ratio (Le Marne 1968b). Its radio emission is centrally peaked and its radio spectrum is flat (Mills et al. 1978, Dickel et al. 1994). It lacks infrared emission (Werner et al. 1978) as well as H109α emission (McGee et al. 1974). This rules out that it is a highly obscured HII region. Its SNR nature was unambiguously established when a strong X-ray source was detected (Long and Helfand 1979) with a spectrum indicating synchrotron emission (Clark et al. 1982). An optical identification proposed by Danziger et al. (1981) consisted of two bright [SII] patches. SNR shocks were also seen in the strong forbidden lines of [OI], [FeII], and [FeIII] and in the broad wings in the [OIII] and Hβ lines. The optical identification does not, however, coincide with the radio or X-ray source (Mathewson et al. 1983a, b).

Judging from the stellar content of LH99, the progenitor of SNR 30 DorB may have been a very massive early O-type star. The association LH99 contains three O3 stars (Schild and Testor 1992) and three WR stars (Lortet and Testor 1991), of which one, Brey73, is an aggregate of 11 components (Testor et al. 1988). These stars are powerful ionizing sources and the surrounding HII region may be expected to show enhanced [OIII] emission. The [OIII] line can therefore not be considered for the identification of the SNR. Chu et al. (1992) identified, in an echellogram in Hα + [NII] through the brightest of the two [SII] patches, a feature indicating an irregular expanding shell. The largest expansion velocity of the shell reached about 190 km s^{-1} in both the receding and the approaching side. The brightest [SII] patch coincided with the eastern edge of this shell. Further analyses of the echellograms and additional information from Fabry–Perot images led to the estimate of the size of the SNR as almost 90″ × 70″. Previous estimates range from $\leq$ 30″ to $\geq$ 76″ (Mills et al. 1978, Gilmozzi et al. 1983, 1984).

The expansion velocity of the optical line-emitting region in the SNR 30 DorB is uncertain. Previous estimates of the velocity spreads in the Hβ and [OIII] lines are 450 km s^{-1} (Danziger et al. 1981), 630 km s^{-1} and 780 km s^{-1} (Gilmozzi et al. 1983, 1984) for the [OIII] line. This should be compared with the just-quoted echelle observations of the Hα and [NII] lines of only $\sim$ 190 km s^{-1} (Chu et al. 1992). Possibly, the latter are not sensitive to broad faint wings of a profile or the slit positions are such that the high-velocity material has been missed. As the instrumental profiles of the previous observations are not well known, the question of a very-high-velocity material has to be left open.

The outlined properties of 30 DorB suggest that it may be a giant Crab-type SNR. It has a bright core and an extended plateau covering $\sim$ 25 × 18 pc. The radio map at 843 MHz (Mills et al. 1984) shows two components within the SNR boundary, a point source near the NW corner and an extended source over the whole SNR. If the point source represents the location of a pulsar, it must have a large proper motion to have moved from where the SN explosion occurred. The expansion velocity of 30 DorB is much slower than that of the Crab Nebula and its size is much larger (the Crab is $\sim$ 3 × 4 pc). The optical emission from 30 DorB is clearly from the blast wave, whereas there is additional pulsar powered emission in the Crab Nebula. The age of 30 DorB may be determined from its size and expansion velocity using Sedov's (1959) similarity solution for an adiabatic shock, i.e. $0.4(radius/V_{exp})$. With

9.3 Supernova remnants in the Clouds

the large expansion velocity the age becomes $\approx$ 10 000 yr. Even then it is an order of magnitude older than the Crab Nebula, aged 940 yr. With the more likely expansion velocity 30 DorB has an age of $\approx$ 24 000 yr. The two SNRs have, however, surprisingly similar radio fluxes.

It is possible that there are several SNRs inside the 30 Doradus nebula. Its complex core makes it difficult to identify them.

N 49 appears optically as an incomplete shell $\sim$ 1 arcmin in diameter. It is evolving in a cloudy, heterogeneous ISM. Line-of-sight velocity dispersions of individual line profiles range from 5 to 130 km s^{-1} (Shull 1983b). The smaller dispersions are associated with a photoionized shell which is at rest with respect to the LMC. It is probably ionized by UV photons from SNR shocks. The larger velocity dispersions, $\geq$ 30 km s^{-1}, are associated with smaller, shock-ionized regions which form a spherically expanding system. The projected expansion velocities, $\sim$ 280 km s^{-1}, indicate an age of at most 16 000 yr for N49. CO observations by Hughes et al.(1989) indicate that the SNR is physically related to a molecular cloud; the average velocities agree extremely well. The CO cloud is located near the brightest X-ray emission region, where the optical emission is also strong.

N63A was first noted as a possible SNR on the basis of its non-thermal radio spectrum by Mathewson and Healey (1964b) and confirmed as such by the presence of strong [S II] lines (Westerlund and Mathewson 1966). It is seen as three bright knots about 0'.5 across, embedded in the large HII region N 63, which contains a stellar association, LH83 $\equiv$ NGC 2030. The western knot is photoionized by an O5–9 main-sequence star; the north-eastern and southern knots form the SNR. The average reddening of the stellar association is $A_V = 0.6$ mag; the age of the association is $\sim$ 4 Myr (Laval et al. 1986). The velocities in the SNR are in the range 100–300 km s^{-1}, rather low in comparison with other SNRs in the LMC. No motion of the gas was detected in the surrounding diffuse nebula. The mean velocity of N63 is 298 $\pm$ 7 km s^{-1}, close to the HI velocity at that position, 308 km s^{-1}.

The large-scale kinematic and structural properties of N63A have been described by Shull (1983a). The SNR evolves in a clumpy ISM. Line-of-sight velocity dispersions of three types have been found. (i) The narrow bands, of high surface brightness, are produced by dense clouds that have been shocked by the SN blast wave and have been accelerated to projected speeds of 20–50 km s^{-1}. (ii) The broad bands have lower surface brightness and may arise from shocks passing through tenuous clouds; their projected expansion velocities are 100–140 km s^{-1}. Extreme velocities in these clouds indicate that the kinematic age of the SNR is 7000 or 20 000 yr, depending upon the assumed expansion geometry. (iii) The spikes arise from two HII regions in the immediate vicinity of N63A; both are photoionized nebulae. They may be fossil regions, produced by the UV/soft X-ray burst from the SN explosion or by the SNR progenitor star.

N63A has also been observed with the Australia Telescope at 4786 MHz and 8638 MHz with the 3 and 6 km arrays, including complete polarization data (Dickel et al. 1993). The derived radio spectral-index map confirms that the two knots are parts of the SNR with typical non-thermal spectra. The integrated emission has a spectral index of $\alpha = -0.6$ ($S_p \sim v^\alpha$). The polarization fraction reaches over 15% at both frequencies in the faint parts of the shell, but is less than 2% at the bright peaks. This is attributed to internal Faraday rotation in the dense knots. It is concluded that N63A is a shell-type SNR with a thick asymmetric shell. The radio shell has a diameter of about 65″, corresponding to a linear size of 15.8 pc at 50 kpc distance.

N132D was first optically identified as a SNR by Westerlund and Mathewson (1966). It lies in the eastern end of the LMC Bar, in a group of small HII regions, N132A–J. DEM186 encompasses N132D and N132H and has the dimensions $7' \times 4'$, with the major axis going N–S. The small HII region detected by Morse et al.(1995), $2'$ south of N132D and coinciding with the molecular cloud mapped by Hughes et al.(1989), may be identical to N132H. The surroundings of N132D suggest that it is in a region of active star formation, with high-mass stars and dense clouds nearby.

N132D belongs to the same class of remnants as Cassiopeia A in our Galaxy. It exhibits optical emission lines of oxygen corresponding to expansion velocities of ~ 2250 km s^{-1}, implying an age of 1300 yr and a large overabundance of oxygen. It has a complex morphology with several components: a high-velocity oxygen-rich ring, with a diameter of ~ 6 pc, a larger limb-brightened shell of normal composition and a diameter of ~ 21 pc, and, towards the north, a still larger diffuse emission region (Hughes 1987, 1994b). The oxygen abundance in the inner ring is 10–100 times the abundance in the outer shell.

N132D is the brightest LMC SNR in the X-ray band (0.2–4 keV) with a strong OVIII 0.645 keV emission, which contributes a significant fraction to the observed X-ray flux. Morphologically, its X-ray structure is similar to the optical with the same dual morphology, but the diffuse emission region does not appear in the X-ray image. It is also unique in that the central material, presumably stellar ejecta, contributes $\sim 10\%$ of its total X-ray emission (Hughes 1987).

Hughes also proposed, from an analysis of the X-ray surface variations in the outer shell, that the supernova explosion had occurred in a low-density cavity in the ISM formed by the HII region of the precursor star, which then has to be of a spectral type later than B0. The observed characteristics indicate that the ejecta have not yet interacted with a significant amount of circumstellar matter.

New optical CCD/interference filter imagery and *IUE* spectroscopy of N132D have been presented by Blair et al.(1994). Contrary to Hughes they find that a few optical features have X-ray counterparts but that there is, in general, little correlation between X-ray and optical features. The generally weak UV emission relative to the optical disagrees with the general character of shock model predictions and shows that photoionization is the dominant excitation mechanism for the UV/optical emission. The derived abundances of the emitting material indicate a precursor near 20 M$_\odot$.

Morse et al.(1995) have carried out [O III] $\lambda 5007$ Fabry–Perot observations of N132D and low-dispersion spectrophotometry from [OII] λ 3727 to [SII] $\lambda 6724$. By combining the kinematic and spectral characteristics of the line-emitting features with X-ray and radio observations they have constructed a model of N132D. Many of the high-velocity, O-rich knots appear to be distributed along the surface of a shell expanding at a velocity of ~ 1650 km s^{-1}. This leads to an expansion age of ~ 3150 yr. This is higher than that derived by Lasker (1980) and Sutherland and Dopita (1995), and by Hughes (1987) under the assumption of a cavity, but is less than the Hughes' estimate applying the Sedov solution. The fastest moving O-rich knots appear to lie outside the expanding shell; if they are allowed to define the expansion velocity the kinematic age increases to ~ 2450 yr.

N186D fulfils the criteria given above for a SNR. It is of particular interest

9.3 Supernova remnants in the Clouds

because of its location in the northern part of a shell-like nebula, N186E. The radial velocity pattern over N186D is complex, one zone with [O III] emission shows highly recessional velocities ($RV = 340$ km s^{-1}), another with bright [SII] filaments shows high approaching velocities ($RV = 150$ km s^{-1}; the systemic velocity, $\equiv$ to the velocity of N186E, is 238 ± 5 km s^{-1}). The latter may be understood as the result of the interaction between the SNR shock and the nebula N186E (Laval et al. 1989). The faint components of N186E have a pattern compatible with expansion from a centre that does not coincide with the geometrical centre of the nebula, where no early-type star is seen. However, at the presumed centre an extended X-ray source exists (Rosado et al. 1993a). It is thus possible that this nebula is a fossil SNR below detection in present radio surveys

0509−67.5, 0519−69.0, and N103B are among the smallest, and presumably also among the youngest, in the LMC, ≤ 1500 yr old (Hughes et al. 1995). X-ray spectra of the three, obtained with ASCA, show strong Kα emission lines of Si, S, Ar and Ca, with no evidence for corresponding lines of O, Ne, or Mg. The dominant feature in the spectra is a broad blend of emission lines around 1 keV, attributed to L-shell emission lines of iron. The elements detected are the major products of nucleosynthesis from type Ia SNe. Two of the SNe, 0509−67.5 and 0519−69.0, have been classified as of type Ia because of the Balmer-dominated SNRs (see above). N103B consists of several small bright knots with emission-line spectra typical for most SNRs and abundances similar to those in e.g. N63A and N49. Its X-ray image has a diameter of only 6 pc with a centrally peaked morphology. Its classification as being of type 1a is thus fully dependent on the X-ray spectrum. The authors conclude that roughly one-half of the SNRs produced in the LMC during the last ~ 1500 yr came from type Ia SNe; a conclusion that may not deviate too much from that drawn by Mathewson et al. (1983) in considering the small numbers and the uncertainties.

0540−69.3 is a Crab-like SNR in the LMC. It contains a pulsar, PSR 0540−69, with a period of 50 ms, a characteristic age of 1660 yr, and is losing rotational energy at a rate of 1.5×10^{38} erg s^{-1} (Seward and Harnden 1994). The SNR has a patchy outer shell with a diameter of 55″ accounting for 20% of the X-ray emission (ROSAT). The shell also emits at radio and optical wavelengths. The radio and X-ray structures are very similar, and the optical patches bracket the brightest X-ray region in the shell. 20% of the X-rays from the central region are due to the pulsar. The steady component may originate in the pulsar or in a surrounding synchrotron nebula or from the 8″ diameter [O III] shell.

1 E 0102.2−7219 is the brightest SNR in soft X-rays in the SMC. It is an oxygen-rich remnant belonging to the same class as N132D (Dopita et al. 1981). It is superposed on the HII region N76, centred to the southwest of the SNR position. Its oxygen-rich material covers a velocity range from -2500 to $+4000$ km s^{-1}. An undecelerated expansion of 3250 km s^{-1} implies an expansion age of 2150 yr (Blair et al. 1989). The SNR may, however, be appreciably younger, of the order of ~ 1000 yr (Tuohy and Dopita 1983, Amy and Ball 1993).

In the X-rays 1 E 0102.2−7219 shows a strongly limb-brightened shell with a radius of ~ 5.8 pc (distance 60 kpc). A high-resolution radio map shows a strongly limb-brightened shell with about the same dimensions. There is evidence of a relatively strong compact radio source near its centre. A ROSAT HRI X-ray image reveals a ring-like component and a spherically symmetric shell component with roughly 39%

of the HRI emissivity coming from the ring, 50% from the shell and 11% from a clumpy component. Most of the clumps appear to be associated with the ring-like component. The optical emission shows a local peak at the inner edge of the X-ray ring and then a gradual decrease further out. At the outer edge of the X-ray ring the optical emission has reached a minimum value. Beyond the blast wave (extreme extent of X-ray emission) there is an optical halo, presumably photoionized by the precursor star, the UV flash of the SN, or the UV radiation of the SNR (Hughes 1994b).

The correlation between the X-ray and radio data is not particularly good in detail. The radio shell is somewhat asymmetric to a level not present in the X-rays. The brightest part of the radio shell is towards the northeast, where the X-ray shell shows its minimum brightness. A weak X-ray clump is seen near the compact central radio source. This could signal the presence of a pulsar in the remnant but could equally well be a bright spot in the SNR shell projected onto the SNR centre.

The UV spectrum of 1 E 0102−7219 has been studied by Blair et al. (1989). They detected emission lines of OI, [OII], OIII], OIV], CIII], CIV, [NeIV] and MgII. By comparing their intensities with optical lines and calculating shock models and models of photoionization by the remnant's X-ray emission, they concluded that a combination of shock heating and photoionization was necessary to explain all the relative line intensities. The mass of the precursor star was probably slightly larger than 15 $M_\odot$.

0101−7236 was detected as a soft X-ray source by Seward and Mitchell (1981). Mathewson et al. (1983) showed it to be an incomplete Hα shell with the overall dimensions $98'' \times 77''$ but less than $12''$ thick. Radio emission appeared to come from a shell appreciably thicker than that seen in Hα. The source has been re-detected in several X-ray surveys. Hughes and Smith (1994) found it to be point-like and suggested that it may be associated with a Be star rather than with the SNR.

Ye et al. (1995) have observed 0101−7236 with the Australia Telescope Compact Array at 1378 and 2380 MHz and with the ROSAT HRI in the energy range 0.1–2 keV. The radio observations show a thick, somewhat filamentary, shell probably producing all the emission. The X-ray radiation in the region is dominated by a point-like source about $3''$ from the known Be star (at $1^h 3^m 14^s, 4; -72°9'12''$ (J2000)) It is most likely produced by the binary system containing the Be star. Support for this conclusion comes from the fact that the X-ray source is variable. No pulsed radio signals from the system have been detected.

No diffuse X-ray emission is observed from SNR 0101−7236. The upper limit for its X-ray emission is at least a factor of three below that of similar SNRs in the SMC. Its optical and radio structure shows the highest brightness in the SW part of the shell. This may indicate that the SNR is bounded in this direction by a region of high-density material which has decelerated the shock to the point where the gas temperature is below the X-ray emitting temperature. In the opposite direction the diffuse hot gas generated by the SNR may have escaped through an opening in the shell. It would then have contributed to producing the superbubble DEM 124.

SNR 0101−7236 is thus the only SNR in the SMC without associated X-ray emission. Its SNR character is otherwise established by its shell radio and Hα morphology and its non-thermal radio spectrum

9.4 SN 1987A

SN 1987A, the nearest visible supernova in 400 yr, has been extensively studied since its detection on 23 February, 1987, by astronomers in all fields. Numerous papers have been written on the supernova itself and its surroundings. Among the most exciting observations were those of neutrinos (see *ESO Conference and Workshop Proceedings* No. 26, 1987) and of γ-rays (see Ramana Murthy and Wolfendale 1993, p.38). Here, only some results concerning the progenitor and the surrounding ISM will be considered. The story of the supernova itself, the remnants of its progenitor, and the SNR, now under formation, will continue for a long time to come.

The surroundings of SN 1987A are of great interest as they reveal the interactions between winds from the progenitor's previous two important evolutionary phases, as a red and a blue supergiant with mass loss and stellar winds. The supernova will soon have reached the stage where we can talk about SNR 1987A. This will happen when the expanding ejecta from the explosion catch up with and interact with the rings formed during the progenitor days and produce a thermal X-ray source with a temperature of a few 10^7 K (Masai and Nomoto 1994). Thermal X-ray emission, resulting from shocks, has already been observed (see below), and so has radio emission, but no significant amount of mass from the expanding supernova has yet reached the ring. Dramatic changes in the ring structure are still to be expected.

SN 1987A is in a star cloud, including the association LH 90. UBV colours have been presented for the 39 brightest stars within 30" from it (Walker and Suntzeff 1990). From 23 of the stars, with very blue colours, the reddening in the region was determined as $E_{B-V} = 0.17 \pm 0.02$ mag. The reddening in the field appears to be uniform and may therefore also apply to SN 1987A.

The progenitor, Sanduleak $-69°202$, B3 Ia, had a complex history of mass loss up to the time of explosion (Plait *et al.* 1995). Dense circumstellar material, highly nitrogen-enhanced, implies that the gas had undergone CNO processing. The star must have been a red supergiant in a previous phase. The circumstellar material then ejected was subsequently swept up by the blue supergiant wind and appears now as an extensive nebulosity in the SN environment. The SN is at the centre of a bipolar nebula, shaped as an hourglass with a bright waist. Higher resolution observations show this waist as an elliptical ring of emission centred on the SN. It is suggested that the extreme-UV flash of the SN excited the gas in the ring, which is now cooling through recombination, emitting the narrow lines observed. If the nebula was formed in the same way as bipolar PNe, it was $\sim 10^4$ yr old when the SN flash occurred.

The progenitor star was probably in motion with respect to its circumstellar environment. The supernova is located at the top of a cocoon-shaped dark bay. Wang *et al.* (1993) suggest that the bay was produced by the shock interaction between the ISM and the energetic wind of the progenitor in its first blue supergiant (BSG) stage. In the following stage, as a red supergiant (RSG), its mass loss was greater but the stellar wind velocity was lower. The star ran through the shocked interstellar bubble and the slow RSG wind interacted with the ISM to form the feature called 'Napoleon's Hat nebula'. Interactions between the RSG and the second BSG winds eventually formed the symmetrical nebular loops, independent of the surrounding ISM.

The circumstellar material near SN 1987A forms a clumpy elliptical ring (Plait *et al.* 1995). *HST* FOC observations were obtained in the [OIII] and Hβ lines between

August 1990 and October 1993 and, with the PC, in the continuum and the [N II] line between April 1992 and May 1993. In [OIII], the ring had a semimajor axis of 6.4×10^{17} cm and a semiminor axis of 4.6×10^{17} cm $\pm 8.1 \times 10^{15}$ cm (distance 50 kpc), implying an inclination of $44° \pm 1°$ if the ring is an inclined circle. The apparent width of the ring was 9.0×10^{16} cm $\pm 1.6 \times 10^{16}$ cm. The position angle of the major axis was $89° \pm 3°$. The radial profile of the ring was not symmetric, with the inner boundary sharper than the outer. The filling factor in the ring was small, the average path length through the emitting gas was only $\sim 4.0 \times 10^{15}$ cm. The ring faded by a factor of ≈ 3.7 in [OIII] between 1990 and 1993. The surface brightness around the ring varied by a factor of 2 in the summed [OIII] image; there were large differences in the path lengths around the ring.

Observations in the UV and in the optical [OIII] $\lambda 5007$ line with the *COSTAR* corrected FOC on the *HST* in January 1994 (Jacobsen *et al.* 1994), 2511 and 2533 days after the outburst, revealed a well-resolved symmetrical expanding envelope which could be traced out to a radius of $\simeq 275$ mas in the near UV. The apparent diameter of the ejecta had grown to 255 ± 2 mas in the near UV and 167 ± 5 mas in the visible, in close agreement with previously inferred expansion rates. A structure in the shape of an incomplete shell-like crescent had started to form within the inner $\simeq 60$ mas core of the nebula.

Copious hard X-rays from Compton down-scattering of γ-rays from the ^{56}Co decay were detected from SN 1987A early on. Soft X-rays from two point sources were observed with the *ROSAT* PSPC in the direction of SN 1987A in 1991 and 1992 (Gorenstein *et al.* 1993). The brighter source was identified with the SN. The emission was interpreted as arising from the interaction of supernova ejecta with the existing blue giant wind. The second source may be a Be/X-ray binary, so far unidentified, or an X-ray echo from the supernova outburst. The latter interpretation requires that the supernova emitted $\sim 10^{47}$ erg in a burst of soft X-rays.

Detection of soft X-rays from SN 1987A also occurred in a deep *ROSAT* PSPC exposure in 1992 by Beuermann *et al.* (1994); a marginal detection had occurred one year earlier. The luminosity in the 0.5–2.0 keV band was $\approx 10^{34}$ erg s^{-1}. The authors favour an origin from the shock interaction of the SN blast wave with the termination shock of the fast wind from the progenitor star. Another possibility is that the X-ray emission originated in the central object or in the shock-heated ejecta and circumstellar material.

No evidence of a companion MS star at the site of SN 1987A has been found so far in the mass range from 2.5 M$_\odot$ up to the mass of the progenitor (Crotts *et al.* 1995).

Spatially resolved profiles of the [OIII] 5007 Å emission line from the three elliptically shaped rings of ionized gas surrounding SN 1987A were obtained by Meaburn *et al.* (1995) in October 1994 using the Manchester echelle spectrometer on the AAT. The morphology and kinematics of the rings were consistent with a model in which two toroidal, circular rings travel at ≤ 25 km s^{-1} in opposite directions along the surfaces of a bi-polar cone centred on the SN. The third, central ring is also in the form of a torus but expands radially at 8.3 km s^{-1} in a plane perpendicular to the common axis of the cones. All the rings were created $2-3 \times 10^4$ yr prior to the SN explosion. They represent the circumstellar environment of the progenitor star in the stage before the explosion.

9.4 SN 1987A

From prime focus photographs, taken over three years with the AAT, Spyromilio et al. (1995) have attempted to derive the three-dimensional structure of the reflecting material in front of SN 1987A. Features complicating the interpretation are bright rims from the nebula N157C and a large wind-blown bubble (Lortet and Testor 1984). The bubble cannot be the source of the expanding light echoes. Possibly a bright, diffuse portion of N157C may be associated with one of the echoes. The other echoes may form either on the front and rear faces of a single supergiant shell or on two shells (scales of 100–500 pc), one inside the other and both containing the supernova.

10

The 30 Doradus complex

30 Doradus is the most luminous HII region in the Local Group of Galaxies and is outstanding in many ways. Its structure is fascinatingly complex, with a large number of bright arcs and apparently dark areas in between, and it is also eloquently called the Tarantula Nebula. Its precise boundary has been difficult to establish. In the Henize (1956) catalogue a large nebulosity, N135, ~ 2 deg^2, encompasses 30 Doradus, SGS LMC 2 and SGS LMC 3, and would, seen at a greater distance, be identified as a single supergiant HII region (Fig. 10.1). Its central portion, N157$\equiv$ DEM263 (Davies et al. 1976, DEM), contains in its NW part N157A, 30 Doradus $\equiv$ NGC 2070, and N157B $\equiv$ NGC 2060. Henize estimated the diameter of N157A to be $\sim$ 30'; often values as low as 15' may be seen for 30 Doradus.

30 Doradus provides a most spectacular O association for which detailed information is now available (Fig. 10.2). Its brightest object, R136 $\equiv$ HD 38268, now known to be a cluster of clusters in itself, was for a while considered to be a supermassive star by many astronomers. Several of the brighter members in the immediate surroundings of the dense cluster core belong to the rare O3If*/WN-A class (Melnick 1985, Walborn 1986) and are among the most massive stars known.

The angular sizes of the major components of the 30 Doradus complex are presented in Table 10.1 as seen at true (50 kpc), galactic (2 and 8 kpc) and large (1 Mpc) distances. The dimensions at 2 and 8 kpc are for comparisons with galactic objects of similar type (Sect. 10.5).

The 30 Doradus region, including SGS LMC 2 and an area south of it, is extremely rich in HII, HI, CO and dust. Also the diffuse X-ray emission in the LMC is brightest in this region (see Sect. 9.1.2). Its distribution is very similar to the optical picture.

It has been difficult to find an explanation for the enormous concentration of gas and dust in this relatively limited part of the LMC. The tidal interaction between the SMC and the LMC has produced no concentration of gas in the 30 Doradus region (see Sect. 3.1), so it has to have arrived there by some other means. A reasonable solution may be found in the Fujimoto and Noguchi (1990) model, in which a gas cloud moved to this part of the LMC following the near collision between the SMC and the LMC 0.2 Gyr ago. The cloud does not necessarily remain in the plane of the LMC. As the HI gas in the disk does not show a void in the 30 Doradus region, the 30 Doradus association, as a strong source of UV radiation, has to be at least 250–400 pc above the plane. This is at about the same height as the HI L-component (see Sect. 8.1.1).

The Hα luminosity in the DEM263 region is measured as 5.7×10^{39} erg s^{-1}.

Table 10.1. *Dimensions of the components of the 30 Doradus complex*

Feature	Diameter	Angular at			
		50 kpc	2 kpc	8 kpc	1 Mpc
30 Dor nebula	200 pc	14′	5°.7	1°.4	40″
30 Dor association	45 pc	3′	1°.2	19′	9″
R136 cluster	2 pc	8″.3	3′.4	52″	0″.4
R136a cluster	0.5 pc	2″.0	52″	12″.9	0″.1

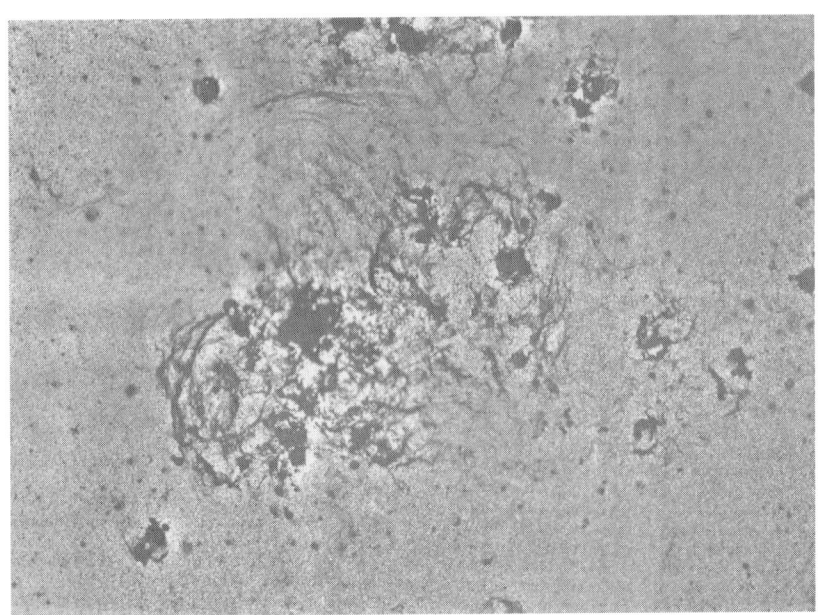

Fig. 10.1. Hα + [NII] photograph of the 30 Doradus region (Meaburn 1979). North is up, east is to the left. N135 encompasses all the nebulosity seen in the lower half of the figure. N157 is the central nebula (overexposed) and contains the 30 Doradus cluster. The filamentary structures to the east and to the west of N157 are SGS LMC 2 and SGS LMC 3, respectively. The long filaments towards the north extend up to the southern end of SGC LMC 4, the shell structure of which is very evident.

The measured fluxes for a subset of diameters are given in Table 10.2 (Kennicutt *et al.* 1995a). The total flux of the region is exceptional: 30% of the LMC Hα flux comes from a region of 30′ radius centred on 30 Doradus, and more than half from a region with 90′ radius ($\sim$ 1.3 kpc).

10.1 The extinction in the 30 Doradus region

The complex structure in the 30 Doradus region may be caused by the distribution of the ionized material only, or it may be partly due to a significantly varying extinction. Comparisons of Balmer line isophotes and 408 MHz radio fluxes (Le Marne 1968b) have suggested a visual absorption of $A_V = 2.0$ mag in the centre of the nebula with a value of 0.8 mag near the outskirts. In a recent comparison of an Hα image

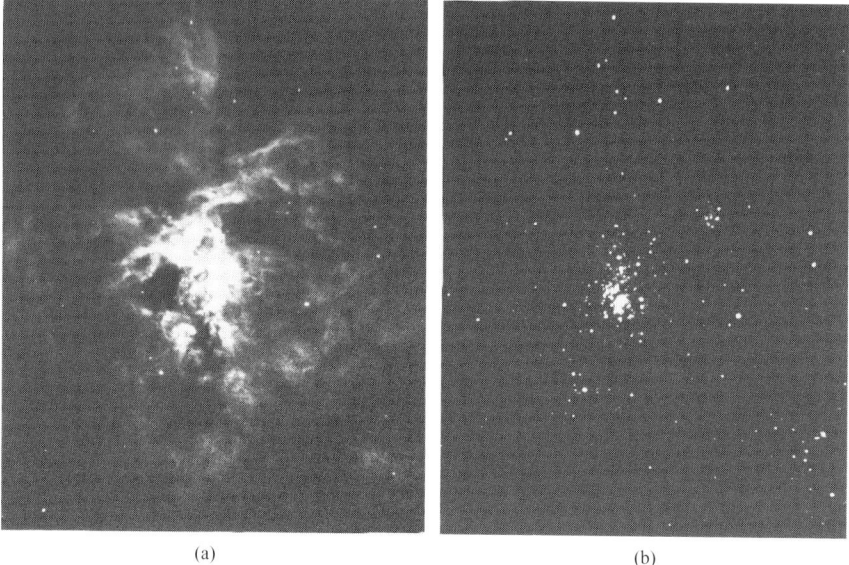

Fig. 10.2. Narrow-band photographs of the 30 Doradus nebula: (a) in the [OIII] 5007 Å-line (half-width 30 Å), and (b) in an emission-free part of the continuum, centred at 5350 Å(half-width 110 Å). (Mount Stromlo observatory 180 cm telescope.)

Table 10.2. *Hα fluxes in 30 Doradus*

Diameter: (arcmin)	1.67	3.0	5.0	10	20	40	60	90
$F(H\alpha)$: (10^{-11} erg cm^{-2} s^{-1})	92	278	460	870	1370	2130	2820	3730

At larger radii other bright HII regions, such as N158, N159, N160, will be included.

and a 13 cm radio map, with a resolution of 7".5, little variation was found in the radio/optical emission ratio over the $\sim 7' \times 7'$ studied (Dickel et al. 1994). The extinction, on a scale of 1.8 pc (distance 50 kpc), was considered constant over this area with a mean extinction at Hα of 1.1 mag. The Hα image delineates the structure well. Very compact HII regions may, however, exist, embedded in large amounts of dust, opaque in the optical and smeared below detection in the radio map.

Other investigators find a significantly varying absorption over the region. Parker and Garmany (1993) estimate the range to be $0.3 \leq E_{B-V} \leq 0.6$. The mean reddening in the central region is $E_{B-V} = 0.44 \pm 0.20$. In the northern field, well separated from the nebula, $E_{B-V} = 0.27$.

Several determinations of the extinction law in the 30 Doradus region have been carried out. Among these, the extensive investigations by Fitzpatrick and Savage (1984) and Fitzpatrick (1985a) have led to useful models.

Fitzpatrick and Savage studied the properties of the UV interstellar extinction in and near the core of the 30 Doradus nebula. Reddened stars within $5'$ (~ 73 pc

10.1 The extinction in the 30 Doradus region

at a distance of 50 kpc) of the core were compared with unreddened LMC stars of similar spectral types. All the 30 Doradus stars were reddened by $E_{B-V} \approx 0.12$ with an extinction law similar in wavelength dependence to those derived in previous investigations. Several of the stars, including R136a, R145, and R147 (the two latter are outside the 30 Doradus core) were found to be additionally reddened by $E_{B-V} \approx 0.18$ with an extinction law qualitatively similar in wavelength dependence to the law found for the Orion region. The simplest explanation of the extinction properties is a two-component model with a layer of 'LMC foreground dust', which affects all the stars, and a deeper layer of 'nebular dust', which affects some of the stars. The 2200 Å bump is present in both curves. The strengths of the bumps, relative to E_{B-V}, are 20–30 % weaker than for the Milky Way curve. The wavelength positions of the bumps and their profiles are indistinguishable from the galactic bump. The far-UV extinction properties of the 'nebular dust' are quite different from those of the foreground material. It is much weaker per E_{B-V} and per E_{Bump}.

The material making up the 'LMC foreground dust' probably extends far from the 30 Doradus central region and is most likely to be the material sampled previously (Nandy *et al.* 1981, Koornneef and Code 1981).

Fitzpatrick found essentially identical extinction properties towards 12 stars located within about a 500 pc projected distance of the 30 Doradus core. The extinction curve possesses a weak 2200 Å bump and a steep far-UV rise. This is similar to the curves derived in previous studies and has been considered to be representative of 'the average LMC extinction law' (Fig. 10.3). It is, however, biased towards the 30 Doradus region but does not include the 'nebular dust'. Six reddened stars, located *outside* the 30 Doradus region, yield extinction curves which are significantly weaker in the far-UV than in the 30 Doradus curve. They all have about the same far-UV strength but different 2200 Å bumps. These 'typical' LMC extinction curves do not differ as much from the galactic curves as is inferred from those derived from stars in the 30 Doradus region. Interstellar grain processing may have occurred in the dynamically active 30 Doradus region, rendering its extinction properties atypical of the LMC.

During the *Astro I* Spacelab mission the Ultraviolet Imaging Telescope (UIT) was used to derive images in four passbands with effective wavelengths between 1615 and 2558 Å (Hill *et al.* 1993). 314 stars were observed in a $9'.7 \times 9'.7$ field centred on 30 Doradus and their spectral types and extinctions were determined. 35 % of the stars appear to be MS stars earlier than B1. The extinction of the more reddened stars follows the Fitzpatrick curve with the 'nebular dust' curve from Fitzpatrick and Savage added (Table 10.3).

Most stars within the high surface brightness area near R136 have, perhaps surprisingly, a low 'nebular dust' extinction, $E_{B-V} \leq 0.03$, while the corresponding value for stars outside this area is typically $E_{B-V} \leq 0.30$ mag. The grains producing the 'nebular dust' curve must have been expelled from the central region by stellar winds. (Fitzpatrick and Savage found that R136 was affected by this dust; R136 is not considered by Hill *et al.*) The extinction caused by this dust varies considerably over the region. Hill *et al.* suggest that the differences between the extinction curves observed in the LMC and in the Galaxy are due to the lower cumulative nucleosynthesis in the LMC and the long time necessary for the component producing the 2200 Å bump (carbon?) to reach the ISM in the form of grains.

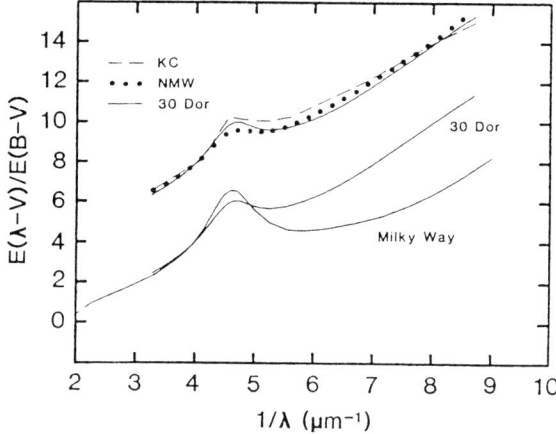

Fig. 10.3. Comparison of the 30 Doradus average extinction curve (Fitzpatrick 1985a) with the galactic average curve (Seaton 1979). In the upper part the 30 Doradus curve is compared with the 'average LMC' extinction curves derived by Nandy et al. (1981, NMW) and Koornneef and Code (1981, KC). The average LMC extinction law (non-30 Dor) derived by Fitzpatrick runs 0.8 and 1.4 units above the galactic curve at $1/\lambda = 7$ and 8, respectively.

Table 10.3. *Extinction in the 30 Doradus nebula*

Extinction	E_{B-V} mag
30 Dor extinction	0.11 ± 0.02
30 Dor 'nebular dust'	0.17 ± 0.17
Galactic foreground	0.10
Total extinction	0.38

Data from Hill et al. (1993)

The extinction towards R136 has also been determined by de Marchi et al. (1993) using the Fitzpatrick and Savage (1984) and Fitzpatrick (1985a) approach (Table 10.4).

The interstellar extinction in the direction of 30 Doradus has also been determined with the aid of the line ratios P_γ/H_δ and $P_\delta/H\varepsilon$ (Greve et al. 1991) for a slit position N–S close to the central cluster. The average extinction derived is $A_V = 1.35 \pm 0.10$, in good agreement with most other averaged data. For two stars in the 30 Doradus nebula, close to this slit position, $A_V = 1.49$ mag was derived (Greve et al. 1990).

10.2 The morphology of the 30 Doradus nebula

The mass of the ionized gas in the 30 Doradus nebula has been estimated to $\leq 10^5$ M$_\odot$ in the central 50 pc (Faulkner 1967). The electron densities are around 2–8 $\times 10^2$ cm^{-3} in the nebula, with a peak of 2×10^3 cm^{-3} in front of R136 (Feast 1961). The excitation source is the rich cluster in the core of the nebula. It has been questioned whether the energy it produces is sufficient to generate the observed large and unusual nebular motions (see Sect. 10.3).

10.2 The morphology of the 30 Doradus nebula

Table 10.4. *The extinction in the R136 cluster*

Extinction	E_{B-V}	R	A_V
LMC foreground	0.16	3.1	0.50
LMC 'nebular dust'	0.18	3.7	0.67
Galactic foreground	0.07	3.1	0.22
Total extinction	0.41		1.39

The galactic foreground extinction is taken from other observers (see Chap. 2).

A radio model of the 30 Doradus region was presented by Mills *et al.* (1978), following a mapping of the region at 1415 MHz with the Fleurs Synthesis Telescope with a beamwidth of 50″. Comparing their data with radio observations at 408 MHz and 5 GHz, they concluded that the emission from 30 Doradus is entirely thermal. Localized nonthermal sources (e.g. N157B, a SNR) were identified. No widespread non-thermal component was found apart from the background radiation from the LMC itself. It was suggested that the nebula consists of a core and a halo with masses of $\sim 6 \times 10^4$ and $\sim 6 \times 10^5$ M$_\odot$, respectively. The visual extinction over the central part of the nebula was determined to be ≈ 2 mag, which implies a dust to gas ratio of $\sim 1\%$.

The morphology of the 30 Doradus complex was studied by Elliott *et al.* (1977) with the aid of a series of photographs in large, medium and fine scale in Hα, Hβ, [OIII], [NII] and the visual continuum (5672 Å). Comparisons with the radio continuum structure at 5000 and 408 MHz showed exact coincidences between the thermal radio peaks and the two regions of maximum brightness in Hα. They appear to be two separate cores, unobscured by dust (Fig. 10.4), and representing two separate centres of ionization. The 30 Doradus cluster is to the south of the brightest nebular region. The two cores appear to be on opposite sides of a ridge of HI. In order to be directly connected with 30 Doradus, the ridge has to be part of the HI L-component (see above); if it is a D(disk) feature its effect is only apparent. The former possibility appears to be the most likely.

Elliott *et al.* concluded that the 30 Doradus HII nebula is produced by ionization fronts eating into neutral material. The small-scale nebulous filaments in the core of the nebula are bright rims on the interfaces between the neutral and ionized matter. Many of the bright emission knots in the nebula have relatively high [NII]/[OIII] ratios and may be bright rims around neutral globules. The [NII] filaments are, as a rule, offset from the Hα filaments away from the central core.

Further out, the scale of the filaments increases, and eventually faint arcs, 1 kpc long, become evident. Most of these filaments belong to either SGS LMC 2 or SGS LMC 3. Some structures, due north of 30 Doradus, are more difficult to ascribe to either shell and to a well-defined source of ionization (Fig. 10.1). A possibility is that they are ejectae from supernova explosions in the core region.

Higher-resolution images of parts of the 30 Doradus nebula have now become available. A 40 pc region of the 30 Doradus nebula, centred about 1 arcmin north of R136, has been imaged in Hα, [SII], and in the continuum with the WFC on

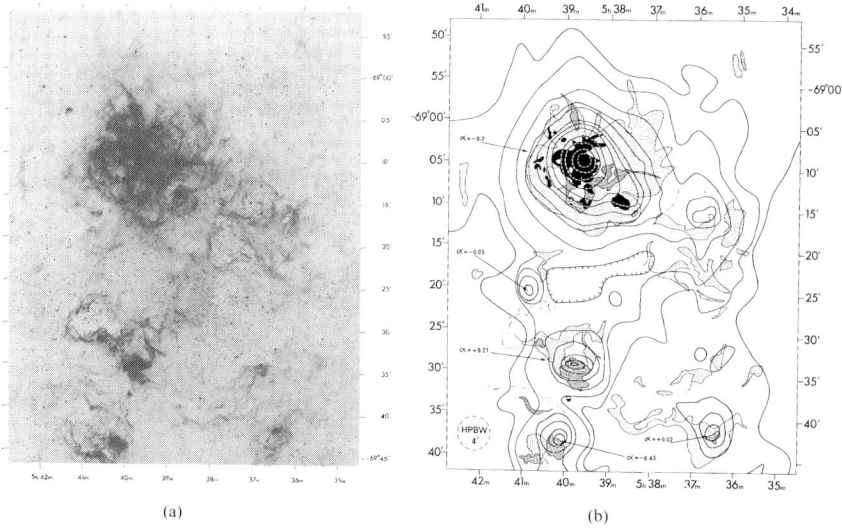

Fig. 10.4. The region of 30 Doradus in (a) Hα and (b) 5000 MHz radio continuum (Elliott et al. 1977). Radio spectral indices are given to indicate thermal and non-thermal sources.

the *HST* on a scale of 0.025 pc per pixel (distance 50 kpc). A ground-based echelle spectrum in Hα and [NII] of the region was also secured (Hunter et al. 1995). The [SII] image shows a sharper, more filamentary appearance than the Hα image, and the filaments are generally offset in [SII] away from R136 in the same fashion as the [NII] filaments. These [SII] filaments locate the ionization fronts and mark the boundaries of the neutral gas in 30 Doradus. Some of them mark the edges of extinction regions. Certain [SII] filaments, 3–13 pc long, appear to be unusually strong. Nothing in particular is seen at their base.

The bright emission knots appear to be wind-blown bubbles centred on O stars. One knot is resolved in a [SII] image into two shells with diameters ~ 0.5 pc, each centred on one or two stars. One of the stars involved is classified as O V; it is in the Walborn and Blades' (1987) 'Knot 1'. If the O star has a wind energy of $L_w = 1-4 \times 10^{36}$ erg s^{-1}, then the bubble would reach a diameter of 0.5 pc in $\sim 10^4$ yr with reasonable assumptions about the gas density, n. (The time is equal to $1-2 \times 10^4\ n^{1.5}$; Weaver et al. 1977.)

The overall appearance of the ionized gas is one of bubbles and filaments embedded in a diffuse, but modulated, emission. R136 and the brightest stars sit in regions of lower surface brightness gas emission and are surrounded by intense ridges and filaments. Directly north of R136 is a feature which may be either a 15 pc diameter bubble or the chance alignment of several ionization fronts (Clayton 1987). To the NE of R136 is a bright emission ridge, about 15 pc long, which bifurcates into broad ridges. The filament is also seen in [NII] (Elliott et al. 1977) and it marks the northern edge of an X-ray bubble (Wang and Helfand 1991a). In the Hα image, extinction patches and lanes are seen: There is a patch to the NW of R136 about 7.3×4.2 pc, and to the W–SW is an 8.8 pc extinction lane.

10.2 The morphology of the 30 Doradus nebula

Table 10.5. *Protostars in the 30 Doradus nebula*

Object	RA(1950)	Dec(1950)	K_0	$(K-L')_0$	M_{bol}
P1	$05^h39^m07.5^s$	$-69°06'36''$	11.0	2.68	-6.5
P2	05 39 01.7	-69 05 28	11.6	2.81	-6.1
P3	05 38 55.3	-69 07 32	11.0	2.64	-6.5
P4	05 39 08.4	-69 05 48	11.8		-5.5

Some strange features are seen in the *HST* Hα and [SII] images. There is a parabola-shaped structure with a length of 0.47 pc, opening away from R136. It is brighter in Hα than in [SII]. In its focus, 0.13 pc from the arc, is a small non-stellar knot that is stronger in [SII] than in Hα. In its immediate neighbourhood is a luminous candidate protostar (Hyland et al. 1992) which could be associated with the knot. The cloud may have been compressed by the wind from the R136 cluster, and the protostar then represents a star formation event caused by the energy output of the cluster into the surrounding ISM.

In the near-infrared (nIR), in JHK, the overall appearance of 30 Doradus agrees with the optical morphology though several highly reddened objects are found (Rubio et al. 1992). Some of them correspond to previously noted protostars (Hyland and Jones 1991, Hyland et al. 1992, see Table 10.5); others seem to be tracers of recent massive star formation.

A complex of four nIR sources is seen about 90″ north of R136, in the vicinity of 'Knot 2' (Walborn and Blades 1987, identified in Fig. 10.2), and within the error box of a H_2O maser position (Whiteoak and Gardner 1986). One of the sources, visible only in the K-image, could be associated with the water maser. The whole nIR complex is projected towards the FIR peak (Werner et al. 1978) and CO emission (Booth and Johansson 1991), suggesting that star formation takes place in this molecular cloud of which 'Knot 2' is the optical counterpart.

A chain of nIR knots is seen to the SW of R136 along an arc-like feature which coincides with one of the more prominent structures of the 30 Doradus nebula. One of the knots is identified with the protostar candidate P3 (Hyland and Jones 1991). In the chain is also the object called 'Knot 3' by Walborn (1991). The CO emission follows roughly the shape of the extended nIR emission and the chain of knots. The feature is also seen at FIR wavelengths (Werner et al. 1978). All these facts point towards an event in which interstellar gas and dust have been compressed by strong stellar winds, possibly SN events, from the central cluster so that a second epoch of star formation has been triggered. The knots are then caused by radiating young stellar objects along the IR nebulosity, seen also in [OIII] but not in the blue continuum. Under the assumption that the O and WR stars in R136 are ~ 2–3 Myr old and that their energetic winds cause a shock velocity of ~ 200 km s^{-1} in the chain, 10.5 pc away, the second epoch of star formation may have started 0.2–0.3 Myr after the initial burst.

High sensitivity (0.02 K in radiation temperature) mapping in the CO (1 → 0) line of a region of $1°.0 \times 0°.75$ centred near the 30 Doradus nebula (Garay et al. 1993) has shown that most of the emission comes from six well discernable molecular clouds with radii between 60 and 160 pc and luminosities between $1-6 \times 10^4$ K km s^{-1} pc^2. The CO brightness temperature and luminosity of these clouds are significantly weaker than those of galactic clouds of similar dimensions. The proportionality factor between the total column-density of gas, N(H$_2$+2H), and the velocity-integrated CO intensity for the 30 Doradus halo clouds is $\sim$ 13 times larger than for molecular clouds of similar size in the inner Galaxy and $\sim$ 2 times larger than the average value for clouds in the LMC.

The strongest CO emission is associated with the 30 Doradus nebula. The bulk of the molecular gas is at the heliocentric velocity 265 km s^{-1}. This is about equal to the average velocity of the stellar component and the ionized gas in the complex, but also to that of the HI L-component in the immediate neighbourhood of 30 Doradus. This result, together with its low brightness temperature, suggests that the CO emission arises from small molecular clumps in a volume dominated by ionized gas and with some neutral gas; the CO is mostly dissociated there. CO emission is also found to be redshifted with respect to the main component. This gas has been accelerated by the stellar winds from the hot stars in the R136 cluster and the surrounding O associations.

Cohen et al.(1988) draw attention to a shell of molecular gas expanding from the 30 Doradus region. It appears as a clumpy ring about 500 pc in diameter with a mass of $\sim 5.6 \times 10^6$ M$_\odot$. A high-velocity component (20 km s^{-1} higher than the ring velocity) is seen in its centre with a mass of $\sim 3.8 \times 10^6$ M$_\odot$. The shell may be expanding at $\sim$ 17 km s^{-1} and with a kinetic energy of $\sim 3 \times 10^{52}$ erg. Its expansion is probably caused by the stellar winds from the young, hot stars in the 30 Doradus central cluster and/or by supernova explosions. It is interesting to note that SN 1987A is in the region of this cloud.

The areas of the 30 Doradus nebula discussed on the basis of presently available images are restricted to its central portions. The ionizing stars (at least 14 WR stars and $\geq$ 200 luminous O stars with $M_V \leq -4$ mag and a lifetime of $\leq$ 5 Myr) have energy outputs somewhere between 2000 and 4000 $\times 10^{36}$ erg s^{-1}. The overall gas density in the region is difficult to estimate. Hunter et al.(1995) arrived at ~ 2.3 cm^{-3} from the HI (McGee and Milton 1966) and CO data (Cohen et al.1988). The combined effect of the stars would lead to a bubble of a radius of about 220 pc and an expansion velocity ≥ 25 km s^{-1}. What then remain in the inner $r = 40$ pc field, studied in the *HST* WFC images, are 'left overs', higher-density spots surrounded by low-density areas and containing stars and, more importantly, protostars. This is indicative of a younger edition of SGS LMC 4, where star formation has already been completed in the 'left overs' in the SSPSF process (see Chap. 6).

10.3 The kinematics of the 30 Doradus nebula

Large-scale line splitting over 30 Doradus was first detected by Smith and Weedman (1972). Observations of the profiles of the [OII] lines in the bright nebular core to the north of R136 show that the peaks in brightness of most of the knots and filaments are displaced 2.3″ northwards as compared with the [OIII] peaks (Cantó et al.1980). In some positions, up to seven velocity components are observed. A typical profile of the [OII] doublet shows distinctly separated velocity components with heliocentric velocities $\sim$ 185, 225, 275, and 300 km s^{-1}. Very similar velocities were obtained in

10.3 The kinematics of the 30 Doradus nebula

the [OIII] line, and are also seen in the CaII 3933 Å line as absorption caused in the LMC (Feast 1961).

Interstellar CaII lines in the spectra of R136, R139, and R145 have revealed the existence of several clouds in these directions with three principal components with heliocentric radial velocities between 230 and 300 km s^{-1} (Blades 1980). A high signal-to-noise spectrum of R136 showed CaII absorption over a (heliocentric) velocity range from 10 to 315 km s^{-1}, with the continuum visible only near 105 km s^{-1} (Blades and Meaburn 1980). The 18 km s^{-1} component (Fig. 10.5) is attributed to local gas, the 70 and 149 km s^{-1} components to gas in the galactic halo, the 284 and 300 km s^{-1} components to gas in 30 Doradus. There remains a weak and broad absorption extending as a wing blueward from the main LMC component from $\sim$ 270 km s^{-1} down to $\sim$ 150 km s^{-1} with an equivalent width of 0.17 Å. As the velocity of R136 is $\sim$ 260 km s^{-1} this feature has large negative velocities relative to R136. It could, consequently, be generated by stellar winds arising from the cluster or by successive supernova explosions in it. The numerous O stars in the cluster, with a total luminosity of $\sim 5 \times 10^7$ L$_\odot$, may produce a stellar wind with $L_w = 1-3 \times 10^{39}$ erg s^{-1}. This may be supplemented by the kinetic energy from supernova explosions, each initially releasing $\sim 5 \times 10^{51}$ erg. The expansion velocity will also depend upon the density of the surrounding interstellar medium. However, the position of 30 Doradus relative to the main HI plane of the LMC must also be taken into consideration. The CaII K feature may originate in a gas cloud well separated from 30 Doradus. There is widespread interstellar gas in the general vicinity of the nebula. It should, however, produce rather narrow interstellar lines and hardly the wide feature seen here.

UV observations of SiIV, CIV and AlIII (Fig. 10.5) were interpreted by de Boer *et al.* (1980) as showing that a corona surrounds the whole LMC. The observed CaII K interstellar lines could be interpreted as showing that extensive halos exist around the Galaxy and around the LMC and that material is also scattered between the two galaxies.

Feitzinger *et al.* (1984) found that the Magellanic interstellar components fell into four well-separated groups in the direction of R136. From a large number of velocity determinations in HI, Hα, NaI, CaII, [OII], [OIII] and UV lines the average values given in Table 10.6 were derived. The data indicate that the inner shell is expanding with a velocity of 24 km s^{-1}.

High-resolution NaI and CaII spectra in the 30 Doradus–SN 1987A region (thus of stars mostly between the HI L-component lobes) show at least five interstellar components in several lines of sight. The most intense absorptions are at 280–290 km s^{-1} and may be considered as components in the LMC galactic disk (Vladilo *et al.* 1993). The most intense HI emission in the field is at ~ 270 km s^{-1}. At this velocity the NaI and CaII absorptions are weak, so that this HI gas may lie beyond the observed stars or have a high degree of Na ionization. The velocities of the observed stars are between 242 and 316 km s^{-1} with a mean velocity of 272 km s^{-1} (12 stars). The majority of the stars would then be in the HI disk. As the Na and Ca clouds have to be in front of the stars they will be moving towards the plane.

Meaburn (1984), using an echelle spectrograph on the Anglo-Australia Telescope, obtained profiles of the [OIII] emission line along three parallel lines, slightly north of the core, over the 30 Doradus nebula. Many sheets, 10–50 pc across, were found, each with a separate radial velocity. Superposed on these motions are those produced

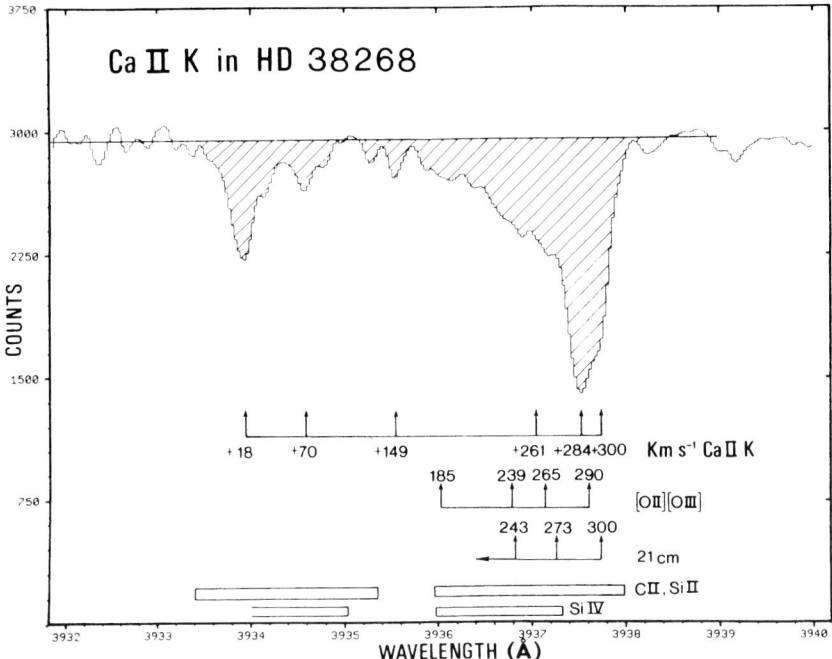

Fig. 10.5. The interstellar CaII K line in the direction of R136 (Blades and Meaburn 1980). The K line velocities marked are those from Blades (1980). The [OII] and [OIII] data are from Cantó et al. (1980) and the UV interstellar lines from de Boer et al. (1980). The HI 21 cm velocities are from McGee and Milton (1966). (Reprinted by permission of Blackwell Science Ltd.)

Table 10.6. *Average heliocentric radial velocities of interstellar-line components in the direction of R136a (Feitzinger et al. 1984)*

Emission km s^{-1}	Absorption km s^{-1}	Source
220		Outer shell
243	241	Principal inner shell structure around R136
262	265	Bulk velocity of hot cavity around R136
290	291	Principal inner shell structure and over-lapping large shells, expanding and contracting
	311	material above the plane (upper branch of LMC rotation curve)

The emission and absorption lines at 290, 291 each show two not fully separated components, 288 and 297, and 286 and 295 km s^{-1}, respectively.

by a number of shells, ~ 15 pc in diameter, expanding at 200 km s^{-1}. They may be pressure driven bubbles, $\leq 2 \times 10^4$ yr old, caused by wind from WR stars. A large number of small shells, ~ 2.5 pc in diameter, are also present. They are $\sim 4.8 \times 10^4$ yr old and were probably formed by low-power winds from O and B stars. In the halo of 30 Doradus, shells with diameters ≤ 100 pc and expansion velocities of ~ 35 km s^{-1} may exist. They may be driven by stellar winds from WR stars in the 30 Doradus halo and SN explosions, or by flows from the powerful compact cluster R136a and WR stars in the surroundings.

Further observations of the 30 Doradus giant HII region underline its complex kinematics. Its outer regions are characterized by a smooth velocity field but its turbulent velocity, 30–40 km s^{-1} FWHM, is considerably higher than in most smaller HII regions (Chu and Kennicutt 1994). In the central 9' (130 pc) core, multiple velocity components are observed at most positions. A large number of expanding structures dominate the velocity field; they range in size from 1 to 100 pc and have expansion velocities of 20–2100 km s^{-1} and are often organized into large hierarchial networks. Integrating these expanding structures produces a simple profile with a Gaussian core and faint extended wings. Several fast-expanding shells, with diameters of 2–20 pc, expansion velocities of 100–300 km s^{-1}, and kinetic energies of 0.5–10 $\times$ 10^{50} erg have been identified. The network and the fast-expanding shells are coincident with extended X-ray sources. They are probably associated with SNRs embedded in shells produced by the combined effects of stellar winds and SNe from the OB associations.

The expanding shells contain roughly half of the kinetic energy in the 30 Doradus complex; this energy is several times higher than the gravitational binding energy in the region. Extrapolation of the current energy injection rate in the nebula over the lifetime of the OB complex suggests that 30 Doradus will evolve into a supergiant shell of the type already exisiting in the LMC.

The high-dispersion spectrum obtained by Hunter *et al.*(1995) of the 30 Doradus nebula immediately north of the R136 cluster indicates a heliocentric velocity in its western half of 273 km s^{-1} and in the eastern half of 253 km s^{-1}. To the west an expanding bubble about 1.6 pc in diameter is observed with a velocity offset from the main Hα gas with $+48$ km s^{-1}. In another portion of the spectrum a low-intensity component is seen, offset -80 km s^{-1} from the main Hα gas. This coincides with one of Chu and Kennicutt's fast shells. Several other velocity offsets occur.

The velocity pattern in the outer parts of the 30 Doradus nebula, the halo, has been studied by Cox and Deharveng (1983), using Hα Fabry–Perot interferograms. They have in particular studied the regions where the radio continuum, at 1415 MHz, and the Hα images show holes in the brightness distribution (Fig. 10.6). These holes are due to a lack of ionized material and not to absorption in the foreground, so that the halo of the 30 Doradus nebula may be formed from empty ionized shells. Their results for 'Shell 1' show two principal sheets of ionized gas in the direction of the centre of the shell, with radial velocities of the order of 235 and 295 km s^{-1}. In the surrounding bright filaments no line splitting is seen.

10.4 The stellar content of the 30 Doradus region

The first CM diagram of stars in the central cluster in the 30 Doradus nebula was presented by Westerlund (1964b) from photographic narrow-band photometry in a pseudo-Strömgren system. A main sequence was found with a number of WR stars and supergiants, classified by the Radcliffe astronomers (Feast *et al.* 1960), and topped by the composite star R136. The total mass of the cluster stars inside a radius of 2'.7 was estimated to be 2×10^4 M$_\odot$.

214 The 30 Doradus complex

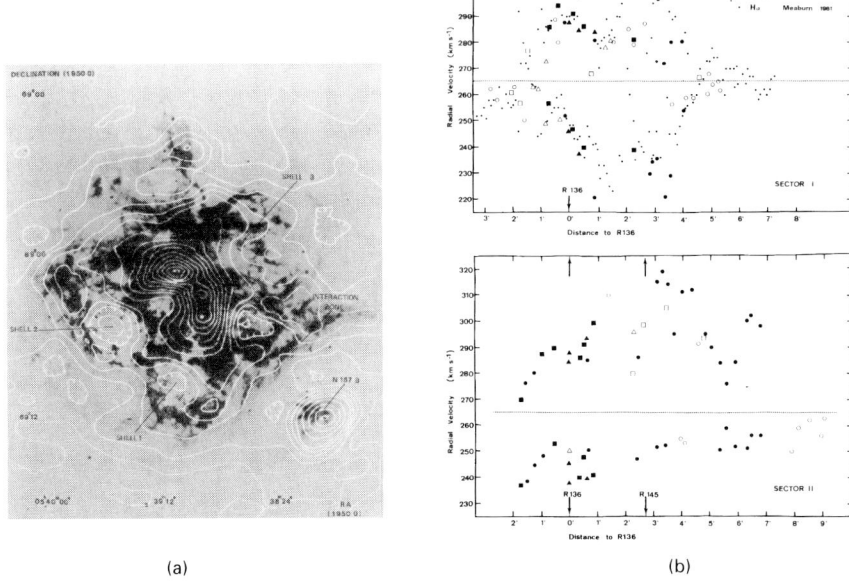

(a) (b)

Fig. 10.6. (a) The 30 Doradus nebula in Hα light with superposed brightness temperature contours at 1415 MHz. A few typical shells are identified. (b) Variation in radial velocity vs the distance to R136 for two sectors containing 'Shell 1' and 'Shell 2'. The 'new' results are those of Cox and Deharveng (1983).

For a full understanding of the composition of this cluster, or association core, photometry with high spatial resolution is necessary, combined with medium resolution spectroscopy of the brightest stars, mostly of very early type, for an accurate classification. Over the past decade extensive knowledge has been gathered. The cluster, $\sim 3'$ in diameter, contains O3 and WN stars, very massive objects in an early evolved state. The luminous central object R136, classified WN5 + OB, $\sim 8''$ across, has been shown to be a dense cluster with dozens of massive components. In and near the bright filaments of the nebula several early O stars have been found embedded in dense nebular knots. They represent young massive objects emerging from their protostellar cocoons. This idea is supported by the discovery of luminous infrared protostars in the same regions (Hyland and Jones 1991, Hyland et al. 1992, Rubio et al. 1992) indicating sequential events.

Within 25 pc from R136, 40 O stars, 12 WR stars, 8 B supergiants and one M supergiant were classified by Melnick (1985). Of the early O stars three were found to be O3V, three O4V and the remainder O5–O9. Contamination by nebular lines of the stellar HeI lines caused many stars to be classified as O(n)V only, leaving the subclass open (for classification criteria see Melnick 1985, Walborn 1993 and references there). Virtually all stars of spectral types O9 and later were found to be supergiants. Three O3If and three O4If stars were identified. Several stars appeared to be transitions between Of and WN-A stars.

R136 was resolved by Feitzinger et al. (1980) into three components, R136, a, b, and c. They were classified by Melnick as follows: R136a, WN4.5 + OB; R136c,

10.4 The stellar content of the 30 Doradus region

WN7 + OB. R136b was too contaminated by light from the other components for a classification. All three were later shown to be multiple (see below).

The 30 Doradus region is extremely rich in WR stars (see e.g. Breysacher 1981, Melnick 1985, Moffat et al. 1987, Morgan and Good 1985, 1987). Moffat et al. list ~ 15 genuine WR stars within $2'.5$ (36 pc) of the core of 30 Doradus, including four in R136a. 75% are of types WN6 and WN7. Only two are of type WC: Brey83 and 88, both multiple systems with the dominating components classified O(4) + WC5? and WC5 + O4, and WC5 + WN4 and WN4.5, respectively. There are also 7 WR/Of stars in the central region. In the surroundings of the nebula, from a radius of $7'.5$ to $45'$ only 2 of 28 WR stars are of late WN. The fraction is still lower in the remainder of the LMC.

In 1990, spectral types were known for 111 OBA stars in the 30 Doradus ionizing cluster (Walborn 1991). 42 of the stars were classified O3–6. In addition 11 WR stars, outside R136, and 13 late-type spectra had been classified. Most of the evolved B- and A-stars are probably not members of the cluster. The R136 content was, at that time, estimated to be at least 26 stars brighter than $V = 16$, of which two were WR stars. The lack of O5–O8 supergiants in the cluster indicates an age of about 2 Myr.

The stellar content of an area of 50 arcmin^2 in the 30 Doradus central region has been studied by Parker (1993) and Parker and Garmany (1993). UBV (CCD) observations were obtained of 1438 stars and BV observations of 912. Spectroscopy of 54 stars, including 24 O stars, was secured. The catalogue is complete (excluding the R136 cluster) to $V = B = 18$, $U = 17$ mag. The positions of the stars in the HR diagram were determined with the aid of the Q-method and available relations between T_{eff} and $(B-V)_0$ and BC and $\log T_{eff}$. The WR stars had to be omitted because of lack of calibrations. The stars brighter than $V = 18$ cover the range $-10.5 \leq M_{bol} \leq -4.0$ around spectral type B2V.

The IMF of the 30 Doradus association has slopes in the range of $\Gamma \approx -1.5 \pm 0.2$, with the flatter slopes being the more likely. The IMF shows a marked curvature even for massive bins ≥ 12 M$_\odot$, where the photometry should be complete. The most massive bins may be underpopulated for a number of reasons: (i) WR stars are omitted and belong to the most massive bins. Adding them will make a considerable impact since there are 14 WR stars in the catalogue and few stars in the most massive bins. (ii) O stars with temperature determinations from the photometry only, may have been placed in too low mass bins. The effect on the IMF of unresolved binaries would, on the other hand, work in the opposite sense: as the binary fraction increases, the observed IMF becomes flatter.

There are indications that two regions of 30 Doradus have different IMF slopes: In the SE region the IMF is well fit by the slope $\Gamma = -1.7$, whereas in the NW region (including R136), where most of the present star formation activity occurs, the curvature problem dominates and the question of incompleteness in the most massive bins arises. The activities in the 30 Doradus region do probably vary from spot to spot, but the trend is obvious, sequential star formation may occur.

The inner region of 30 Doradus has a significant population of M supergiants intermingled with the well-known WN and OB stars (Hyland et al. 1978, from a 2 μm survey). This suggests that two major bursts of star formation have occurred in the last 10 Myr.

10.4.1 *Multiple stellar systems in the 30 Doradus nebula*

The question about the possible existence of supermassive stars in the Magellanic Clouds was solved when R136 was resolved into numerous stars of normal mass. Instead, a multiplicity problem was introduced. A number of the most luminous stars in the LMC have been investigated with modern techniques and most of them have been shown to be multiple. In ground-based imaging a resolution of $0''.3$ has now been reached (see Heydari-Malayeri *et al.* 1993). With the *HST* $0''.1$ appears to be achieved. Presently available results for objects in the 30 Doradus nebula will be considered here.

Two fields SW of the 30 Doradus nebula, SNR 30 DorB/NGC 2060 and 30 DorC/NGC 2044, have been studied extensively with ground-based telescopes. Both fields contain young SNRs whose progenitors were very massive judging from the present stellar content of the systems (Chu and Kennicutt 1988, Chu *et al.* 1992). In addition to the very massive stars detected in these fields, a number of partially resolved compact multiple systems have also been identified (Lortet and Testor 1984, Testor *et al.* 1988, Schild and Testor 1992, Heydari-Malayeri *et al.* 1993). Walborn *et al.* (1995) have used the *HST* to derive *UBV* images for three of these massive, compact multiple systems. In the rich stellar cluster NGC 2060, where one of the WR stars (Brey73) had been resolved by Testor *et al.* into 12 components, the *HST* images showed the two brightest components to consist of four components each. In the 30 DorC region a similar resolution is achieved. In NGC 2044 'West' the star Heydari-Malayeri No. 5, WN3–4 + OB, is resolved into five components and the Of star, No. 9, into two. In NGC 2044 'East' several of the 'individual' O and B stars are resolved into two components. Of particular interest is the B + K star Schild and Testor No. 62 which is clearly resolved in the *HST* images. Tentative HR diagrams indicate that some of the stars may still be composite: their luminosities and masses are too high for their spectral types. Still higher resolution is thus needed before the luminosities of the individual WR, Of, O and B stars can be established with complete certainty. This is needed for the interpretation of their evolution and for the determination of the upper stellar mass limit and of the IMF.

10.4.2 *The core of the ionizing cluster in 30 Doradus*

The first observations of R136 with the *HST* Faint Object Camera (FOC) were obtained in 1990 (Weigelt *et al.* 1991). R136a (Fig. 10.7) was resolved into a very compact cluster of more than eight stars within a $0''.7$ diameter. Continued *HST* observations have led to further resolution of the objects in R136 into stellar components. High angular resolution observations with a Fine Guidance Sensor on the *HST* (Lattanzi *et al.* 1994) established that the central object in R136a is a triple star, a1, a2 and a third component $\Delta V = 1.1$ mag fainter than a1, $\sim 0''.08$ away. The present mass of a1 was estimated to be between 30 and 80 $M_\odot$; its progenitor must have been a 60–160 $M_\odot$ object.

Observations with the *HST* PC led to the identification of 214 stars within R136, a box with the dimensions $7''.3 \times 8''.7$, and 800 stars in a square with $67''$ sides (Campbell *et al.* 1992). Renewed observations increased the number to 213 stars within $3''.3$ (0.8 pc; distance 50 kpc) of the centre of R136a, and over 800 stars in a $35'' \times 35''$ area (Malumuth and Heap 1994). Reddening corrections were determined with the aid of the Q-method because of the large scatter in the $(U - B)/(B - V)$ diagram.

10.4 The stellar content of the 30 Doradus region

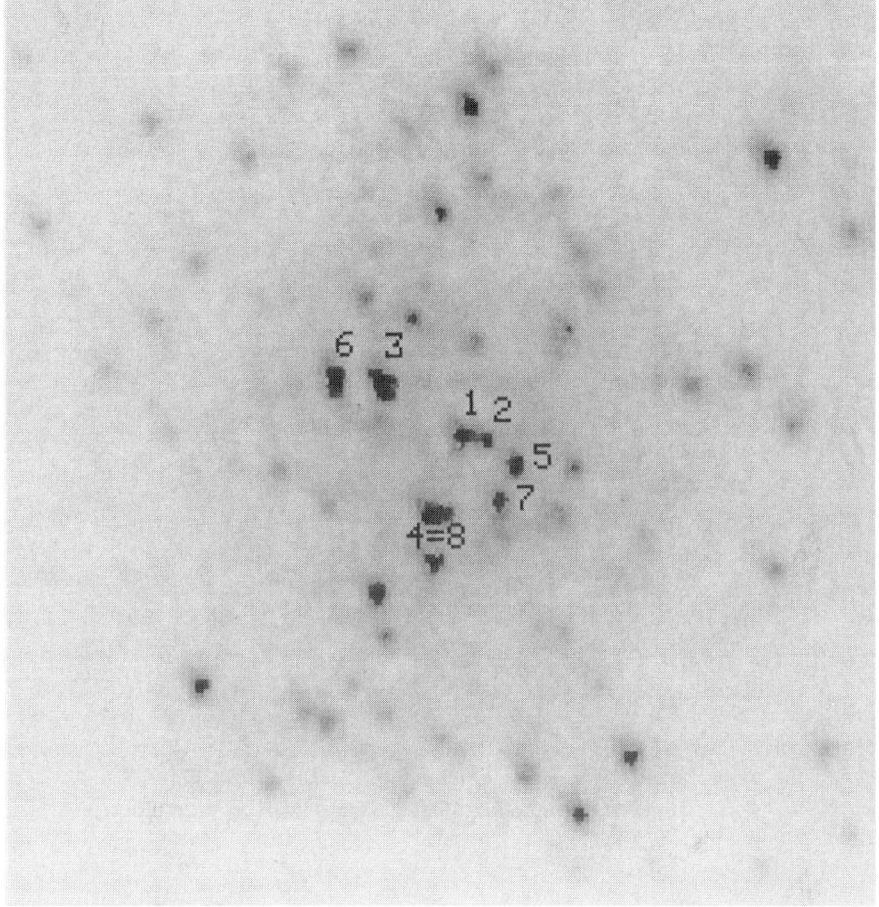

Fig. 10.7. The R136a cluster with the Weigelt *et al.* components identified. The distance between a2 and a5 is 0″.16.

The limits for detection are $B = 17.5$ in R136a and about 2 mag fainter outside the core; the limits for completeness are $B \sim 16$ and 18 mag, respectively.

The LFs of the outer and inner regions have the slopes $\alpha = -0.83$ and -0.66, respectively, where $\log \Phi(L_B) \propto -0.4\alpha M_B$, and M_B is the absolute B magnitude. There is, apparently, a higher number of luminous stars in the inner region than in the rest of the (central part of the) cluster. The corresponding slopes, Γ, of the IMF are $\Gamma = -1.82 \pm 0.41$ and -0.90 ± 0.38 for the outer and inner cluster region, respectively. The slope within R136a is evidently much flatter than outside. The former agrees well with Γs for young associations in the LMC (see Table 5.6). The reason for the flat IMF inside R136a is found in an overabundance of luminous stars in the core. For 45 stars in the central $2'' \times 2''$, $\Gamma = 0.0$ (Heap *et al.* 1992). If these objects are disregarded in deriving the inner LMF, a slope of $\Gamma = -1.61$ is obtained, in agreement with that valid for the outer part.

The surface *mass* density, Σ, profile as a function of radius is well fitted by the surface projection of a King (1966) model with $\log R_t/R_c = 2.06$ and $R_c = 0.24$ pc

(0″.96). A four times smaller value was derived by Campbell *et al.* (1992), using the surface *brightness* profile. The latter is not a scaled version of the former: The central 1″ region contributes 21% of the luminosity but only 10% of the mass (Malumuth and Heap 1994).

The global properties of the 30 Doradus ionizing cluster within a radius of 17″.5 (4 pc) are:
total mass $\geq 1.7 \times 10^4$ M$_\odot$,
total $L_{bol} = 7.8 \times 10^7$ L$_\odot$,
total N$_{ion} = 3.3 \times 10^{51}$ photons s^{-1},
central density $= 9.5 \times 10^3$ M$_\odot$ pc^{-3},
central surface density $= 1.9 \times 10^4$ M$_\odot$ pc^{-2}.

The number of ionizing photons, N_{ion}, is sufficient to account for the Hα emission observed by Kennicutt and Hodge (1986) out to a radius of 200″.

A UV colour–magnitude diagram (1300 Å and 3400 Å) of the R136 cluster was derived by de Marchi *et al.*(1993) from *HST* FOC observations. A total of 165 stars more massive than 10 M$_\odot$ were found inside a 3″ radius corresponding to a space density of ~ 125 stars pc^3. The most probable age of the R136 cluster was found to be 3 ± 1 Myr from the MS turn-off and the spectral properties of the stars: WNL stars exist but no red supergiants. The most massive stars may have initial masses of the order of 50 M$_\odot$.

10.4.3 Individual stars

The star Mk42 ($\equiv$ Brey77), classified O3 by Melnick and O3f/WN by Walborn *et al.*(1992), has been observed with the GHRS on the *HST* and the CASPEC on the ESO 3.6 m telescope by Heap *et al.*(1991). A weak HeI λ 4471 line was seen in the optical spectrum and used, with Hγ and HeII λ 4542, and NLTE-model photospheres to determine the stellar parameters: $T_{eff} = 42\,500 \pm 2000$ K, log $g = 3.5 \pm 0.15$, $N(He)/N(H) = 0.10$–0.15, and $v_{rot} = 200$–250 km s^{-1}. Using $V_0 = 11.40$ and $(m-M)_0 = 18.57$, they derived a stellar radius $R = 29$ R$_\odot$ and a $L_{bol} = 2.3\ (\pm 0.4) \times 10^6$ L$_\odot$. The luminosity was also derived by comparing the reddening corrected UV and visible continua to models with the above T_{eff}. The best fit was obtained for the following colour excesses: $E_{B-V} = 0.06$ for the foreground, 0.04 for the LMC 'halo', 0.10 for the 30 Doradus region, and 0.20 mag for the 'nebular dust' (in principle following the distribution discussed above (Sect. 10.1, and $L_{bol} = 2.5 \times 10^6$ L$_\odot$ derived, in agreement with the value given above. The T_{eff} and L_{bol} suggest, with Maeder's (1990) tracks for stars of low metallicity ($Z = 0.005$), a ZAMS mass for Mk42 of 120 M$_\odot$ and a present mass of 110 M$_\odot$. From the spectroscopic gravity and radius a mass of $M_{grav} = 90\ (+37, -26)$ M$_\odot$ results. Mk42 is obviously one of the most massive stars known.

The GHRS spectrum is dominated by stellar wind features. Lines identified with P Cygni profiles are NV λ1240, OV λ1370, CIV λ1549, HeII λ1640, and NIV λ1718. Heap *et al.* derived $v_\infty = 3000$ km s^{-1} and $\beta = 0.7$–1.0 in the wind velocity law: $v(r) \approx v_\infty\ (1 - R_*/r)^\beta$. The mass-loss rate was estimated in a number of ways and eventually given as 4×10^{-6} M$_\odot$ yr^{-1} with a large uncertainty. Spectral analysis further indicated that Fe and O are reduced by a factor of 4 relative to the Sun, carbon is more strongly depleted and nitrogen is about solar.

Ultraviolet spectra of the two components R136a5 and R136a2 in the very dense R136a cluster have also been obtained with the aid of the GHRS on the *HST* (Heap *et al.* 1994). (There are nearly 50 stars within a region 0.5 pc across.) R136a2 is 0".16 from R136a5 and only 0".08 from R136a1 (see Fig. 10.7) and thus is unavoidedly contaminated. The spectrum obtained was estimated to be two parts R136a2 and one part R136a1. R136a2 is classified WN4-w (w indicating that the emission lines are weak) and R136a5 O3f/WN. The UV spectrum of the latter shows many similarities to that of Mk42. The major differences between the two stars are that R136a5 is one magnitude fainter and has a higher wind velocity, $V_\infty = 3200\text{–}3400$ km s^{-1}. The following properties of R136a5 were derived: $T_{eff} = 42\,500$ K, $R = 16.4$ R$_\odot$, $L_{bol} = 8 \times 10^5$ L$_\odot$ and $M \approx 50$ M$_\odot$, typical for a main-sequence O star. A very high mass-loss rate is indicated by the HeII 6140 Å emission: $\dot{M} = 1.8 \times 10^{-5}$ M$_\odot$ yr^{-1}. This is about an order of magnitude higher than is assumed by current evolutionary models and will drastically alter the evolutionary path of the star.

If the Langer *et al.*(1994) evolutionary sequence for massive stars of galactic composition: O ⇒ H-rich WN ⇒ LBV ⇒ H-poor WN ⇒ H-free WN ⇒ WC ⇒ SN, is applicable to LMC stars, both R136a2 and R136a5 will be on the main sequence with the WN4 star, R136a2, being the initially more massive, and, therefore, now the more evolved star. Neither star is a fully developed WN star, the emission lines are too weak and N IV] 1485 Å missing. They are transition objects in the possible scenario for massive stars.

There appear to be no stars in R136a brighter than $M_V = -7$. This may be due to pulsation-enhanced mass loss: massive O stars evolve directly into H-rich WN stars like R136a2, and the following LBV stage is very brief.

10.5 Comparison of the R136 cluster with galactic clusters

NGC 3603 appears to be the densest concentration of early-type massive stars known in the Galaxy and has often been compared with the 30 Doradus central cluster. It is also the most massive visible HII region in the Galaxy. Walborn (1984) considered its spectacular HD 97950 system as a close-up view of R136. In a volume with diameter 0.25 pc (corresponding to 1" at the distance of the LMC) there are six stars, all of which may be double in turn. The luminosities of the brightest components were estimated as $M_V \sim -7.9$. Recently obtained *HST* PC images of NGC 3603 (Moffat *et al.* 1994) confirm the remarkable similarity between it and 30 Doradus. Certain differences are worth noting. All known WR stars, three, in NGC 3603 are within 0.1 pc of its centre, only 20%, two, of those in 30 Doradus are equally close. The WR stars in NGC 3603 are, on the average, 1 mag brighter than those in R136. It might be the consequence of the older age of NGC 3603; the earliest spectral type is O4; in R136 it is O3. The many bright WR stars in 30 Doradus, outside R136 confirm that R136 is a young central cluster in an association.

Moffat *et al.*(1994) ask why NGC 3603 type clusters are so rare in the Galaxy while R136/30 Doradus is merely the youngest example of the LMC's numerous populous clusters. There are, however, clusters in the Galaxy that are only slightly more evolved than NGC 3603 and not too different in composition. The two dense clusters Sher 1 and Westerlund 2, both containing one WR star, of types WN4 and WN7, respectively, belong to this category. Moffat *et al.*(1991) noted the similarity between the latter cluster and NGC 3603 but also that it is less compact. Other differences are that its WN 7 star is outside the cluster core and its most luminous stars are of type O7 with $M_V(\text{Sp}) \approx -5.2$. This cluster may well be an evolved version of NGC 3603 and R136. It also shows similarities to the 30 Doradus cluster: it is the core of the HII region RCW 49 and contains an X-ray source which may be a SNR or a wind-powered X-ray bubble (Goldwurm *et al.* 1987). It is heavily reddened, with the extinction varying over the cluster area. Its most luminous stars may be double.

differences are that its WN 7 star is outside the cluster core and its most luminous stars are of type O7 with $M_V(\text{Sp}) \approx -5.2$. This cluster may well be an evolved version of NGC 3603 and R136. It also shows similarities to the 30 Doradus cluster: it is the core of the HII region RCW 49 and contains an X-ray source which may be a SNR or a wind-powered X-ray bubble (Goldwurm *et al.* 1987). It is heavily reddened, with the extinction varying over the cluster area. Its most luminous stars may be double.

Ultraviolet spectra of the two components R136a5 and R136a2 in the very dense R136a cluster have also been obtained with the aid of the GHRS on the *HST* (Heap et al. 1994). (There are nearly 50 stars within a region 0.5 pc across.) R136a2 is 0″.16 from R136a5 and only 0″.08 from R136a1 (see Fig. 10.7) and thus is unavoidably contaminated. The spectrum obtained was estimated to be two parts R136a2 and one part R136a1. R136a2 is classified WN4-w (w indicating that the emission lines are weak) and R136a5 O3f/WN. The UV spectrum of the latter shows many similarities to that of Mk42. The major differences between the two stars are that R136a5 is one magnitude fainter and has a higher wind velocity, $V_\infty = 3200$–3400 km s^{-1}. The following properties of R136a5 were derived: $T_{eff} = 42\,500$ K, $R = 16.4$ R$_\odot$, $L_{bol} = 8 \times 10^5$ L$_\odot$ and $M \approx 50$ M$_\odot$, typical for a main-sequence O star. A very high mass-loss rate is indicated by the HeII 6140 Å emission: $\dot{M} = 1.8 \times 10^{-5}$ M$_\odot$ yr^{-1}. This is about an order of magnitude higher than is assumed by current evolutionary models and will drastically alter the evolutionary path of the star.

If the Langer et al. (1994) evolutionary sequence for massive stars of galactic composition: O ⇒ H-rich WN ⇒ LBV ⇒ H-poor WN ⇒ H-free WN ⇒ WC ⇒ SN, is applicable to LMC stars, both R136a2 and R136a5 will be on the main sequence with the WN4 star, R136a2, being the initially more massive, and, therefore, now the more evolved star. Neither star is a fully developed WN star, the emission lines are too weak and N IV] 1485 Å missing. They are transition objects in the possible scenario for massive stars.

There appear to be no stars in R136a brighter than $M_V = -7$. This may be due to pulsation-enhanced mass loss: massive O stars evolve directly into H-rich WN stars like R136a2, and the following LBV stage is very brief.

10.5 Comparison of the R136 cluster with galactic clusters

NGC 3603 appears to be the densest concentration of early-type massive stars known in the Galaxy and has often been compared with the 30 Doradus central cluster. It is also the most massive visible HII region in the Galaxy. Walborn (1984) considered its spectacular HD 97950 system as a close-up view of R136. In a volume with diameter 0.25 pc (corresponding to 1″ at the distance of the LMC) there are six stars, all of which may be double in turn. The luminosities of the brightest components were estimated as $M_V \sim -7.9$. Recently obtained *HST* PC images of NGC 3603 (Moffat et al. 1994) confirm the remarkable similarity between it and 30 Doradus. Certain differences are worth noting. All known WR stars, three, in NGC 3603 are within 0.1 pc of its centre, only 20%, two, of those in 30 Doradus are equally close. The WR stars in NGC 3603 are, on the average, 1 mag brighter than those in R136. It might be the consequence of the older age of NGC 3603; the earliest spectral type is O4; in R136 it is O3. The many bright WR stars in 30 Doradus, outside R136 confirm that R136 is a young central cluster in an association.

Moffat et al. (1994) ask why NGC 3603 type clusters are so rare in the Galaxy while R136/30 Doradus is merely the youngest example of the LMC's numerous populous clusters. There are, however, clusters in the Galaxy that are only slightly more evolved than NGC 3603 and not too different in composition. The two dense clusters Sher 1 and Westerlund 2, both containing one WR star, of types WN4 and WN7, respectively, belong to this category. Moffat et al. (1991) noted the similarity between the latter cluster and NGC 3603 but also that it is less compact. Other

11

Chemical abundances

Knowledge about the chemical abundances of objects of various ages in the Magellanic Clouds is essential for our understanding of their evolution in the past. The abundances of their lighter elements have been derived from spectra of HII regions, and those of the heavier elements from spectra of supergiants. An extensive study was carried out by Pagel et al. (1978). A summary of the composition of the HII regions was presented by Dufour (1984). Little has changed since then. The summary is still often used as a reference for comparison with stellar abundances, and it is therefore reproduced in Table 11.1.

A problem in interpreting the many results derived for the metallicity of cluster and field stars is the model dependence. The early works pointed towards an underabundance in the cluster stars relative to the field stars, but, as will be seen below, there is a clear tendency towards more and more agreement in metallicity between cluster and field stars. Differences have also frequently been found between the chemical abundances of the ISM and the stars in the Clouds, though warnings against overinterpreting the observational results have been given (Pagel 1993). The uncertainties in the derived data, for the stars as well as for the ISM, are still considerable.

For the study of the stellar abundances in the Magellanic Clouds only the most luminous and most extreme supergiants (A-type, $M_V \leq -9$) were available for high-dispersion studies up to about 1975. The existing models were insufficient for fine analysis work because of the complicated physics of the atmospheres of these stars, which, anyhow, cause severe difficulties for fine analysis work: their atmospheres are very extended (no plane parallel solutions); NLTE effects are important in spite of the low temperature ($T_{eff} \sim 9000$ K) because of a low log g (~ 0.7–0.9); and large-scale hydrodynamic motions exist so that hydrostatic models become inadequate. Early analyses of A-type supergiants must therefore be regarded with caution.

The new large telescopes and the modern spectroscopic equipment and detectors have made it possible to use higher dispersion and to study MS stars in addition to or, perhaps better, instead of the supergiants. Presently available equipment permits high-resolution spectroscopy to $V \sim 15$ with reasonable exposures, so that B stars that are only slightly evolved can be reached in both Clouds.

11.1 Element abundances in the interstellar medium

The abundances, [M/H], of the Cloud HII regions relative to the solar neighbourhood are given in Table 11.1 according to Dufour (1984); and in Table 11.2 they are given

Table 11.1. *Abundances in Magellanic Cloud HII regions relative to the Sun*

	[He]	[C]	[N]	[O]	[Ne]	[S]	[Ar]
LMC	−0.10	−0.75	−0.99	−0.44	−0.41	−0.38	−0.37
SMC	−0.13	−1.49	−1.50	−0.85	−0.83	−0.74	−0.79

Table 11.2. *Abundances in Magellanic Cloud HII regions relative to Local HII regions*

	[M$_1$/H]	[M$_2$/H]	[He/H]	Ref.
LMC/Local	−0.26	−0.58	−0.07	Feast (1989b)
	−0.31	−0.36	−0.06	Russell and Dopita (1992)
SMC/Local	−0.61	−1.20	−0.10	Feast (1989b)
	−0.57	−0.77	−0.09	Russell and Dopita (1992)

M$_1$ refers to the elements O, Ne, S, Cl, and Ar which have about the same values of [M/H]; M$_2$ refers to C and N with a distinctly greater under- abundance.

according to Feast (1989b). The latter table also gives mean values from Russell and Dopita (1992). Details will be found below.

24 components of interstellar CaII, 13 of NaI, 3 of KI, 4 of CaI and 3 of LiI ($\lambda 6707$) were identified by Vidal-Madjar et al. (1987) in high-resolution spectra of SN 1987A; it was the first detection of interstellar Li in an external galaxy. Of the components, those with velocities $235 \leq V_{hel} \leq 280$ km s^{-1} and $1 \leq W($NaI D$_2)/W($CaII K$) \leq 2$* were considered to have formed in LMC gas. The 7 CaII components in this range represent the resolved components of features seen toward stars in and around the 30 Doradus nebula. SN 1987A must then be located in or behind the LMC main plane, and at least at the distance of 30 Doradus. The unsaturated components of CaII and NaI give $N($CaII$) \sim 2 \times 10^{11}$ cm^{-2} and $N($NaI$) \sim 1 \times 10^{11}$ cm^{-2}. A comparison with the HI 21 cm column densities at the two velociites 247 and 268 km s^{-1}, which match some of the CaII and NaI components, gives $N($CaII$)/N($HI$) \sim 2.2 \times 10^{-10}$ and $N($NaI$)/N($HI$) \sim 1 \times 10^{-10}$. This is in agreement with an overall underabundance of metals in the LMC as compared with the Galaxy. The only KI component in the LMC velocity range gives, on the other hand, $N($KI$) \simeq 9 \times 10^{10}$ cm^{-2}, which is larger than the galactic value.

The lithium column value is found to be N(Li) $\sim 6.7 \times 10^7$ cm^{-2}. Comparisons with $N($KI$)$ and $N($HI$)$ indicate an underabundance of Li in the LMC with a factor of ~ 2 relative to the Galaxy.

11.2 Element abundances in clusters

Many young MC clusters are embedded in rich unbound stellar coronae. Such halos may be the remnants of the associations in which these clusters were born (van den Bergh 1991), but, as most young clusters are members of superassociations, the apparent coronae may in reality only be a part of the superassociation field (see

---
* W = equivalent width in mÅ.

11.2 Element abundances in clusters

Table 11.3. *[Fe/H] values for LMC and SMC clusters*

Cluster	LMC Log Age yr	[Fe/H]	Cluster	SMC Log Age yr	[Fe/H]
NGC 1984	6.84	−0.15	NGC 330	6.84	−1.4
NGC 2209	8.84	−1.0	NGC 152	8.90	−0.8
NGC 1978	9.30	−0.4	NGC 411	9.26	−0.7
NGC 2173	9.48	−0.6	NGC 416	9.40	−1.6:
NGC 2155	9.78	−1.5	L 113	9.70	−1.4
ESO 121-SC03	9.95	−0.9	K 3	9.90	−1.2
NGC 2210	≥10.08	−2.0	L 1	10.00	−1.2
NGC 2257	≥10.08	−1.8	NGC 121	10.08	−1.4

Slightly different values may be found in Chap. 4 for some of the clusters. The cluster showing the greatest variations in [Fe/H] is NGC 330. Note e.g. the difference between [Fe/H] for NGC 330 given here and in Table 11.5.

Sect. 6.1.1). There is thus little reason to expect anything but minor differences in abundances between cluster and field stars. The last generation clusters are sometimes found to be younger, sometimes older than the surrounding field. As a rule, field supergiants are older than the clusters and associations in the same superassociation. They may thus be less metal-rich than the associations unless they have formed out of the same gas but on different time scales. It is of interest to establish how small a difference in age is needed to produce a noticeable difference in metallicity.

Metallicities for Cloud clusters of all ages may be found in the tables in Chap. 4. A comparison of [Fe/H] data for a number of LMC and SMC clusters of similar age is presented in Table 11.3. Various techniques have been used to determine the metallicities: isochrone fittings, photometry or spectroscopy of individual stars. Details may be found in the compilations by Feast (1989), Seggewiss and Richtler (1989), and Westerlund (1990).

In the LMC the oldest clusters are the most metal deficient, whereas in the SMC an appreciable scatter is seen. This may be due to problems in interpreting the observations or in the isochrone fittings. It may also be real and due to the effects of inhomogeneities or gradients in the ISM from which the clusters have formed. The clusters scatter over much of the galaxy and are kiloparsecs apart. If the clusters in the two Clouds are compared age by age, the differences are mostly negligible. As the uncertainties in the determinations are still great, definite conclusions should not be drawn.

Among individual clusters that have attracted much interest are NGC 330 in the SMC and NGC 1818 in the LMC.

Numerous attempts to determine the metallicity of NGC 330 have been reported. CCD Strömgren photometry (in v, b, y) has been carried out for cluster stars and nearby field stars by Grebel and Richtler (1992). They found that the supergiant field stars are systematically more metal-rich than the probable cluster members. The mean metallicity, [M/H], of the former is −0.74 dex; the cluster stars have [M/H] = −1.26 dex (assumed reddening $E_{B-V} = 0.03$). It is difficult to understand why the young cluster should be less metal-rich than the field stars of about the same age. With the peculiar structure ascribed to the SMC, spatial variations may of course exist. There are, however, also clusters in the LMC where similar effects have been observed. The metallicity difference between stars in NGC 1818 and in its surrounding field is even larger.

Hilker et al.(1995) have carried out CCD Strömgren photometry of NGC 1866 and NGC 330 and their surroundings in order to determine the metallicities of the clusters and some field stars. Because of the vague character of a metallicity determined by this method, the emphasis is put on the difference between the two clusters rather than on absolute values. NGC 1866 is thus more metal-rich than NGC 330 by 0.50 ± 0.06 dex. If the (spectroscopic) metallicity of NGC 330 is −1.1 dex, then NGC 1866 has −0.6 dex.

Reitermann et al.(1990) have analysed two B stars with $M_V \sim -5$, one, D12, in the field of the NGC 1818, the other,330/3, in the field of NGC 330.* Both stars are relatively slow rotators. For both stars cluster membership is in doubt. D12 is located at about five times the radius of the cluster core from the centre of NGC 1818 and may well be a field star. NGC 330/3 has been suggested to be a non-member (Wolf and Reitermann 1989) because of the measured underabundance of ~ -0.7 dex. This is in good agreement with that of late-type field stars in the SMC, but differs from the underabundance in the cluster of more than -1.0 dex derived for a late-type supergiant (Spite 1989).

The Reitermann et al. analyses differ from the previous ones in several important aspects: the temperatures are derived on continuum fits (including IUE continua); the reddening of NGC 330/3 is taken as $E_{B-V} = 0.12$, instead of 0.04; and the galactic star HR 3663, B3III, is used for comparison instead of ι Her. (Within the error bars there is agreement between the abundances of HR 3663 and the Sun.) The abundances given in 1990 are presented in Table 11.4, relative to the Sun, via HR 3663, for a microturbulent velocity $\xi = 10$ km s^{-1}. Also shown are the abundances relative to the Sun,via HR 2618 (a B-type giant), determined by Jüttner et al.(1993). The model atmosphere parameters of the two MC stars have changed slightly: For NGC 330/3 T_{eff} from 18 700 K to 18 800 K and log g from 3.00 to 2.91; for NGC 1818/D12 the corresponding changes are 17 900 K to 19 600 K and 2.85 to 3.13. Also the reddening used for NGC 330/3 is lower, $E_{B-V} = 0.03$. These changes, if significant for the abundance calculations, should affect all elements. As, however, only a few elements, C, Mg, Al, and Fe, suffer pronounced changes, the change of comparison star may have the most important effect.

The differences between the two investigations may be considered acceptable with regard to the accuracy achieved. Only in the case of carbon is there a significant discrepancy.

The following conclusions may be drawn from Table 11.4 when the 1992 results are given highest weight.

(i) The He abundance is nearly solar in both stars and agrees with that of the HII regions in the Clouds.

(ii) Carbon is depleted relative to the solar composition. This is in accord with the analyses of F supergiants (see Sect. 11.4). It is compatible with the upper limits

---

* The stellar numbers are from Robertson (1974) for the LMC star and from Arp (1959b) for the SMC star.

11.2 Element abundances in clusters

Table 11.4. *Differential element abundances in NGC1818/D12 and NGC 330/3 compared with HII region abundances*

NGC Element	1818/D12 (1990)	1818/D12 (1992)	330/3 (1990) $\Delta \log \epsilon$	330/3 (1992)	HII LMC	HII SMC
He	± 0.0	− 0.13	− 0.2	− 0.12	− 0.10	− 0.13
C	− 1.1	− 0.62	− 1.6	− 0.82	− 0.75	− 1.49
N	− 0.2	− 0.46	− 0.8	− 1.08	− 0.99	− 1.50
O	− 0.7	− 0.76	− 0.8	− 1.11	− 0.44	− 0.85
Mg	− 1.1	− 0.77	− 1.2	− 0.90		
Al	− 1.1	− 1.02	< − 1.1	− 0.91		
Si	− 1.3	− 0.97	− 1.5	− 1.00		
S	− 0.5	− 0.29	< − 0.9	− 0.41	− 0.38	− 0.74
Fe	− 0.8	− 0.79	− 0.9:	− 0.79		

$\Delta \log \epsilon = \log \epsilon(MC) - \log \epsilon(Sun)$; $\xi = 10$ km s^{-1}. The accuracy (1990) in the element abundances is estimated to ± 0.2 for the LMC star and ± 0.3 for the SMC star.

found for two evolved B supergiants in the LMC (Kudritzki *et al.* 1987). The carbon abundance in the LMC star agrees with that in the HII regions (contrary to what was concluded from the 1990 data), whereas it is probably overabundant in the SMC star relative to the SMC HII regions.

(iii) Nitrogen is in excess by 0.5 in the LMC star relative to the mean of the heavier elements, whereas in the SMC star the abundance of N and of the heavier elements is about the same. In both stars the N abundance exceeds the abundance in the HII regions by ∼ 0.5. This may be interpreted as evidence for mixing of products from the CNO burning to the surface, though the stars are not too evolved, or it may be due to a different evolution of the metal-poor Cloud stars. (The Clouds have a different global evolution as compared with our Galaxy.).

(iv) Oxygen is about as much depleted as the heavier elements. Compared to the HII regions, O is slightly depleted in both the LMC star and the SMC star. The F supergiants in the SMC behave in nearly the same way.

(v) In the LMC star the elements, Mg, Al, Si, Fe are considerably less abundant ($\Delta \log \epsilon = -0.9$) than in the F supergiants in the field with $\Delta \log \epsilon = -0.3$ (see Sect. 11.4). The red supergiant B12 in NGC 1818 has the iron peak (Ca to Ni) elements deficient by -0.8 to -1.0 (Richtler *et al.* 1989), whereas Al is deficient by -0.6 and Si only by -0.2. The B star, B30, in the blue globular cluster NGC 2004 yields a similar deficiency ($\Delta \log \epsilon = -0.6$) for the heavier elements (Jüttner *et al.* 1993). On the other hand the two B supergiants studied by Kudritzki *et al.* (1987) have Si, Mg and Al abundances close to those of the present BIII stars.

In the SMC star the mean deficiency of Mg, Al, Si and Fe, $\Delta \log \epsilon = -0.9$, is larger by 0.25 dex than the abundance of the heavier elements in the F supergiants in the field, $\Delta \log \epsilon = -0.65$ (see Sect. 11.4). For another member of the NGC 330 cluster, the cool supergiant 330/A7, $\Delta \log \epsilon = -1.3$ for the iron peak elements and -1.2 for oxygen (Spite *et al.* 1986), which is relatively close to the result presented for NGC 330/3.

(vi) Of the listed heavier elements, only S is also observed in the HII regions. Its abundance in the ISM agrees with that in the LMC star and probably also with that in the SMC star.

The Jüttner (1993) and Jüttner et al. (1993) investigations give differential abundances for a total of nine BIII stars, six LMC stars (four in clusters, two in the field) and three SMC stars (2, 1), relative to HR 2618. A relatively safe conclusion of their analysis appears to be that there is no obvious difference in abundances between the cluster stars and the field stars *in the LMC*. The two cluster stars *in the SMC*, on the other hand, display a mean underabundance of 0.2 dex relative to the field stars. Whether this is significant will be revealed when the spectra of more MS stars have been analysed. It may then also be obvious whether spatial inhomogeneities exist in the chemical composition of the Clouds.

Differences between the metal content of the youngest clusters and the youngest field stars have, however, been found in both Clouds by Spite et al. (1991, 1993). The metal abundance in NGC 330 is, according to them, about -1 dex; in the SMC field population it is about -0.6 dex. The difference is even stronger in the LMC: -0.9 dex and -0.6 dex for NGC 1818 and NGC 2004, respectively, but -0.3 dex in the young LMC field population. This is in contrast with the results just quoted for the LMC.

A medium-resolution spectroscopic investigation of field stars in the SMC and the LMC (Thévenin and Jasniewicz 1992) confirmed the Spite et al. results for the field population: the mean abundances of [Fe/H] were found to be -0.46 dex and -0.19 dex, respectively. The stars studied are luminous, in the SMC ($m-M = 18.9$) between $-8.8 < M_V < -5.7$, in the LMC (18.5) $-6.2 < M_V < -5.4$. The metal abundances of young Magellanic Cloud clusters should thus be lower than the average field abundances. The differences may, however, be considered negligible (Pagel 1993).

Recent investigations by Jasniewicz and Thévenin (1994, J&T) tend to support the similarity in composition between field and cluster stars in the SMC also. Medium-resolution spectra of stars in NGC 330 in the SMC and in eight LMC clusters were obtained and analysed with the aid of synthetic spectra. Abundances were derived for NGC 330 and five of the LMC clusters; some of the data are given in Table 11.5. The [Fe/H] value for NGC 330 is compatible with the value for field SMC stars, -0.46 dex ± 0.11. NGC 1818 appears marginally underabundant, the field LMC stars have [Fe/H] $= -0.19$ dex and thus are not as underabundant as was found by Richtler et al. (1989). The composition of NGC 1850 agrees reasonably well with that of NGC 1818 and the field stars. The result for NGC 2004 agrees within the error bars with those derived by Jüttner et al. (1989) and Jüttner (1993). The carbon is significantly depleted. The abundances derived for NGC 2100 agree with those for NGC 1818.

The chemical abundances of the LMC clusters show a great dispersion; the same result has been found for the LMC field supergiants (Luck and Lambert 1992) (see below).

The agreement between the data in columns 6 and 8 may be considered acceptable in the case of NGC 2004 and NGC 2100 considering the precision of the analyses, but the discrepancy is rather large for NGC 330 and NGC 1818. The analysis by Meliani et al. (1994) of five red supergiants in NGC 1818 with the aid of medium-resolution spectroscopy tends to support the Jüttner et al. data. The mean metallicity for the five stars is [Fe/H] ≈ -0.9. Carbon abundances have been obtained by computing

Table 11.5. *Abundances of MC clusters. Given are log [M/H] with respect to solar values (J&T 1994)*

cluster	Age Myr	C/H	Mg/H	Ca/H	Fe/H	No. of stars	Fe/H 1992
NGC 330		−0.60	−0.48	−0.54	−0.55	10	−0.80
NGC 1818	14	−0.41	−0.36	−0.43	−0.37	5	−0.79
NGC 1850	38	−0.40	−0.10	−0.15	−0.12	2	
NGC 1854	34	−0.70	−0.20	−0.45	−0.50	1	
NGC 2004	8	−1.00	−0.53	−0.63	−0.56	3	−0.71
NGC 2100	10	−0.44	−0.35	−0.44	−0.32	4	−0.37

The precision of the J&T abundances is estimated to 0.4 dex in C, 0.3 dex in Mg and Ca, and 0.2 dex in Fe. The Fe/H values in column 8 are from Jüttner *et al.* (1993).

synthetic spectra in the G-band region. The mean carbon abundance is $\epsilon(C) = 7.9$ for four stars, and a lower value of $\epsilon(C)^* = 7.2$ for star B26, for which a convective mixing, or exchange of material with a companion, may have occurred.

A K supergiant ($V = 14.4$) in the LMC association NGC 1948, $\sim$ 7–25 Myr old, on the periphery of SGCS LMC 4, has been shown to have a metallicity [Fe/H] $\approx$ $= -0.4$ dex (Spite *et al.* 1993). Its C and N abundances and the ratio $^{12}C/^{13}C = 9$ $\pm$ 5 indicate that the star has undergone an extensive internal mixing. The derived abundances relative to iron are [C/Fe] $= -0.2$, [N/Fe] $= +0.4$ and [O/Fe] $= -0.25$. As expected, no Li is visible in the stellar spectrum.

If the anomalies in the C and N abundances are due to mixing of CN-cycled products, the sum of these abundances ought to be equal to the sum of the abundances of these elements in the material from which the star has formed. Here log N_{C+N} = 8.31. In the LMC HII regions log $N_{C+N} = 7.94$ (Dufour 1984). The excess in this K star is the same as for the Luck and Lambert stars.

Spite *et al.* find, further, that the iron peak elements in the K star are deficient relative to the Sun by a factor of about 2, as is the C + N abundance. The rare earths heavier than Zr are enhanced relative to iron by a factor of $\sim$ 2.5. If the star is representative of the association, the association has the same abundance characteristics as the young LMC field stars.

11.3 Element abundances in hot field stars and Cepheids

From an extensive programme of NLTE analyses of OB supergiants in the LMC Kudritzki (1988) reported for two LMC stars that the higher metals are markedly underabundant, Mg by a factor of $\sim$ 10, whereas He and N are strongly enriched (even in comparison with LMC HII regions). The two supergiants are very evolved, exhibiting CNO-cycled matter at their surfaces.

The difficulties encountered in dealing with supergiant stars increases the importance of studying unevolved young stars, such as main-sequence OB stars, whose chemical composition can be expected to reflect that of their parent interstellar medium. One drawback is that the B stars are frequently fast rotators. If $v >$ 100 km s^{-1} the star is not suitable for fine analysis.

* $\epsilon(C) = \log N(C) = 12 + \log N(C)/N(H)$.

Table 11.6. *Abundances in Cloud supergiants*

Element	SMC [M/H]	σ	N	LMC [M/H]	σ	N
C	−1.02	0.28	11	−0.02	0.26	19
N	−0.27	0.35	7	+0.37	0.47	7
O	−0.70	0.19	8	−0.23	0.30	9
Na I	−0.42	0.37	4	−0.14	0.07	2
Mg I	−0.45	0.12	7	−0.22	0.36	7
S I	−0.41	0.15	6	+0.06	0.05	3
Fe I	−0.53	0.12	7	−0.34	0.35	6
Fe II	−0.49	0.06	5	−0.10	0.41	2

Data are from Luck and Lambert (1992).

The main-sequence SMC star AV304, B0.5V, $V = 14.98$, is exceptionally sharp-lined and therefore well suited for a detailed analysis. Dufton *et al.*(1990) have used model atmosphere techniques to derive its atmospheric parameters and chemical composition. They find a general heavy element depletion (relative to τ Sco) of ≈ -0.6 to -0.8 dex, with nitrogen being more severely depleted (< -1.1 dex) and oxygen possibly less depleted (-0.45 dex). Helium has an approximately normal abundance.

Most of the chemical-abundance estimates for Cepheids have been made from overall blanketing effects and measure mainly [Fe/H]. The mean abundances of LMC and SMC Cepheids are -0.09 and -0.65 dex, respectively (Harris 1983).

Luck and Lambert (1992) have derived element abundances for 14 Cepheids and non-variable supergiants (seven in each Cloud) along with four galactic comparison objects. Their CNO data suggest a disparity between LMC and SMC supergiants. Both Clouds show a range of CNO abundances larger than that of galactic supergiants.

(i) They find that only HV 5497, in the LMC, shows any detectable Li ($\log\epsilon(\text{Li}) = 2.5$). This star is unique among the analysed objects in that it is metal-rich, [Fe/H] = +0.2, which may explain its Li-richness.

(ii) The [Fe/H] ratios are consistent with the analysis errors dominating the abundance dispersion in the SMC, while within the LMC a real dispersion in [Fe/H] is possible.

(iii) The carbon abundances of the stars in the LMC and especially the SMC are higher than those of the HII regions – new analyses of the HII regions are needed.

(iv) The oxygen abundances for the SMC and the LMC behave differently. The SMC is consistent with a constant [O/Fe] ratio ($= -0.18 \pm 0.19$) while the LMC ([O/Fe] $= -0.01 \pm 0.23$) is not. The [O/Fe] ratios of the SMC stars are lower than those found in galactic stars of equal metallicity. The [O/Fe] ratios of both Clouds are compatible with the [O/Fe] ratios found in galactic supergiants.

(v) The α elements of the LMC and, to a lesser extent, the SMC, are overabundant relative to galactic supergiants. However, the elements Na and Al are normal in the LMC relative to galactic supergiants but deficient in the SMC. The galactic supergiants show enhancements of both elements relative to solar-type dwarfs.

(vi) The light s- and r-process elements (Sr, Y, Zr) show galactic ratios with respect to Fe in both Clouds, but the heavy s- and r-process elements (Ba–Sm) are enhanced by $\sim +0.4$ dex. The galactic supergiants show enhancements of these elements over solar-type dwarf stars.

If star formation occurred in a single short burst some Gyr ago and was followed by a long period of quiescence, then the ISM will continue to be enriched in Fe produced in SN of type Ia. Oxygen will not be replenished as it is produced in SN of types Ibc and II (with massive short-lived progenitors). Thus, the O/Fe ratio in the ISM will decrease with time (Gilmore and Wyse 1991). Marginal support is found in the work by Luck and Lambert.

11.4 Element abundances in cool field stars

A high-resolution spectroscopy of eight F-type supergiants in each Cloud has been carried out by Russell and Bessell (1989). The reddening and the photometric temperatures and gravities were determined with the aid of Strömgren uvby and Cousins BVRI photometry. The stars have luminosities between -5.5 M_V and -8.6; most of them are less luminous than previously investigated supergiants in the Clouds. Abundances were derived for over 20 elements in both Clouds. Large microturbulent velocities in the F giants makes the Strömgren m_1 index less useful as a metallicity measure.

Spite (1989), Spite et al. (1989a,b), and Spite and Spite (1990) have analysed rather high-resolution spectra of F and G supergiants in the Magellanic Clouds. The following conclusions are in common to them and to Russell and Bessell:

(i) The metallicity in the SMC field is [Fe/H] $= -0.65 \pm 0.2$ dex and that in the LMC field [Fe/H] $= -0.30 \pm 0.2$ dex.

(ii) The pattern of the elemental abundances is essentially the same in the Magellanic Cloud field stars as in Canopus, except for the rare earth elements, which are enhanced in the former.

(iii). The large underabundance of carbon, suggested from analyses of the HII regions in the Magellanic Clouds, is not observed in the cool stars. For other elements, including oxygen, good agreement is found with the HII regions.

Russell and Bessell also found that the microturbulent velocities of the Magellanic Cloud stars were very similar to those of galactic F supergiants. The physical parameters for the two most luminous stars in their program, G0Ia supergiants in the SMC, were difficult to determine. Much higher microturbulent velocities were now derived for these stars, thus explaining the previous rather high estimates of their metallicities.

The suggestion of a scatter of abundances in the Magellanic Cloud stars of about -0.2 dex (in addition to the observational uncertainties) may be interpreted as a moderately incomplete mixing of the ISM in each Cloud (Russell and Bessell 1989). Spite et al. suggested instead that the apparent inhomogeneities are due to a low metallicity in the halo and a progressively increasing metallicity in the disk. Metal-poor clusters, such as NGC 1818 and NGC 330, discussed above, may then have formed outside the disks of the Clouds, with various delays in time, from unprocessed material that escaped mixing. A similar suggestion has been made by Feast (1989b): NGC 330 may have formed far from the other young objects in the SMC, maybe in the Magellanic Stream. However, as discussed above, NGC 330 may have the same composition as the surrounding field. This tends to support the idea of an incomplete mixing of the ISM. Both Clouds are rather fragmented (as discussed in Chap. 3), so that it may not be surprising to find some relatively young clusters which have formed in less processed material than others.

High-resolution spectroscopy of nine F supergiants in the field of the LMC has been carried out by Hill *et al.*(1995). Their LTE-analysis showed that all the stars are metal-deficient, -0.34 dex $\leq$ [Fe/H] ≤ -0.14, confirming the Russell and Bessell and Spite *et al.* results. The iron abundance is surprisingly uniform. For all stars [C/Fe] $\leq$ 0 but exceeds the value observed in LMC HII regions by $+0.2$ dex. The [O/Fe] ratio is very similar to the one found for the galactic supergiant Canopus. Sodium does not seem to be enhanced. Among α elements Si, S, Ca and Ti are enhanced but Mg is slightly underabundant (with respect to Fe) when compared with the solar value. Heavy s- and r-process elements are overabundant by $+0.3$ dex on average. This overabundance could be somewhat smaller for Eu, a pure r-process element. Lithium could not be detected in any of the stars, in general agreement with the galactic warm supergiants.

Colours, magnitudes and radial velocities for a proper-motion selected sample of SMC halo giants in a field near NGC 121 have been determined by Suntzeff *et al.*(1986). For ten stars with good spectrophotometric indices they found $\langle[\text{Fe/H}]\rangle = -1.63 \pm 0.31$ (s.d.). From 30 of the selected stars they derived $\langle[\text{Fe/H}]\rangle = -1.56 \pm 0.32$ (s.d.) by fitting their colours and magnitudes to the Sandage–Roques composite cluster HR diagram, using $m - M = 18.85$. They concluded that the spread in metallicity is real and similar in shape, width and mean metallicity to that of globular clusters and RR Lyraes of the galactic halo, despite the fact that the mass of these systems is very different (see their Fig. 9). The metallicity found in the NGC 121 field is appreciably lower than that found for F supergiants, [Fe/H] $= -0.65 \pm 0.2$ dex (see above),

11.4.1 Abundances in RR Lyrae stars

In the SMC the field RR Lyraes appear more metal-poor, [Fe/H] $= -2.0 \pm 0.2$, than the oldest clusters. NGC 121 has [Fe/H]$= -1.4$ and an estimated age of 12 ± 2 Gyr (Walker 1991).

In the LMC the field RR Lyraes are more metal-rich, [Fe/H]$= -1.8$, than some of the oldest clusters (NGC 1786, 1841), [Fe/H]$= -2.2$. The ages of the true globulars may be as high as 16–18 Gyr (e.g. NGC 1841 and the Reticulum cluster $\equiv$ GLC 0435-59), the same age as M 92 in the Galaxy.

The metal abundance of GLC 0435-59 was determined (Walker 1992a) from:
(i) the positions of the RRab Lyrae stars in the period–amplitude diagram: [Fe/H] $= -1.71 \pm 0.04$;
(ii) by comparing the mean period of the RRab Lyrae stars, P_{ab}, with clusters of known abundances: [Fe/H] $= -1.7$;
(iii) by comparing the giant branch colour with those of galactic clusters: [Fe/H] $= -1.69 \pm 0.08$.

The RR Lyraes in NGC 2257 have the same metallicity as the field RR Lyraes but are 0.2 mag brighter; the cluster is then ~ 4 kpc in front of the field (Nemec *et al.* 1985). The youngest cluster in this age group may be H 11, > 10 Gyr, but without RR Lyraes (see also Chaps. 4 and 7).

Table 11.7. *Metallicities from RR Lyrae stars in clusters*

Cluster	Age gyr	[Fe/H]	Ref.
NGC 121	12 ± 2	−1.4	Walker 1991
NGC 2257		−1.8 ± 0.3	Nemec et al. 1985
		−1.8 ± 0.1	Walker 1989
NGC 1841	16 − 18	−2.2 ± 0.2	Walker 1990
GLC 0435-59	16 − 18	−1.7 ± 0.1	Walker 1992a
NGC 2210		−1.9	Walker 1985

The ages of the LMC globulars are frequently given as ≥ 12 Gyr.
For NGC 121 ∼ 10 Gyr is often used.

11.5 Element abundances in planetary nebulae

All PNe in the Magellanic Clouds are overabundant in nitrogen as compared with HII regions (Dennefeld 1989). This is believed to be the result of CN processing in the parent star; the products are brought to its surface during the first dredge-up. The nitrogen abundance is even larger in the Peimbert type I nebulae because of the intensive hot-bottom burning of CNO (Renzini and Voli 1981). The N abundance in the PNe corresponds roughly to the C + N + O abundance in the HII regions. In the type I PNe the He overabundance is also significant.

The carbon abundance should be notably down in type I PNe as a consequence of the efficient processing of CNO at lower metallicities, and is seen to be so. Similarly, N/O has its highest value in the SMC type I PNe. Carbon is instead overabundant in the non-type I PNe, most in the SMC, as compared with HII regions. This is the result of the third dredge-up bringing the results of the He burning to the surface. This process is most efficient in low metallicity surroundings.

The most carbon-poor type I PNe in the SMC is SMP28. It was found by Meatheringham *et al.* (1990) to have a carbon abundance < 1/450th solar; its electron temperature is very high (∼ 25 000 K). The high temperature is ascribed to the lack of efficient cooling by carbon. Nitrogen is overabundant relative to oxygen by a factor of 1.57 in relation to the mean abundance in SMC PNe. It is believed that SMP28 has evolved from a massive progenitor, with a main-sequence $M_{init} > 5$ M$_\odot$, possibly > 7 M$_\odot$, which has undergone both a second and a third nuclear dredge-up, as well as very efficient hot-bottom burning. These processes raised the surface abundances of He and N, depleted O and reduced C drastically. SMP28 is well represented by a model in which the central star has $T^\star = 1.8 \times 10^5$ K and radius $R^\star = 0.09$ R$_\odot$. The nebular mass is estimated to be 0.71 M$_\odot$ and the central star has $M^\star \sim 0.65$–0.71 M$_\odot$.

11.5.1 Metallicity gradients?

Is there a metallicity gradient in the disk of the LMC? Kontizas *et al.* (1993b) found a gradient by dividing the clusters into an inner and an outer system; the latter contains older clusters. This effect is not seen by Olszewski *et al.* (1991), who, using roughly the same material, found that the mean abundance in the outer clusters is nearly the same as in the inner clusters. A critical enrichment period lies between 3 and 10 Gyr

232 *Chemical abundances*

Table 11.8. *Abundances in the ISM of the LMC and the SMC relative to the ISM in the Solar Vicinity for elements with* $Z \leq 12$ *(Russell and Dopita 1992)*

Object	[He]	[C]	[N]	[O]	[Ne]	[Na]
LMC	−0.06	−0.29	−0.43	−0.35	−0.29	+0.57:
SMC	−0.09	−0.60	−0.94	−0.67	−0.63	−0.62
Vicinity	+0.01	−0.29	−0.37	−0.14	−0.09	+0.23

'LMC' and 'SMC' stand for 'MC − solar vicinity', 'vicinity' for 'Solar vicinity − Sun'. The abundances are derived from HII regions and SNRs except for C and Na which are derived from F supergiants.

Table 11.9. *Abundances in the ISM of the LMC and the SMC relative to the ISM in the Solar Vicinity for elements with* $12 \leq Z \leq 18$ *(Russell and Dopita 1992)*

Object	[Mg]	[Al]	[Si]	[S]	[Cl]	[Ar]
LMC	−0.10		−0.18:	−0.36	−0.40	−0.13
SMC	−0.59	−0.30:	−0.600	−0.47	−0.46	−0.61
Vicinity	−0.03	+0.20	+0.050	−0.21	−0.09	−0.18

The abundances are from HII regions and SNRs except for Mg, Al, Si, which are from F supergiants.

and $-1.5 <$ [Fe/H] < -0.8. The metallicity gap correlates with the age gap found in the age distribution of LMC clusters.

The constant ratio C star/M star contradicts a gradient and so do the Cepheid abundances. No gradient is seen in the latter, neither east–west nor radially (Harris 1983).

11.6 Summary of abundance determinations

The studies presented above show that the Magellanic Clouds are moderately metal-deficient. Certain inhomogeneities appear to exist in the ISM. It is important to determine whether it is a halo–disk effect or whether it is due to a real fragmentation. The latter appears at present most likely.

The [Fe/H] data of the various age groups discussed above are illustrated in Fig. 11.1. Both Clouds show a roughly exponential increase in metal abundance with time over the last 10 Gyr (Feast 1989b). For young objects (age ≤ 0.2 Gyr) the mean [Fe/H] is ~ -0.2 and -0.5 for the LMC and the SMC, respectively.

The abundances representative of the Magellanic Clouds are summarized in Tables 11.8, 11.9, and 11.10, following Russell and Dopita (1992).

Helium has a largely cosmic origin and may be left out in the consideration of abundances. Relative to the Local ISM all elements show about the same relative abundances, whereas relative to the Sun at least carbon among the lighter elements appears underabundant (see Table 11.8; the table also shows that the C and O

11.6 Summary of abundance determinations

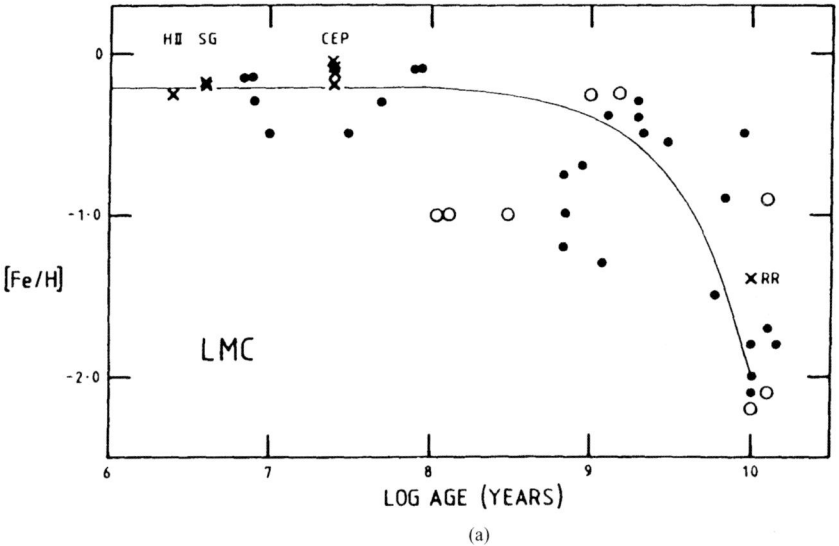

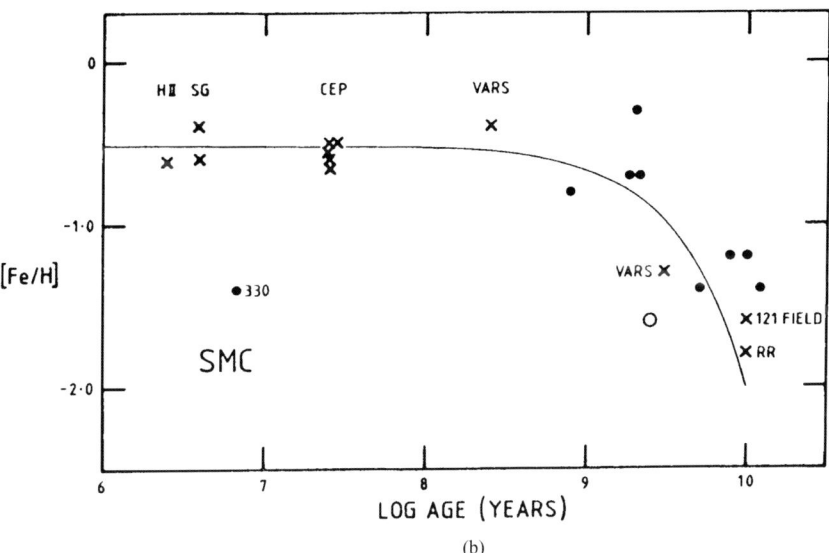

Fig. 11.1. Relation between [Fe/H] and age in (a) the LMC and (b) the SMC. Clusters are plotted as dots or circles (less certain), Crosses identify HII (regions), SG (supergiants), CEP (heids), RR (Lyrae) and VARS (short-period variables) as marked. In (b) the SMC cluster NGC 330 is identified as well as the field around NGC 121. Note that there are determinations of [Fe/H] for NGC 330 that will move the cluster appreciably upwards in the diagram. (Reprinted from Feast (1989b), Figs. 3 and 4, by permission of Springer-Verlag.)

Table 11.10. *Abundances in the ISM of the LMC and the SMC relative to the ISM in the Solar Vicinity for elements with $Z > 18$ (Russell and Dopita 1992)*

Object	[Ca]	[Cr]	[Mn]	[Fe]	[Ni]	[Sr]	[Ba]
LMC	−0.23	−0.10	−0.19	−0.19	−0.21	−0.31	−0.23
SMC	−0.43	−0.47	−0.37	−0.58	−0.41	−1.45:	−0.94
Vicinity	−0.25	−0.11	−0.14	−0.11	−0.00	−0.15	−0.00

The heavy element abundances are from F-type supergiants. Of heavy elements not included in this table vanadium may be of particular interest as it appears overabundant in the Clouds, +0.20 in the LMC, −0.34 in the SMC.

abundances in the Local galactic ISM are less than those of the Sun despite the fact that the Sun is ~ 4 Gyr old).

The stellar carbon abundances in the Clouds (relative to iron) appear entirely normal; it is therefore the HII regions that must be overdeficient in this element.

The s-process appears to have been less effective, and the r-process more effective in the Magellanic Clouds at producing heavy neutron capture elements, when compared with the Galaxy.

The ISM of the LMC has a mean metallicity 0.2 dex lower than the local ISM; the metallicity of the SMC is 0.6 dex lower. The ISMs of both the Magellanic Clouds *and* the Galaxy have significantly non-solar elemental ratios.

How accurate are the results derived to date? Appreciable changes in the metallicities can be noticed as the models used evolve or change. Sensitivity to reddening and temperatures derived from intrinsic colours play a role. The differences between available data are still too large for definite conclusions regarding individual objects or classes of object. This is particularly true for the intermediate-age clusters. Evident is that nitrogen and carbon are underabundant in the HII regions, and that the stellar abundances of these elements are higher than those in the HII regions. All planetary nebulae are overabundant in nitrogen and carbon in the non-type I PNe. This is evidently a result of the nuclear processing in their progenitors.

12

The structure and kinematics of the Magellanic System

As discussed in Chap. 3, the LMC, the SMC and the Galaxy form an interacting system. At times the interactions have had severe effects. Features observed in the LMC and the SMC, including radial velocities of individual objects, are therefore not necessarily determined solely by their rotation and motions.

When the first HI 21 cm-line radial velocities were measured, it was concluded that both Clouds were rotating systems, flattened and tilted and with extensive spiral structures (Kerr and de Vaucouleurs 1955a,b). The rotational motions were at first looked for using the optical centres, 5^h24^m, $-69°.8$ for the LMC and $0^h51^m, -73°.1$ for the SMC. In order to obtain symmetrical rotation curves it was, however, necessary to introduce radio centres of rotation: in the LMC, at 5^h20^m, $-68°.8$, appreciably displaced from the optical centre, and in the SMC 1^h10^m, $-73°.25$. The centre of rotation in the LMC has since been a source of much discussion (see Sect. 3.4.4). In the SMC the problems are of another nature.

12.1 The structure and kinematics of the LMC

Images of the LMC in most wavelength regions are dominated by radiation from its Extreme Population I constituent (stellar associations, supergiants, etc.) or the connected gas (HII regions, HI complexes, molecular clouds) and dust which display the regions of recent star formation as an asymmetric pattern, not completely at random but with some structure. The interpretations of this structure differ appreciably, though stochastic self-propagating star formation (SSPSF, Gerola and Seiden 1978) is often considered an important active mechanism, replacing in some ways the density-wave phenomenon in spiral galaxies.

12.1.1 *The youngest population*

Many kinematical studies of the LMC have been carried out over the past three decades, generally based on surveys of the central system ($r < 3$ kpc). As those dealing with Population I objects, supergiants and emission nebulae, as a rule derive rotation curves in good agreement with that of neutral hydrogen, they will not be considered in detail here (see Prévot *et al.*(1989) for references). Most analyses have led to the conclusion that the LMC is dominated by a flat disk with a solid body motion in its central parts and signs of differential rotation in the outer regions but with extensive irregularities of the velocity field. The neutral hydrogen has a more global coverage of the LMC than the other Population I components. This, and the frequently appearing multiple peaked profiles, suggests that it may give the most vital information. Much information about the fragmentation of the LMC in the line of sight is also found in the multiple components frequently seen in the interstellar lines. This has been extensively discussed in Sect. 8.7 and will not be repeated here.

McGee and Milton (1966) have presented very useful 21 cm HI line data for the LMC which have played an important role in interpretations of its structure and kinematics. They proposed a model consisting of a principal plane containing two very large spiral arms, and another plane, inclined to the former by about 20°, containing two shorter arms emanating from each end of the stellar Bar. The best radio centre of rotation was found to be 5^h20^m, $-69°.0$. Kerr (1971) considered that too much regularity was attributed to the system in this model and saw the LMC gas as having a complex shape, with many localized concentrations, some of which are well away from the mean plane of a rather flattened system. In some regions there may be a tendency towards a rudimentary spiral pattern. He also speculated that the LMC may not be massive enough for a density-wave phenomenon to be established. (In the SMC there are no signs of any density-wave fragments at all.) It has been shown that density waves do not contribute to the structure in the LMC (Dixon and Ford 1972). The distribution of the stellar age-groups in the 'major spiral arm' contradicts star formation in that way. More recently, Feitzinger (1980) has also concluded that the 'spiral' filaments are not connected with density waves. The part of the 'arm' investigated by Dixon and Ford has in fact the stellar distribution expected in a superassociation where SSPSF has been active (Westerlund 1989).

Recent determinations of the HI rotation curve of the LMC have been carried out by Rohlfs et al. (1984) and by Luks and Rohlfs (1992).

Rohlfs et al. interpreted the roughly parallel isovelocity lines of the intensity-weighted mean galactocentric radial velocity perpendicular to a major axis at $PA = 208°$ as the signature of a rotating flat disk. The central region, within $r \leq 1.5°$, was found to be strongly perturbed by non-circular motions so that only the outer parts could be used to determine a point of symmetry as the centre of rotation. There is a kink in the northern part of the rotation curve, about 3° from the centre with no counterpart on the southern side. Their maps show that the kink is connected with the decrease in weighted mean radial velocity of about 15 km s^{-1} in the SGS LMC 4 area, where an HI void is seen.The disturbances in the central region may well be due to another HI void and a corresponding supergiant shell, which may be more difficult to outline as part of it may be superposed on the Bar.

The large gas cloud embedding the 30 Doradus complex and extending south for about 2° is visible as a strong perturbation of the velocity field in the Rohlfs et al. 21 cm line map with a second component of lower radial velocity. A similar component is seen at about the NW end of the Bar. These components were interpreted as being connected with a warp of the LMC disk. They correspond to the tilted plane in the McGee and Milton model.

Over the past 30 years numerous attempts have been made to interpret the observations of the youngest population as showing a spiral-arm structure similar to the one displayed by the hydrogen gas and the youngest stars in our Galaxy. Two lines of thought regarding the structure of the LMC appear to have developed independently of each other.

In one, (1), the radio centre is given the role played by the nucleus in our Galaxy, i.e. as the centre of rotation and mass for all classes of object (see Freeman et al.

12.1 The structure and kinematics of the LMC

1983, Meatheringham *et al.* 1988). It finds its strongest support in the rotation curve(s) derived from the HI line profiles.

In the other, (2), the 30 Doradus nebula is considered the mildly active nucleus of the LMC and the origin of the spiral arms (Schmidt-Kaler and Feitzinger 1976, Feitzinger *et al.* 1987). Its support is found in the thermal sources seen in the radio continuum and in the young associations seen particularly well in the ultraviolet.

(1). The major component of the two HI features found by Luks and Rohlfs (1992) in their HI maps (see Sect. 8.1) has a symmetric rotation curve with no obvious influence of the Bar. Its kinematics can be modelled by a flat disk ($i = 33^{\circ}$; position angle of the line of nodes $= 162°$) in differential rotation with its centre at $5^h 12^m 48^s - 69°.6$ (1950), provided the tangential velocity of the LMC is considered (see Table 3.1). The rotation curve is linear for $r \leq 0.7$ kpc from the centre, corresponding to an angular rotation of 50 km s^{-1} kpc^{-1}; for $r \geq 1.4$ kpc the curve flattens. A turnover occurs at $r = 3.5$ kpc in the southern branch with $V_{rot,max} \sim 70$ km s^{-1}.

The centre of rotation found for the disk component by Luks and Rohlfs, 1.2° from the centre of the Bar, 0°.1 off its centreline, differs appreciably from that normally used in rotation analyses of components of various ages in the LMC. Their results underline Kerr's (1971) conclusion that the LMC gas has a complex shape. The question may be asked if equally disturbing irregularities do exist in all generations, though smoothed out, and affect all determinations of LMC kinematics. It is, anyhow, somewhat surprising that relatively good correlations are suggested to exist between HI rotation solutions and solutions for much older objects, considering the existing uncertainty in the HI structure.

The L-component, consisting of two large complexes linked by the 30 Doradus nebula (see Sect. 8.1), was assumed to be in front of the disk as no absorption of the disk gas could be detected. As absorption has since been observed, the lobes and 30 Doradus are now assumed to be behind the main HI disk (see Sect. 6.1.2).

The velocity difference of the L-component and the disk is at most parts of the order of 20–30 km s^{-1} except at the eastern border where the L-component has a kinematical warp reaching up to 60 km s^{-1}. Also, the L-component shows signs of differential rotation; its inclination and line of nodes are, however, not known.

(2). A number of well separated sources in the 1.4 GHz radio continuum map form long ridges, suggested to originate in the 30 Doradus nebula (Feitzinger *et al.* 1987). The ridges correlate well with a series of blobs in the 100 μm emission in the *IRAS* maps and with the UV brightness distributions derived by Smith *et al.* (1987) for hot stars (at 1500 Å (Fig. 12.2) and 1900 Å). Most outstanding is the chain of OB associations and HII regions defining 'a bright and sharply defined spiral' through the Bar towards the NW. The chain may, however, be part of one or more supergiant shells (Chap. 6).

Feitzinger *et al.* (1987) acknowledge the SSPSF as having been active in the Constellation III area, in the way suggested by Dopita *et al.* (1985c), and consider it as an isolated space–time cell. Nevertheless, two of their long ridges of star formation loci, F and E2 (their Fig. 2), include as dominating parts major portions of the star forming shell in Constellation III. Similarly, their ridges A and B enclose much of SGS LMC 2.

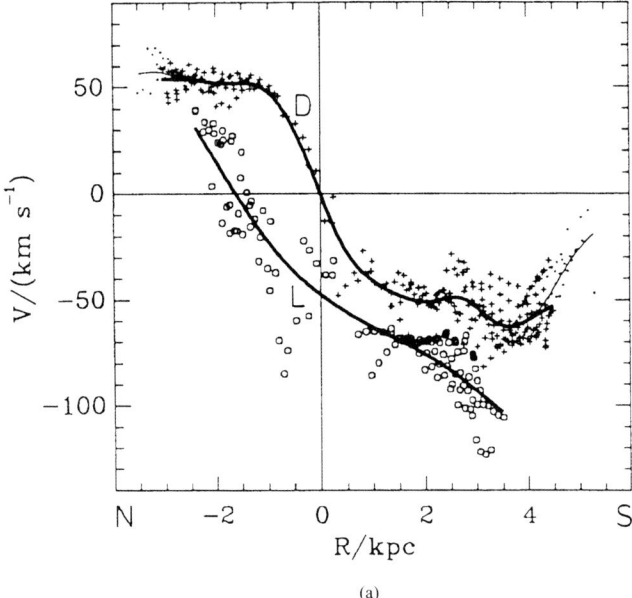

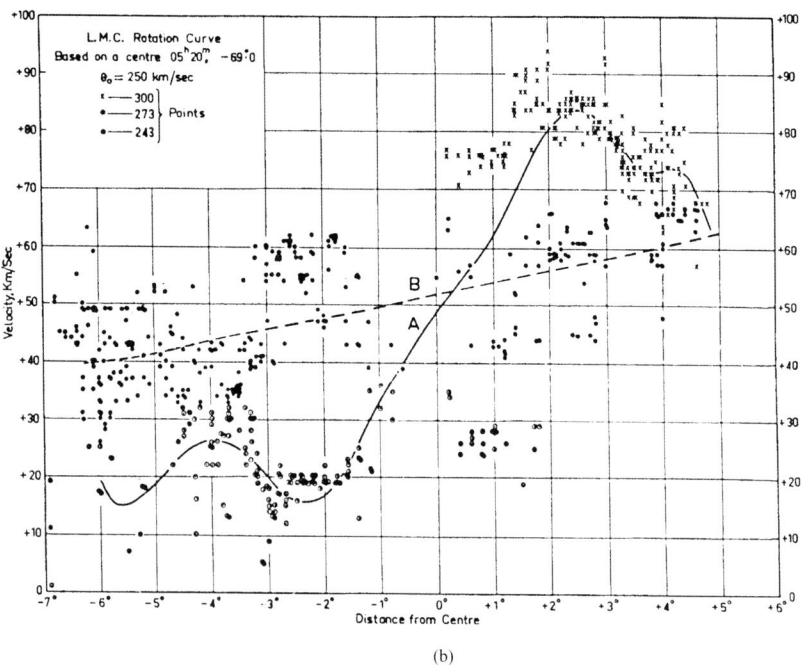

Fig. 12.1. (a) Rotation curves for the disk (+) and L (∘) components in Luks and Rohlfs' (1992) model of the HI gas in the LMC from a sector of ± 20° from the line of nodes. (b) The McGee and Milton (1966) rotation curves modelled on two spiral arms. The velocity points are in a sector ± 10° from a line of nodes in $PA = 171°$. Curve 'A' is fitted to +300 and +243 km s^{-1} points, curve 'B' to +273 km s^{-1} points. Note that '−' indicates North in (a), South in (b).

12.1 The structure and kinematics of the LMC

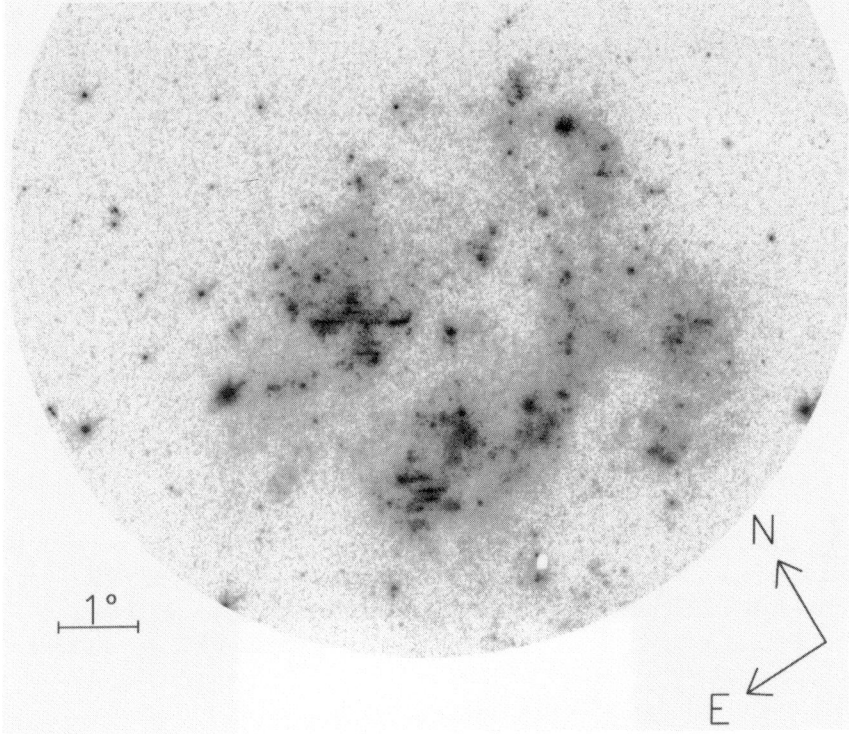

Fig. 12.2. LMC at 1500 Å (Smith *et al.* 1987).The passband has a FWHM of 200 Å.

It would be surprising if the thermal radio sources did not coincide well with the HII regions which, as sites for the most recent star formation, are also very rich in UV radiating stars and, presumably, in dust. It may be considered more remarkable that the HI complexes also follow this pattern so well (McGee and Milton 1966). In the LMC many HII regions and their associations are apparently still embedded in, or surrounded by, neutral hydrogen and, possibly, molecular clouds.

Feitzinger *et al.*(1987) have presented a SSPSF model which simulates the observed distribution of stars, gas and dust and considers molecular cloud formation, three gas phases and detailed gas distributions in neighbouring cells and perpendicular to the disk around the star forming sites. The energetics of the stochastic percolating star formation processes come from the differential rotation and the energy input of supernovae, WR, O, and B stars. In the model star formation migrates in a systematic manner. The role of the origin of all the ridges, the 30 Doradus nebula, is, however, not explained.

The maximum extent of the UV emission observed by Smith *et al.* (Fig. 12.2) corresponds to a circle of about 5.5 kpc in diameter. A deterministic self-propagating star formation (DSPSF) model was developed in order to describe the 'bright and sharply defined spiral' as a star formation front. In their preferred solution the main spiral feature is fitted to the front. Several of the most prominent features, such as Shapley Constellations I, III, and IV, cannot be included in the model, and the computed front does not reach the centre of 30 Doradus.

It is difficult to accept models presenting the most recent star formation as migrating through the LMC in a systematic manner as the ages of the objects in widely separated stellar associations are virtually identical. A more reasonable explanation is that bursts of star formation occurred rather simultaneuously in a number of areas in the central region of the LMC (Westerlund 1964c). A more or less simultaneous burst of star formation over much of the central regions of the LMC is also required for the formation of the HI sheet seen above and below the LMC plane (Meaburn *et al.* 1987).

Following the outbursts, the SSPSF process (though modifications are needed, see below) led to the formation of large superassociations with diameters of over 1 kpc (Westerlund 1989); some, or parts, of them, are delimited by the SGSs (Chap. 6). The bursts responsible for the formation of the Extreme Population I in the central region of the LMC may have been caused by the close encounter between the LMC and the Galaxy around 50 Myr ago.

An explanation of the generally highly asymmetric distribution of gases in the LMC has been suggested by Fujimoto and Noguchi (1991): at the approach of the SMC and the LMC ~ 0.2 Gyr ago many gas clouds collided hydrodynamically, shock compressions of various scales were generated and aggregates of various masses were formed. The larger clouds orbited ballistically in the gravitational fields of the LMC, the SMC and the Galaxy, mainly in the vicinity of the LMC but not necessarily in its equatorial plane. The largest was the CO- and H-rich cloud that now contains the 30 Doradus nebula and SGS LMC 2, and is the most active region in the LMC.

12.1.2 The east–west asymmetry in the LMC

The FIR/6.3 cm brightness ratio shows a systematic difference between the eastern and western half of the LMC (Xu *et al.* 1992) with most of the eastern giant HII regions, including 30 Doradus, showing lower FIR/6.3 cm brightness ratios than those in the western part.

The asymmetry could be due to different ages of massive stars in the two parts, which give rise to different non-ionizing to ionizing flux ratios. It could also be due to a difference in the gas-to-dust ratio. The first possibility appears to be the most likely. It is supported by the existence of a similar asymmetry in the diffuse X-ray emission (0.1–3.5 keV range) (Wang *et al.* 1991, see Chap. 9). The star forming regions in the western half of the LMC may be systematically older than those in the eastern half, so that the stars responsible for the ionization of HII regions are less massive and have higher non-ionizing-to-ionizing flux ratios.[*] They will thus heat the dust but not contribute to the thermal radio emission. The age difference between the western and eastern star forming regions need not be large to explain the flux differences. The interaction between the SMC and the LMC ~ 0.2 Gyr ago may have produced star forming clouds in the western part first. In the model we have followed above, the star forming cloud now in the 30 Doradus region took most of the 0.2 Gyr to move to its present position. Also the intermediate-age generations may show some

---

[*] It is obvious that the evolution of the stars is slightly more advanced in the western than in the eastern parts of the LMC: 48% of the supergiants are OB2 stars in the west, 57% in the east, whereas for B7–G the percentages are 32 and 20, respectively (Sect. 5.4). The superassociations in the west are also less gas-rich than those in the east (Chap 6).

12.1 The structure and kinematics of the LMC

Table 12.1. *Rotation solutions for LMC old clusters*

Objects SWB class	N	V_{circ} km s^{-1}	$\Theta(PA)$ deg	V_{sys} km s^{-1}	σ	Comments
HI		79	− 9	44		FIO
I–III	24	81±11	1±5	40±3	15	FIO (2)
IV–V	18	48±15	31±11	25±4	13	FIO (7)
VII	9	119±15	44±6	38±4	16	FIO (9)
V–VII	25	90±9	41±5	26±2	17	FIO (3)
V–VII	28	72±15	31±10	39±4	21	SOSH velocities
New, outer	50	71±8	14±8	44±3	19	SOSH, $V_t = 0$ km s^{-1}
New, outer	50	69±8	−7±10	42±3	17	SOSH, $V_t = 150$
V–VI	17	54±14	−7±14	36±5	17	SOSH, $V_t = 150$
VII	12	66±16	36±11	41±4	23	SOSH, $V_t = 150$
VII; R ≥ 2°	9	51±8	− 9±7	40±2	23	SOSH, $V_t = 250$

V_{circ} is the velocity in the plane of the LMC; it is equal to $V_m/\sin i$, where V_m is the amplitude of rotation solution.
$\Theta(PA)$ is the position angle of the line of nodes of the disk.
V_{sys} is the systemic velocity of the solution and σ is the rms dispersion around the solution.
$V(\theta) = \pm V_m\{[\tan(\theta - \theta_0)\sec i]^2 + 1\}^{-0.5} + V_{sys}$.
The 12 old clusters are those listed in Table 4.3; the three inside $R = 2°$, and not included in the last solution, are NGC 2005, NGC 2019, and NGC 1835.

asymmetry. The SW part of the LMC, outside the young star forming regions, does not contain stars younger than 1 Gyr (see Sect. 4.4).

12.1.3 The old and intermediate-age clusters

Doubts of the LMC as a single, uniformly rotating disk arose from an investigation of the radial velocities of 59 clusters (Freeman *et al.* 1983, FIO). Clusters younger than 1 Gyr (SWB types I–III) were found to have motions similar to the gas in their vicinity and share the rotation solution derived from HI and HII region velocities (Table 12.1). Also the intermediate-age clusters (SWB IV–V) form a flattened system. The oldest clusters (SWB VII, age ≥ 10 Gyr) appeared to lie in a highly flattened disklike system with, according to FIO, a z scale height ∼ 0.5 kpc at $R \approx 3$ kpc. The LMC would then have its oldest clusters rotating in a disk separate from that of the other populations but lacking a spherical halo population (defined by globular clusters).

Schommer *et al.* (1992, SOSH) have analysed the radial velocities of 83 star clusters in the LMC, using individual stellar velocities measured at the calcium triplet. Position angles and radii were measured around the HI rotation centre, 5^h21^m, $-69°17'$; the tilt of the LMC was taken to be $i = 27°$; both quantities are in agreement with FIO. About one-half of the clusters are more than 5° from the centre. They exhibit disk kinematics and are most likely of intermediate age. When corrections are made for the transverse velocity, V_t, of the LMC, the outer clusters and the inner intermediate-age clusters form a disk which is aligned with the inner HI kinematics and the outer LMC isophote major-axis position angle.

A comparison of the solutions for the SWB type V–VII clusters, FIO (3) and SOSH, shows that the discrepancy in V_{sys} between this group and the young population has disappeared with the new velocities.* The systemic velocities for all SOSH groups show good agreement. The odd PA for the line of nodes for the oldest 12 clusters disappears when only the nine outer clusters are considered. The latter show rotation with a small velocity dispersion characteristic of a thick disk or a flattened spheroid. They do not have the kinematics of an isothermal, or slowly rotating, pressure supported halo. In the inner 2° the old clusters exhibit peculiar velocities, as do the CH stars and the old LPVs, possibly due to perturbations from the Bar.

The problems encountered by FIO appear to have been solved. There is only one young and intermediate-age disk in the SOSH solutions, and the outer SWB VII cluster disk shows a PA and a V_{sys} in agreement with that. Its smaller rotation amplitude indicates that the old disk lags the circular velocity in agreement with the increased velocity dispersion of this old component.

The rotation curve does not show signs of a Keplerian falloff out to at least 5–6 disk scale lengths, implying the existence of dark matter associated with the LMC. The mass of the LMC is estimated to 1.5–2.5×10^{10} $M_\odot$ out to 8° radius. No old, isotropic, slowly rotating halo has been found in the cluster population. It may exist; more studies of RR Lyraes and very old metal-poor giants are needed. No strong effects on the stellar velocities in the outer parts of the LMC are seen that may be due to tidal interaction between the Clouds and the Galaxy.

12.1.4 The planetary nebulae

Meatheringham et al. (1988, MDFW) have determined radial velocities for 94 PNe in the LMC, and have derived a rotation solution for the PN population. It is essentially identical with that of the HI, but the vertical velocity dispersion of 19.1 km s^{-1} is much greater than the 5.4 km s^{-1} found for HI. A comparison of the radial velocity of each PN with that of HI in its vicinity shows a poor local correlation between the HI and PN dynamics. Fig. 12.3 shows the velocity differences between the PNe radial velocities and the local HI radial velocities as compared with the rotation solution for HI. Distinct regions of increased dispersion in HI velocities are visible. They correspond closely in position with the superassociations (see Chap. 6), where gas has obviously been severely stirred up. The higher dispersion of the PNe radial velocities is also evident. It is difficult to see any tendency to a rotation solution in the scatter. It appears necessary to analyse the composition of the PN population in the LMC more before drawing conclusions about its behaviour as a single unit. It appears unlikely that the PNe in the LMC should be much different from those in the SMC as far as division into age groups is concerned. There are three distinct groups of PNe also in the LMC: (i) type I PNe with high electron temperatures, (ii) optically thick PNe, and (iii) optically thin PNe. The first group originates from high-mass precursors and may behave differently kinematically from the other groups.

Nevertheless, the large vertical velocity dispersion of the PNe is consistent with it being the result of orbital heating and diffusion over the lifetime of the PNe. Then the bulk of the PNe cannot represent a halo population..

In the HI rotation solution consideration is given to the transverse velocity of the LMC. The best solution is derived with the transverse velocity $V_t = 275 \pm 65$ km s^{-1}

---

* There are some remarkable differences in velocities for some clusters. Most striking is NGC 1841 for which FIO give a galactocentric velocity of 84 ± 37, and SOSH 10 km s^{-1}. The SOSH value has been confirmed by Storm and Carney (1991).

12.1 The structure and kinematics of the LMC

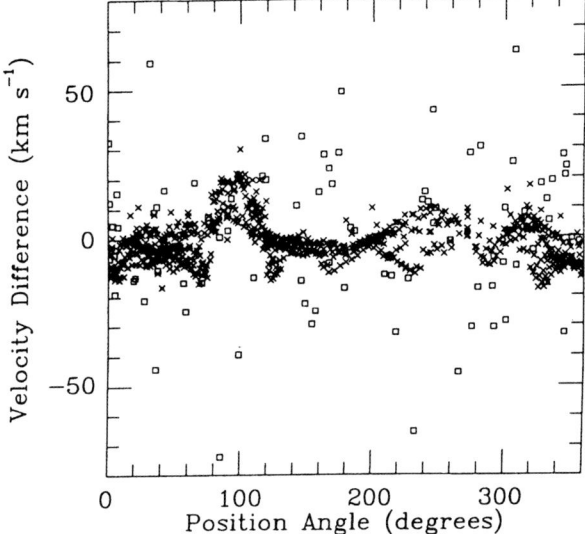

Fig. 12.3. The velocity difference obtained from HI data (X) and the planetary nebulae (squares) in the LMC relative to the HI rotation solution. Major deviations in the HI data are due to star forming regions.

Table 12.2. *Rotation solutions for PNe in the LMC*

Sample	Number	$V_\circ$	$\theta_\circ$
New PNe	11	40	164
MDFW PNe	95	42	170
All PNe	106	42	168
All PNe with $R \geq 4°$	30	46	170

and LMC near perigalacticon. This is consistent with a maximum galactic mass of $\sim 4.5 \times 10^{11}$ $M_\odot$ out to 51 kpc. The rotation curve obtained after correction for this velocity implies a LMC mass of $(4.6 \pm 0.3) \times 10^9$ $M_\odot$ within a radius of 3° or about 6×10^9 $M_\odot$, total.

Vassiliadis et al. (1992) have added velocity data for 11 PNe in the outer fields of the LMC in order to confirm the conclusions drawn by MDFW that the PN population forms a flattened disk with an almost identical rotation solution to that of the HI. Their results, expressed in the position angle of the kinematic line of nodes, $\theta_\circ$ and the galactocentric velocity of the LMC, $V_\circ$, are given in Table 12.2.

Ignoring the velocity dispersions, Vassiliadis et al. find that no significant difference exists between the PNe and the HI kinematics, in agreement with the MDFW analysis.

12.1.5 The oldest stellar field populations

Radial velocities for a group of the oldest long-period variables (LPVs) in the LMC have been determined by Bessell et al. (1986). These stars are similar to the Mira variables in 47 Tuc and the galactic centre. The systemic velocity of these LPVs is indistinguishable from that of the HI gas. The intrinsic line-of-sight dispersion of the LPV velocities about their systemic-velocity mean is 30 km s^{-1}. The authors find this consistent with that expected for stars belonging to a flattened disk with a scale height of ~ 0.3 kpc; they also see marginal evidence that the kinematic system, to which the old LPVs belong, rotates. However, the fact that the material is rather limited and most of the stars are within 1° of the HI centre of rotation, where the HI motions are strongly perturbed by non-circular motions (Rohlfs et al. 1984), makes it difficult to accept this interpretation as the only one possible.

Hughes et al. (1991) have obtained radial velocities for a significant sample of LPVs and have applied a kinematic analysis to a wide range of LMC populations (HI, CO, PN, CH stars, clusters, LPVs). The oldest LPVs (~ 10 Gyr) were found to have a high velocity dispersion and a low rotational velocity, proving that they belong to a flattened spheroid population (maximum height ≤ 2.8 kpc); it may be part of an old disk population (~ 4 Gyr). The dynamics of the LMC is otherwise dominated by a single rotating disk.

The CH stars are considered as good halo tracers in the Galaxy. In the LMC the velocity distribution, derived by Hartwick and Cowley (1991) from radial velocities for 81 CH stars, appears asymmetrical, with a relatively sharp cutoff at the high-velocity end and with an extended low-velocity tail. The overall velocity dispersion for these stars is 24 ± 2 km s^{-1}. In a kinematic halo at least 40 km s^{-1} is expected. Such a halo does not appear to exist in the LMC. The CH star velocity dispersion is comparable with that of LMC planetary nebulae (19.1 km s^{-1}, see above). The kinematical age of the CH stars is suggested to be of the order of a few Gyr. However, two groups of CH stars may exist, one of which may have about the same age as the oldest LPVs and belong to the spheroid population. The other may be associated with a flattened-disk system and contain stars with luminosities up to $M_{bol} = -6$ or brighter. They may belong to a young AGB population, age ~ 0.1 Gyr, or be products of binary mergers, age a few Gyr (see Sect. 7.4).

Further studies show that the velocity dispersion of the CH stars *outside* 5° from the LMC centre is only 10 km s^{-1} (Suntzeff et al. 1993). This is less than that of the intermediate-age populations. The SWB type V–VI clusters (1–3 Gyr) have a velocity dispersion of 17 km s^{-1} (Schommer et al.. 1992), the PNe (1–3 Gyr), as mentioned above, have a velocity dispersion of 19.1 km s^{-1}, and the LPVs (1–5 Gyr) have a velocity dispersion of ~ 18 km s^{-1} (Hughes et al. 1991). Only the youngest populations in the LMC have velocity dispersions as low as these CH stars.

The CH stars *inside* 5° from the LMC centre have an asymmetrical velocity distribution with a tail towards low velocities. This may indicate a scatter of these stars in depth. The mean velocity of the 'tail stars' is about 30 km s^{-1} smaller than that of the disk stars, or, if the Hartwick and Cowley conclusion is used, the old-cluster-like population of the CH stars has a mean velocity ~ 15–20 km s^{-1} smaller than the LMC itself. These values are of the same order as those for the runaway supergiants.

12.2 The structure and kinematics of the SMC

No convincing rotational effects were found by Feast et al. (1961), who concluded that their results could be represented by a random scatter around a mean velocity of 166 ± 3 km s^{-1}.

12.2 The structure and kinematics of the SMC

The early interest in interpreting the SMC as a rotating galaxy with a spiral structure (see de Vaucouleurs and Freeman 1973) has at present been replaced by the question about its extension in depth and about its possible fragmentation.

12.2.1 The extension of the SMC in depth

The SMC was first suspected to have an appreciable extent in depth by Johnson (1961). Hindman (1964) remarked on the complexity of its HI distribution and suggested (Hindman 1967) that in addition to a flattened system seen edge-on, three expanding shells of gas could be distinguished within the main body. In a recent 21 cm line mapping of the SMC with the Australia Telescope Compact Array (Staveley-Smith et al. 1995) a remarkably complex HI structure is revealed with compact knots, filaments, 'bubbles' and supergiant shells. One of the latter overlaps with Hindman's southernmost shell. Some of the features are well outside the optically defined main structure. The dissimilarities between the HI and the optical 'young population' morphology are great. A detailed comparison of the new HI data with images also available in the infrared and the radio continuum may be expected to give extremely important information about the evolution of the youngest generations in the SMC. Its contribution to our knowledge about the spatial structure of the SMC is anticipated with great expectation. Meanwhile, available data have serve.

The A, B, and O supergiants in the SMC have an appreciable spread in distance, up to 7 kpc in spatial depth (Azzopardi 1982). The younger stars have a larger distance on the mean and concentrate in the NE sector. At least two stellar groupings have been found to exist.

Extensive radio, optical and UV observations have made it evident that several velocity groups exist in the SMC, partly overlapping.

Two separate entities in the HI distribution, each with its own stellar and nebular populations, were seen in velocity-space by Mathewson and Ford (1984). They proposed that the SMC was torn apart by the close encounter with the LMC some 0.2 Gyr ago into a low-velocity fragment, the Small Magellanic Cloud Remnant (SMCR), in front of a higher-velocity fragment, the Mini-Magellanic Cloud (MMC).

Confirmation that the higher-velocity system is behind the lower-velocity gas is found in the velocities determined for a number of early-type stars and for the interstellar calcium with the aid of the CaII line at 3933 Å in their spectra (Cohen 1984). The radial velocities of the observed stars correspond as a rule to that of the higher-velocity HI gas, those of the CaII lines to that of the lower-velocity HI gas. In the few cases where two components of CaII are seen, the stellar velocity agrees with that of the higher-velocity gas. The velocity dispersion of the stars with respect to the gas at each point is 13 km s^{-1}, ignoring three stars with vastly deviating velocities. Also the CaII velocities show a good agreement with those of HI at each point.

Using HI velocity and Cepheid data Caldwell and Coulson (1986) found the SMC to consist of a central bar seen edge-on, a near arm in the NE, a far arm in the SW, and a mass of material pulled out of the centre of the SMC and seen in front of the SW arm. They associated the far arm with the lower-velocity HI component, and the near arm and the 'pulled-out' material with the higher-velocity HI component, contrary to Mathewson and Ford (1984) and to the results from the interstellar Ca II line just described.

In an attempt to spatially separate the two HI masses Mathewson et al. (1986) measured the distances of 161 Cepheids in the SMC, using the PL relation in the infrared. The Cepheids were found to extend from 43 to 75 kpc with a maximum concentration at 59 kpc. A rather complete subsample of the younger Cepheids (P $\geq$ 10 days) split into two components, each with a depth of about 6 kpc, and with centres 12 kpc apart.

Torres and Carranza (1987) used Hα interferometric observations and available radial velocities for HI, supergiants, emission and planetary nebulae, and interstellar calcium to derive a four-component velocity pattern for the SMC. The components have RV_{hel} = 105, 140, 170, and 190 km s^{-1} or, on the average, RV_{GSR} = -55, -20, $+10$, and $+30$ km s^{-1}. HI is seen in all four components. The supergiants, the emission nebulae and the PNe are all found in the 140 and 170 km s^{-1} components. The diffuse ionized hydrogen is missing in the 190 km s^{-1} component and the interstellar CaII is seen only in the 140 and 190 km s^{-1} components. Absorption features which may be associated with these four components are also seen in the UV (Fitzpatrick 1985b; Fitzpatrick and Savage 1985).

In a further development of their model Mathewson et al. (1988) measured radial velocities and distances of 61 Cepheids in three fields in the SMC Bar and carried out a high-resolution HI survey of this region. They combined these data with available velocity data for other Population I components. The results confirmed that the SMC has a depth of about 20 kpc and that the NE section of the Bar is 10–15 kpc closer than the southern. There are regions where Cepheids cluster in groups of similar velocities and distances. These groups coincide with groups of F–M supergiants and with particular features on the HI velocity maps. When combined with the interstellar absorption-line data it is seen that the stars can be associated with the HI emission, and the low-velocity components are nearer than the high-velocity components. Certain groups of Cepheids suggest that a correlation exists between velocity and distance: the slope is about 4 km s^{-1} kpc^{-1}. This gives a time scale of expansion of 2.5×10^8 yr, in agreement with the collision model.

Martin et al. (1989) carried out an extensive analysis of the structure and motions of the SMC on the basis of McGee and Newton's (1981) high-quality profiles of the 21 cm line. Their analysis is more extensive than that presented by Mathewson and Ford (1984) and Caldwell and Coulson (1986), as they, instead of using only a projection on the major axis, study the HI velocity distribution along 12 cuts parallel to the major axis. Accurate radial velocities of 307 young stars and 35 HII regions are used together with very high spectral resolution profiles of interstellar absorption lines. They found:

1. The HI in the Bar is distributed in four components: a very low (VL) radial velocity component in the SW (RV_{GSR} = -40 to -50 km s^{-1}); a low (L) velocity major component (-28 km s^{-1}) covering the Southern half of the SMC; a high (H) velocity important component ($+9$ km s^{-1}) seen everywhere except in the SW; and a very high (VH) but weak component ($+30$ km s^{-1}) seen mainly in the NE. The VH and H components may be connected.

Components L and H correspond approximately to the SMCR and MMC in the Mathewson and Ford model. All four components agree with those noted by Torres and Carranza. The VH and H components may be connected.

12.2 The structure and kinematics of the SMC

2. A majority of the young stars can be associated with one of the four HI components described above. However, 53 stars and one HII region have deviating velocities and are not associated with HI clouds. Some of them appear to form physical groups, one of which may be between the H and the VH complexes and another may be in front of the L complex.

3. The main complex H is located behind the L complex. In the south and southwest outskirts of the Bar the depth is particularly large. Several 'L subcomplexes' may exist, one of which may be the nearest known entity.

4. Interstellar CaII is seen at higher and lower velocities than the 21 cm HI emission lines. The higher-velocity gas must be in front of the H and L components (see Fitzpatrick and Savage 1985) and the lower-velocity gas must be in front of the H component.

Martin *et al.* conclude that most of the young SMC stars lie within a depth of < 10 kpc; this corresponds approximately to the lateral diameter of the SMC. This depth is smaller than that found by Mathewson *et al.* (1986, 1988) and Caldwell and Coulson (1986). It is similar to the one deduced by Welch *et al.* (1987), who found that the SMC does not extend beyond its tidal radius, 4–9 kpc, and that it is not in 'the process of irreversible disintegration'. Martin *et al.* consider that there is no proof that the distances of SMC Cepheids are linked to their radial velocities. There is, however, as mentioned above, some support for that correlation in Mathewson *et al.*'s new data.

A still larger extent in depth of the SMC has been observed by Hatzidimitriou and Hawkins (1989) and Hatzidimitriou and Cannon (1993). The intermediate-age population in two regions over 2.5 kpc from the optical centre, one in the NE and one in the SW, were studied. In the NE field a maximum extension of ~ 20 kpc was observed; part of this population is also closer to us than the one in the SW. The distances of a number of RGC/HB stars, determined from their magnitudes (Sect. 7.2), showed a strong linear correlation, with their radial velocities with a slope of 6 km s^{-1} kpc^{-1}, with the more distant stars having the highest radial velocities. This supports the results by Mathewson *et al.* (1988) presented above and leads to the same time scale of expansion, ~ 0.2 Gyr.

12.2.2 Planetary nebulae, carbon stars and metal-poor giants

The PNe in the SMC form an unstructured spheroidal population apparently associated with the Bar and with the centroid at $0^h 49^m 39^s$, $-73°30'$ (1950) and a mean radial velocity $RV_{GSR} = -17$ km s^{-1} (Dopita *et al.* 1985a). No evidence is seen in these data of any organized rotation, nor of any bimodal velocity distribution of the kind suggested by Feast (1968) and also noted by Torres and Caranza (1987).

If the precursors of the PNe are the late AGB stars, i.e. mainly carbon stars, then their distributions should be similar. Radial velocities for the field population of C stars in three regions of the SMC were determined by Hardy *et al.* (1989) with an individual precision of ± 1.8 km s^{-1}. They found that the C star population does not behave kinematically like the extreme Population I. There is no evidence of velocity splitting, nor of the rotation of the main body. As expected, the C stars behave kinematically like the PN system with which they share the mean velocity ($RV_{LSR} \approx 137$ km s^{-1}) as well as a velocity dispersion of ≈ 27 km s^{-1}. The same velocity dispersion is shown by a sample of halo metal-poor giants near NGC 121 (Suntzeff

Table 12.3. *Radial velocities observed in the SMC*

RV_{GSR} km s^{-1}	Components	Ref.
-55	HI, Hα interferometric	Torres and Carranza
-20	HI, I, EM, PN, Hα, CaII	(1987)
$+10$	HI, I, EM, PN, Hα	
$+30$	HI, CaI	
-17	PN, no bimodal distribution	Dopita *et al.* (1985a)
$-40 - -50$	VL, component in SW	Martin *et al.* (1989)
-28	L, major component in S	
$+9$	H, important everywhere except in SW	
$+30$	VH, weak component in NE	

Absorption features associated with the four Torres and Carranza components have been observed in the UV (Fitzpatrick 1985a; Fitzpatrick and Savage 1985).

et al. 1986). C stars and PNe may belong to a spheroidal-like system, though radial velocities for a larger sample of both groups are needed before final conclusions are drawn.

Suntzeff *et al.* also compared the kinematics of halo giants in the NGC 121 field with the SMC in general. By correcting the observed radial velocities of the SMC halo field for the solar apex motion they found a velocity of -29 km s^{-1}. This is significantly smaller than the velocities of $+2 \pm 3$ km s^{-1} for the main body of the SMC and $+21$ km s^{-1} for the K1 region on the eastern side of the SMC Bar (Ardeberg and Maurice 1979). The three values together give the impression of a small gradient in stellar radial velocity roughlyt orthogonal to the HI major axis. However, they more likely show the multiple-peaked velocity structure in the SMC. The halo field population should then be associated with the low-velocity, SMCR, component corresponding to the bulk of the SMC.

The data for PN, C stars, and metal-poor giants just presented appear to show that the near collision with the LMC 0.4–0.2 Gyr ago did leave the older stellar component roughly spheroidal, whereas the gaseous component was drawn out along our line of sight.

Somewhat contradictory results have been derived by Reid and Mould (1990). They found that the radial velocities of the AGB stars in a SMC field just east of the Bar showed a bimodal distribution of the kind known from HI and other components. The narrowness of the giant branch speaks against a depth of more than 5 kpc in this population.

12.3 The motions in the Magellanic Stream and in the Bridge

The radial velocities of the HI clouds in the Magellanic Stream with respect to the galactic centre become increasingly more negative from 0 km s^{-1} at MS I to -200 km s^{-1} at MS VI. There is a sharp discontinuity in velocity of ~ 100 km s^{-1} between the top of the InterCloud region and MS I (Mathewson *et al.* 1987). Other

12.3 The motions in the Magellanic Stream and in the Bridge

components are also seen in HI as overlapping the main Stream. Several of them appear to be typical high-velocity clouds (Morras 1985). Comments on the structures and motions of the Bridge, including the SMC Wing and the Stream, are also given in Chap. 3.

Appendix I

Acronyms and abbreviations used frequently in the text

LMC = Large Magellanic Cloud
SMC = Small Magellanic Cloud

ISM = InterStellar Medium

BC = Bolometric Correction
M_{bol} = absolute bolometric magnitude
L_{bol} = bolometric luminosity
T_{eff} = effective temperature
$\Gamma = -x$ = Slope of the IMF
X, Y, Z = relative content of H, He, and heavy elements (M)
$[M/H] = \log N(M/H)_{star} - \log N(M/H)_{\odot}$
FWHM = Full Width at Half Measure
SSPSF = Stochastic Self-Propagating Star Formation

LF = Luminosity Function
IMF = Initial Mass Function
CMD = Colour Magnitude Diagram
HR diagram = Hertzsprung Russell Diagram
MS = Main Sequence
ZAMS = Zero-Age Main Sequence
AGB = Asymptotic Giant Branch
TP = Thermal Pulse
E-AGB = Early AGB
TP-AGB = Late AGB
RGB = Red-Giant Branch
SGB = Sub-Giant Branch
HB = Horizontal Branch
RGC = Red-Giant Clump
PN = Planetary Nebula
SN = SuperNova
SNR = SuperNova Remnant

Identification of objects:

Stars: AV = Azzopardi, Vigneau 1982; BI = Brunet *et al.* 1975; Brey = Breysacher 1988; Mk = Melnick 1985; R = Feast *et al.* 1960; S = Henize 1956
Clusters: Br = Brück 1975, 1976; H = Hodge 1988c; HS = Hodge and Sexton 1966; K = Kron 1956; LW = Lyngå and Westerlund 1963; SL = Shapley and Lindsay 1963
Stellar associations: LH = Lucke and Hodge 1970
Nebulae: DEM = Davies *et al.* 1976; N = Henize 1956
Dark nebulae: H = Hodge 1974a
Planetary nebulae: WS = Westerlund and Smith 1964; SMP = Sanduleak *et al.* 1978, Sanduleak 1984
X-ray sources: E = *Einstein*; RX = *ROSAT*; CAL =*Columbia Astrophys. Lab.*

Motions:
LSR = Local Standard of Rest
GLR = Galactic Standard of Rest
mas = milliarcsecond

Photometry:

UV = far ultraviolet
U, B, V, R, I = standard photometric passbands: ultraviolet, blue, visual, red and (photographic) infrared
nIR = near InfraRed passbands: J,H,K, L, M at, 1.3, 1.6, 2.2, 3.5, 5 μm
FIR = Far InfraRed passbands, usually IRAS passbands: 12, 25, 60, 100 μm
V, B, L, U, W = passbands in Walraven photometric system

Observatories

ADH = Armagh–Dunsink–Harvard
ESO = European Southern Observatory
CTIO = Cerro Tololo International Observatory

Space instruments:

Einstein Observatory: HRI = High Resolution Imager; IPC = Imaging Proportional Counter
ROSAT = Röntgen Observatory SATellite: PSPC = Position Sensitive Proportional Counter
IUE = International Ultraviolet Explorer
HST = Hubble Space Telescope: PC = Planetary Camera; FOC = Faint Object Camera; GHRS = Goddard High Resolution Spectrograph; WFC Wide Field Camera

Appendix 2

Reviews and proceedings

Problems of the Magellanic Clouds, W. Buscombe, S.C.B. Gascoigne, G. de Vaucouleurs, 1955, *Austr. J. Sci. Suppl.* **17**, No. 3.

Recent Developments in Studies of the Magellanic Clouds, A.D. Thackeray, 1963, *Advances in Astron. Astrophys.* **2**, 264.

The Galaxy and the Magellanic Clouds, IAU/URSI Symp. No. **20**,1964, eds. F.J. Kerr, A.W. Rodgers (Austr. Acad. Sci., Canberra).

The Magellanic Clouds, Symposium at Mt Stromlo Observatory, 1965, eds. J.V. Hindman, B.E. Westerlund, E.R. Wilkie.

Magellanic Clouds, B.J. Bok, 1966, *Ann. Rev. Astron. Astrophys.* **4**, 95.

The Stellar Content of the Magellanic Clouds, B.E. Westerlund, 1970, *Vistas in Astronomy*, **12**, 335.

The Magellanic Clouds, European Southern Observatory Symposium in Santiago de Chile, March 1969, 1971, ed. A. Muller (D. Reidel, Dordrecht).

Die Magellanschen Wolken, Th. Schmidt, 1972, *Mitteilungen der Astron. Gesellschaft Nr. 31*.

Structure and Dynamics of Barred Spiral Galaxies, in particular of the Magellanic Type, G. de Vaucouleurs and K.C. Freeman, 1973, *Vistas in Astr.* **14**, 163.

Galaxien vom Magellanischen Typ, J.V. Feitzinger, 1980, *Space Sci. Rev.* **27**, 35.

Structure and Evolution of the Magellanic Clouds, IAU Symp. No. **108**, 1984, eds. S. van den Bergh, K.S. de Boer (D. Reidel, Dordrecht).

The Magellanic Clouds: their evolution, structure and composition, B.E. Westerlund, 1990, *A&AR* **2**, 29.

The Magellanic Clouds, IAU Symp. No. **148**, 1991, eds. R. Haynes, D. Milne (Kluwer, Dordrecht).

Recent Development of Magellanic Cloud Research, A European Colloquium, 1989, eds. K.S. de Boer, F. Spite, G. Stasińska.

New Aspects of Magellanic Cloud Research, Proc. 2nd European Meeting on the Magellanic Clouds, Heidelberg, 1993, eds. B. Baschek, G. Klare, J. Lequeux (Springer-Verlag, Berlin).

For a listing of catalogues see *Nomenclature in the Magellanic Clouds*, M.-C. Lortet, S. Borde, F. Ochsenbein, 1993, *Dictionary of the Nomenclature of Celestial Objects*.

Bibliography

Abbe C., 1867, *MNRAS* **27**, 262.
Alcaino G., Liller W., 1987, *AJ* **94**, 372.
Alexander J.B., 1960, *MNRAS* **121**, 99.
Allen D.A., 1980, *ApL* **20**, 131.
Alongi M., Bertelli G., Bressan A., Chiosi C., Fagotto F., Greggio L., Nasi E., 1993, *A&AS* **97**, 851.
Alvarez H., Aparici J., May J., 1987, *A&A* **176**, 25.
Alvarez H., Aparici J., May J., 1989, *A&A* **213**, 13.
Amy S.W., Ball L., 1993, *ApJ* **411**, 761.
Andersen J., Blecha A., Walker M.F., 1984, *MNRAS* **211**, 695.
Andersen J., Blecha A., Walker M.F., 1985, *A&A* **150**, L12.
Andersen J., Blecha A., Walker M.F., 1987, *MNRAS* **229**, 1.
Ardeberg A., Maurice E., 1979, *A&A* **77**, 277.
Ardeberg A., Linde P., Lindgren H., Lyngå G.,1985, *A&A* **148**, 263.
Arellano Ferro A., Mantegazza L., Antonello E., 1991, *A&A* **246**, 341.
Arp H.C., 1957, *ApJ* **149**, 91.
Arp H.C., 1958, *AJ* **63**, 273.
Arp H.C., 1959a, *AJ* **64**, 175.
Arp H.C., 1959b, *AJ* **64**, 254.
Arp H.C., 1967, *ApJ* **149**, 91.
Arp H., Thackeray D., 1967, *ApJ* **149**, 73.
Avner E.S., King I.R., 1967, *AJ* **72**, 650.
Azzopardi M., 1982, *Comptes Rendus sur les Journées de Strasbourg* (Observatoire de Strasbourg), p.20
Azzopardi M., 1989, *Recent Developments of Magellanic Cloud Research* (eds. K.S. de Boer, F. Spite, G. Stasinska), p.57.
Azzopardi M., Lequeux J., Maeder A., 1988, *A&A* **189**,34.
Azzopardi M., Vigneau J., 1977, *A&A* **56**, 151.
Azzopardi M., Vigneau J., 1982, *A&AS* **50**, 291.
Baade W., 1963, *Evolution of Stars and Galaxies* (ed. Cecilia Payne-Gaposchkin; Harvard University Press), p. 233.
Bajaja E., Loiseau N., 1982, *A&AS* **48**, 71.
Balona L.A., Jerzykiewicz M., 1993, *MNRAS* **260**, 782.
Barbero J., Brocato E., Cassatella A., Castellani V., Geyer E.H., 1990, *ApJ* **351**, 98.
Barlow M.J. Morgan B.L., Standley C., Vine H., 1986, *MNRAS* **223**, 151
Barnes III Th.G., Moffett Th.J., Gieren W.P., 1993, *ApJ* **405**, L51.
Batchelor R.A., McCulloch M.G., Whiteoak J.B., 1981, *MNRAS* **194**, 911.
Becker S.A., 1981, *ApJS* **45**, 478.
Becker S.A., 1982, *ApJ* **260**, 695.
Becker S.A., Mathews G.J., 1983, *ApJ* **270**, 155.
Bencivenni D., Brocato E., Buonanno R., Castellani V., 1991, *AJ* **102**, 137.
Berman B.G., Suchov A.A., 1991, *Ap. Space Sci.* **184**, 169.
Bertelli G., Bressan A., Chiosi C., 1985, *A&A* **150**, 33.
Bertelli G., Mateo M., Chiosi C., Bressan A., 1992, *ApJ* **388**, 400.

Bertelli G., Bressan A., Chiosi C., Mateo M., Wood P.R., 1993, *ApJ* **412**, 160.
Bessell M.S., 1991, *A&A* **242**, L17.
Bessell M.S., Wood P.R., 1993, *New Aspects of Magellanic Cloud Research* eds. B. Baschek, G. Klare, J. Lequeux; Springer-Verlag, Berlin), p. 271.
Bessell M.S., Wood P.R., Lloyd Evans T., 1983, *MNRAS* **202**, 59.
Bessell M.S., Freeman K.C., Wood P.R., 1986, *ApJ* **310**, 710.
Beuermann K., Brandt S., Pietsch W., 1994, *A&A* **281**, L45.
Bhatia R.K., 1990, *PASJ* **42**, 757.
Bhatia R.K., 1992, *Mem. S. A. It.* **63**, 141.
Bhatia R.K., Hatzidimitriou D., 1988, *MNRAS* **230**, 215.
Bhatia R.K., MacGillivray H.T., 1987, *ESO Conf. and Workshop Proc.* **27** (eds. M. Azzopardi, F. Matteucci), p. 485.
Bhatia R.K., MacGillivray H.T., 1988, *A&A* **203**, L5.
Bhatia R.K., MacGillivray H.T., 1989, *A&A* **211**, 9.
Bhatia R., Piotto G., 1994, *A&A* **283**, 424.
Bhatia R.K., Cannon R.D., Hatzidimitriou D., 1987, *ESO Conf. and Workshop Proc.* **27** (eds. M. Azzopardi, F. Matteucci), p. 489.
Bhatia R.K., Read M.A., Hatzidimitriou D., Tritton S., 1991, *A&AS* **87**, 335.
Bica E., Alloin D., Santos, Jr. J.F.C., 1990, *A&A* **235**, 103.
Bica E., Clariá J.J., Dottori H., 1992, *AJ* **103**, 1859.
Bica E., Clariá J.J., Dottori H., Santos, Jr. J.F.C., Piatti A., 1991, *ApJ* **381**, L51.
Blades J.C., 1980, *MNRAS* **130**, 33.
Blades J.C., Madore B.F., 1979, *A&A* **71**, 359.
Blades J.C., Meaburn J., 1980, *MNRAS* **190**, 59P.
Blades J.C., Barlow M.J., Albrecht R. et al., 1992, *ApJ* **398**, L 41.
Blair W.P., Raymond J.C., Long K.S., 1994, *ApJ* **423**, 334.
Blair W.P., Raymond J.C., Danziger J., Matteucci F.,1989, *ApJ* **338**, 812.
Blanco V.M., McCarthy M.F., 1975, *Nature* **258**, 407.
Blanco V.M., McCarthy M.F., 1983, *AJ* **88**, 1442.
Blanco V.M., McCarthy M.F., 1990, *AJ* **100**, 674.
Blanco V.M., Frogel J.A., McCarthy M.F., 1981, *PASP* **93**, 532.
Blanco V.M., McCarthy M.F., Blanco B.M., 1980, *ApJ* **242**, 938 (BMB).
Böhm-Vitense E., Hodge P., Proffitt Ch., 1985, *ApJ* **292**, 130.
Bok B.J., 1964, *IAU/URSI Symp. No.20*, eds. F.J. Kerr and A.W. Rodgers; Austr. Acad. Sci., Canberra), p. 335.
Bok B.J., Bok P.F., 1969, *AJ* **74**, 1125.
Bolte M., 1987, *ApJ* **315**, 469.
Bomans D.J., Dennerl K., Kürster M., 1994, *A&A* **283**, L21.
Booth R. S., 1993, *New Aspects of Magellanic Cloud Research* (eds. B. Baschek, G. Klare, J. Lequeux; Springer-Verlag, Berlin), p. 26.
Booth R.S., de Graauw Th., 1991, *IAU Symp. No. 148* (eds. R. Haynes, D. Milne; Kluwer, Dordrecht), p. 415.
Booth R., Johansson L.E.B., 1991, *The Magellanic Clouds, IAU Symp. No. 148* (eds. R. Haynes, D. Milne, Kluwer, Dordrecht), p. 157.
Boreiko R.T., Betz A.L., 1991, *ApJ* **380**, L27.
Boroson T.A., Liebert J., 1989, *ApJ* **339**, 844.
Bothun G.D., Thomson I.B., 1988, *AJ* **96**, 877.
Bothun G., Elias J.H., MacAlpine G., Matthews K., Mould J.R., Neugebauer G., Reid I.N., 1991, *AJ* **101**, 2220.
Braunsfurth E., Feitzinger J.V., 1983, *A&A* **127**, 113.
Bregman J.M., 1979, *ApJ* **229**, 514.
Bressan A., Bertelli A., Chiosi C., 1986, *Mem. Soc. Astr. It.* **57**, 411.
Breysacher J., 1981, *A&AS* **43**, 203.
Breysacher J., 1988, *Étude des Étoiles du type Wolf–Rayet dans Les Nuages de Magellan* (Thesis) Université de Paris.
Brown Th., Cordova F., Ciardullo R., Thomson R., 1994, *ApJ* **422**, 118.
Brunet J.P., Imbert M., Martin N., Mianes P., Prévot L., Rebeirot E., Rousseau J., 1975, *A&AS* **21**, 109.
Brück M.T., 1975, *MNRAS* **173**, 327.

Brück M.T., 1976, *Star clusters in the Small Magellanic Cloud, Occasional Reports of the Royal Obs.,* Edinburgh, No.1.
Brück M.T., Hawkins M.R.S., 1983, *A&A* **124**, 216.
Brück M.T., Marşōglu A., 1978, *A&A* **68**, 193.
Burkert A., Livio M., Truran J.W., 1990, *ApJ* **360**, 68.
Burki G., Maeder A., 1976, *A&A* **51**, 247.
Buscombe W., Gascoigne S.C.B., de Vaucouleurs G., 1954, *Aust. J. Sci. Suppl.* **17**, No.3.
Butcher H., 1977, *ApJ* **216**, 372.
Butler C.J., 1971, *IAU Coll. No. 15, Veröff. Remeis-Sternw. Bamberg* **9**, No. 100, p.9.
Butler D., Demarque P., Smith H.A., 1982, *ApJ* **257**, 592.
Byrd G., Valtonen M., McCall M., Innanen K., 1994, *AJ* **107**, 2055.
Byrds G.G., 1976, *ApJ* **208**, 688.
Caldwell J.A.R., Coulson I.M., 1985, *MNRAS* **212**, 879.
Caldwell J.A.R., Coulson I.M., 1986, *MNRAS* **218**, 223.
Caldwell, J.A.R., Coulson I.M.,1987, *AJ* **93**, 1090.
Caldwell J.A.R., Laney C.D., 1991, *IAU Symp. No. 148* (eds. R. Haynes, D. Milne; Kluwer, Dordrecht), p.249.
Caloi V., Cassatella A., Castellani V., Walker A., 1993, *A&A* **271**, 109.
Campbell B., Hunter D.A., Holtzman J.A., et al., 1992, *AJ* **104**, 1721.
Cantó J., Elliott K.H., Goudis C., Johnson P.G., Mason D., Meaburn J., 1980, *A&A* **84**, 167.
Capaccioli M., Della Valle M., D'Onofrio M., Rosino L., 1990, *ApJ* **360**, 63.
Caplan J., Deharveng L., 1985, *A&AS* **62**, 63.
Caplan J., Deharveng L., 1986, *A&A* **155**, 297.
Caplan J., Ye T., Deharveng L., Turtle A.J., Kennicutt R.C., 1996, *A&A* **307**, 403, in press.
Carney B.W., 1990, *ASP Conf. Ser.* **14**, 15.
Carney B.W., Janes K.A., Flower Ph.J., 1985, *AJ* **90**, 1196.
Carney B.W., Storm J., Jones R.V., 1992, *ApJ* **386**, 663.
Cassatella A., Barbero J., Geyer E.H., 1987, *ApJS* **64**, 83.
Castellani V., Chieffi A., Pulone L., 1991, *ApJS* **76**, 911.
Caswell J.L., 1995, *MNRAS* **272**. L31.
Caswell J.L., Haynes R.F., 1981, *MNRAS* **194**, 33P.
Caulet A., Deharveng L., Georgelin Y.M., Georgelin Y.P., 1982, *A&A* **110**, 185.
Cester B., Marsi C., 1984, *Astr. Space Sci.* **107**, 167.
Chevalier C., Ilovaisky S.A., 1977, *A&A* **59**, L9.
Chiosi C., Maeder A., 1986, *ARA&A* **24**, 329.
Chiosi C., Pigatto L., 1986, *ApJ* **308**, 1.
Chiosi C., Bertelli G., Bressan A., 1986, *Mem. S. A. It.* **57**, 507.
Chiosi C., Wood P.R., Capitanio N., 1993, *ApJS* **86**, 541.
Chiosi C., Bertelli G., Bressan A., Nasi E., 1986, *A&A* **165**, 84.
Chiosi C., Bertelli G., Bressan A., Wood P.R., Mateo M., 1992, *ApJ* **385**, 205.
Chiosi C., Bertelli G., Meylan G., Ortolani S., 1989a, *A&AS* **78**, 89.
Chiosi C., Bertelli G., Meylan G., Ortolani S., 1989b, *A&A* **219**, 167.
Chiosi C., Vallenari A., Bressan A., Deng L., Ortolani S., 1995, *A&A* **293**, 710.
Chu Y.H., 1983, *ApJ* **269**, 202.
Chu Y.-H., Kennicutt R.C., 1988, *AJ* **96**, 1874.
Chu Y.-H., Kennicutt R.C., 1994, *ApJ* **425**, 720.
Chu Y.-H., Mac Low M.-M., 1990, *ApJ* **365**, 510.
Chu Y.-H., Kennicutt R.C., Schommer R.A., Laff J., 1992, *AJ* **103**, 1545.
Chu Y.H., Wakker B., Mac Low M.-M., García-Segura G., 1994, *AJ* **108**, 1696.
Ciardullo R., Jacoby G.H., Ford H.C., 1989, *ApJ* **344**, 715.
Clampin M., Nota A., Golimowski D.A., Leiterer C., 1993, *New Aspects of Magellanic Cloud Research* (eds. B. Baschek, G. Klare, J. Lequeux; Springer-Verlag, Berlin), p. 273.
Clark D.H., Tuohy I.R., Long K.S., Szymkoviak A.E., Dopita M.A., Mathewson D.S., 1982, *ApJ* **255**, 440.
Clark G., Doxsey R., Li F., Jernigan J.G., van Paradijs J., 1978, *ApJ* **221**, L37.
Clark G., Li F., van Paradijs J., 1979, *ApJ* **227**, 54.
Clayton C.A., 1987, *A&A* **173**, 137.
Clementini G., Cacciari C.,1989, *IAU Coll. No. 111* (ed. E.G. Schmidt; Cambridge University Press), p.255.

Cohen J.G., 1984, *AJ* **89**, 1779.
Cohen J. G., Rich R.M., Persson S.E., 1984, *ApJ* **285**, 595.
Cohen J.G., Frogel J.A., Persson S.E., Elias J.H., 1981, *ApJ* **249**, 481.
Cohen R.S., Dame T.M., Garay G., Montani J., Rubio M., Thaddeus P., 1988, *ApJ* **331**, L95.
Cohen J.G., Rich R.M., Persson S.E., 1984, *ApJ* **285**, 595.
Conti P.S., Fitzpatrick E.L., 1991, *ApJ* **373**, 100.
Conti P.S., Garmany C.D., Massey P., 1986, *AJ* **92**, 48.
Copetti M.V.F., Dottori H.A., 1989, *A&AS* **77**, 327.
Corbet R.H.D., Mason K.O., Cordova F.A., Branduardi-Raymont G., Parmar A.N., 1985, *MNRAS* **212**, 565.
Corsi C.E., Buonanno R., Fusi Pecci F., Ferraro F.R., Testa V., Greggio L., 1994, *MNRAS* **271**, 385 and *Microfiche MN271/1*.
Courtès G., Viton M., Sivan J.P., Decher R., Gary A., 1984, *Science* **225**, 179.
Courtès G., Viton M., Bowyer S., Lampton M., Sasseen T.P., Wu X.-Y., 1995, *A&A* **297**, 338.
Cowley A.P., Hartwick F.D.A., 1991, *ApJ* **373**, 80.
Cowley A.P., Dawson P., Hartwick D.A., 1979, *PASP* **91**, 628.
Cowley A.P., Crampton D., Hutchings J.B., Helfand D.J., Hamilton T.T., Thorstensen J.R., Charles P.A., 1984, *ApJ* **286**, 196.
Cowley A.P., Crampton D., Hutchings J.B., Remilliard R., Penfold J.E., 1983, *ApJ* **272**, 118.
Cowley A.P., Schmidtke P.C., Crampton D., Hutchings J.B., 1990, *ApJ* **355**, 288.
Cox P., Deharveng L., 1983, *A&A* **117**, 265.
Crampton D., Hutchings J.B., Cowley A.P., 1978, *ApJ* **223**, L79.
Crampton D., Cowley A.P., Hutchings J.B., Schmidtke P.C., Thompson I.B., 1990, *ApJ* **355**, 496.
Crotts A.P.S., Kunkel W.E., Heathcote S.R., 1995, *ApJ* **438**, 724.
Dachs J., 1970, *A&A* **9**, 95.
Da Costa G.S., 1991, *The Magellanic Clouds, IAU Symp. No. 148* (eds. R. Haynes, D. Milne; Kluwer, Dordrecht), p. 183.
Da Costa G.S., 1993, *ASP Conf. Ser.* **48** (eds. G.H. Smith and J.P. Brodie), 363.
Da Costa G.S., Mould J.R., 1986, *ApJ* **305**, 159.
Da Costa G.S., Mould J.R., 1988, *ApJ* **334**, 214.
Danziger I.J., Goss W.M., Murdin P., Clark D.H., Boksenberg A., 1981, *MNRAS* **195**, 33.
Davies R.D., Wright A.E., 1977, *MNRAS* **180**, 71.
Davies R.D., Elliott K.H., Meaburn J., 1976, *Mem. RAS* **81**, 89, DEM.
De Boer K.S., Nash A.G.,1982, *ApJ* **255**, 447.
De Boer K.S., Savage B.D., 1980, *ApJ* **238**, 86.
De Boer K.S., Koornneef J., Savage B.D., 1980, *ApJ* **236**, 769.
de Boer K.S., Morras R., Bajaja E., 1990, *A&A* **233**, 523.
DeGioia-Eastwood K., Meyers R.P., Jones D.P., 1993, *AJ* **106**, 1005.
Della Valle M., 1991, *A&A* **252**, L9.
De Marchi G., Nota A., Leitherer C., Ragazzoni R., Barbieri C., 1993, *ApJ* **419**, 658.
Demers S., Irwin M.J., Kunkel W.E., 1993, *MNRAS* **260**, 103.
Demers S., Grondin L., Irwin M.J., Kunkel W.E., 1991, *AJ* **101**, 911.
Dennefeld M., 1989, *Recent Developments of Magellanic Cloud Research* (eds. K.S. de Boer, F. Spite, G. Stasinska), p.107.
Dennerl K., Kürster M., Pietsch W., Voges W., 1993, *New Aspects of Magellanic Cloud Research*, Lecture Notes in Physics 416 (eds. B. Baschek, G. Klare, J. Lequeux; Springer, Heidelberg), p.74.
De Vaucouleurs G., 1980, *PASP* **92**, 579.
De Vaucouleurs G., 1993, *ApJ* **415**, 10.
De Vaucouleurs G., Freeman K.C., 1973, *Vistas Astron.* (Pergamon Press, Oxford) **14**, 163.
Dickel J.R., Milne D.K., Junkes N., Klein U., 1993, *A&A* **275**, 265.
Dickel J.R., Milne D.K., Kennicutt R.C., Chu Y.-H., Schommer R.A., 1994, *AJ* **107**, 1067.
Dickey J.M., Mebold U., Marx M., Amy S., Haynes R.F., Wilson W., 1994, *A&A* **289**, 357.
Dixon M.E., Ford V.L., 1972, *ApJ* **173**, 35.
D'Odorico S., Molaro P., Pettini M., Stathakis R., Vladilo G., 1987, *ESO Conf. & Workshop Proc. No. 26* (ed. I.J. Danziger), p. 525.
Domgörgen H., Bomans D.J., de Boer K.S., 1995, *A&A* **296**, 523.
Dopita M.A., 1985, *Astrophys. Space Sci. Libr.* **120**, 269.
Dopita M.A., 1987, *IAU Symp. No. 115* (eds. M. Peimbert, J. Jugaku), p.501.
Dopita M.A., 1993, *IAU Symp. No. 155* (eds. R. Weinberger and A. Acker), p.433.

Dopita M.A., Meatheringham S.J., 1991a, *ApJ* **367**, 115.
Dopita M.A., Meatheringham S.J., 1991b, *ApJ* **374**, L21.
Dopita M.A., Meatheringham S.J., 1991c, *ApJ* **377**, 480.
Dopita M.A., Ford H.C., Webster B.L., 1985b, *ApJ* **297**, 593.
Dopita M.A, Mathewson D.S., Ford V.L., 1985c, *ApJ* **297**, 599.
Dopita M.A., Tuohy I.R., Mathewson D.S., 1981, *ApJ* **248**, L105.
Dopita M.A., Ford H.C., Lawrence C.J., Webster B.L., 1985a, *ApJ* **296**, 390.
Dopita M.A., Ford H.C., Bohlin R, Evans I.R., Meatheringham S.J., 1993, *ApJ* **418**, 804.
Dopita M.A., Vassiliadis E., Meatheringham S.J. *et al.*, 1994, *ApJ* **426**, 150.
Dopita M.A., Vassiliadis E., Meatheringham S.J., Bohlin R., Ford H.C., Harrington J.P., Wood P.R., Stecher Th.P., Maran S.P.,1995, *ApJ, submitted.*
Dottori H.A, Bica E.L.D., 1981, *A&A* **102**, 245.
Dufour R.J., 1984, *IAU Symp. No. 108* (eds. S. van den Bergh, K.S. de Boer; Reidel, Dordrecht), p. 353.
Dufour R.J., Duval J.E., 1975, *PASP* **87**, 769.
Dufton P.L., Fitzsimmons A., Howarth I.D., 1990, *ApJ* **362**, L59.
Ebisawa K., Mitsuda K., Inoue H., 1989, *PASJ* **41**, 519.
Einasto J., Haud U., Jôever M., Kaasik A., 1976, *MNRAS* **177**, 357.
Elias J.H., Frogel J.A., Schwering P.B.W., 1986, *ApJ* **302**, 675.
Ellingsen S.P., Whiteoak J.B., Norris R.P., Caswell J.L., Vaile R.A., 1994, *MNRAS* **269**, 1019.
Elliott K.H., Goudis C., Meaburn J., Tebbutt N.J., 1977, *A&A* **55**, 187.
Elson R.A.W., 1991, *ApJS* **76**, 185.
Elson R.A.W., Fall S.M., 1985, *ApJ* **299**, 211.
Elson R.A.W., Fall S.M., Freeman K.C., 1987, *ApJ* **323**, 54.
Elwert G., Hablick D., 1965, *ZfAph* **61**, 273
Elmegreen D.M., Elmegreen B.G., 1980, *AJ* **85**, 1325.
Epchtein N., Braz M.A., Sère F., 1984, *A&A* **140**, 67.
Epstein A., Delvaille J., Helmken H., Murray S., Schnopper H., Doxsey R., Primini F., 1977, *ApJ* **216**, 103.
Faulkner D.J., 1967, *MNRAS* **135**, 401.
Feast M.W., 1960, *Observatory* **80**. 104.
Feast M.W., 1961, *MNRAS* **122**, 1.
Feast M.W., 1964, *MNRAS* **127**, 195.
Feast M.W., 1968, *MNRAS* **140**, 345.
Feast M.W., 1970, *MNRAS* **149**, 291.
Feast M.W., 1972, *MNRAS* **159**, 113.
Feast M.W., 1979, *MNRAS* **186**, 831.
Feast M.W., 1988, *ASP. Conf. Ser.* **4**, 9.
Feast, M.W., 1989a, *Recent Developments of Magellanic Cloud Research* (eds. K.S. de Boer, F. Spite, G. Stasińska), p.75.
Feast, M.V., 1989b, *The World of Galaxies* (eds. H.G. Corwin, L. Bottinelli; Springer-Verlag, Berlin), p.118.
Feast M.W., 1992a, *Highlights of Astronomy* (ed. J. Bergeron; Kluwer, Dordrecht).
Feast M.W., 1993, *New Aspects of Magellanic Cloud Research* (eds. B. Baschek, G. Klare, J. Lequeux; Springer-Verlag, Berlin), p. 239.
Feast M.W., Black C., 1980, *MNRAS* **191**, 285.
Feast M.W., Walker A.R., 1987, *ARAA* **25**, 345.
Feast M.W., Whitelock P.A., 1992, *MNRAS* **259**, 6.
Feast M.W., Glass I.S., Whitelock P.A., Catchpole R.M.,1989, *MNRAS* **241**, 371.
Feast M.W., Thackeray A.D., Wesselink A.J., 1960, *MNRAS* **121**, 337.
Feast M.W., Thackeray A.D., Wesselink A.J., 1961, *MNRAS* **122**, 433.
Feitzinger J.V., 1980, *Space Sci. Rev.* **27**, 35.
Feitzinger J.V., 1987, *IAU Symp. No. 115* (eds. M. Peimbert, J. Jugaku; Reidel, Dordrecht), p.521.
Feitzinger J.V., Schmidt-Kaler Th.,1982, *ApJ* **257**, 587.
Feitzinger J.V., Hanuschik R.W., Schmidt-Kaler Th., 1984, *MNRAS* **211**, 867.
Feitzinger J.V., Isserstedt J., Schmidt-Kaler Th., 1977, *A&A* **57**, 265.
Feitzinger J.V., Schlosser W., Schmidt-Kaler Th., Winkler Ch., 1980, *A&A* **84**, 50.
Feitzinger J.V., Haynes R.F., Klein U., Wielebinski R., Perschke M., 1987, *Vistas Astron.* **30**, 243.
Ferraro F.R., Fusi Pecci F., Testa V. *et al.*, 1995, *MNRAS* **272**, 391.

Fischer Ph., Welch D.L., Mateo M., 1992, *AJ* **104**, 1086.
Fischer Ph., Welch D.L., Mateo M., 1993, *AJ* **105**, 938.
Fitzpatrick E.L.,1985a, *ApJ* **299**, 219.
Fitzpatrick E.L., 1985b, *ApJS* **59**, 77.
Fitzpatrick E.L., 1986, *AJ* **92**, 1068.
Fitzpatrick E.L., Garmany C.D., 1990, *ApJ* **363**, 119.
Fitzpatrick E.L., Savage B.D., 1984, *ApJ* **279**, 578.
Fitzpartick E.L., Savage B.D., 1985, *ApJ* **292**, 122.
Flower Ph.J., 1984, *ApJ* **278**, 582.
Fong R., Jones L.R., Shanks T. *et al.*, 1987, *MNRAS* **224**, 1059.
Fransson C., Cassatella A., Gilmozzi R. *et al.*, 1989, *ApJ* **336**, 429.
Frantsman Ju.L., 1988, *Ap&Space Sci.* **145**, 287.
Frantsman Ju., Shmeld I., 1995, *IAU Symp. No. 164* (eds. P.C. van der Kruit, G. Gilmore; Kluwer, Dordrecht), p. 415.
Freeman K.C., Illingworth G., Oemler Jr. A., 1983, *ApJ* **272**, 488.
Frenk C.S., Fall S.M., 1982, *MNRAS* **199**, 565.
Frogel J.A., 1984, *PASP*, **96**, 856.
Frogel J.A., Blanco V.M., 1983, *ApJ* **274**, L57.
Frogel J.A., Blanco V.M., 1990, *ApJ* **365**, 168.
Frogel J.A., Richer H.B., 1983, *ApJ* **275**, 84.
Frogel J.A., Mould J., Blanco V.M., 1990, *ApJ* **352**, 96.
Fujimoto M., Murai T., 1984, *IAU Symp. No. 108* (eds. S. van den Bergh, K.S. de Boer; Reidel, Dordrecht), p. 115.
Fujimoto M., Noguchi M., 1990, *PASJ* **42**, 505.
Gaposchkin S., 1970, *Smithsonian Astroph. Obs. Spec. Rep.* 310.
Garay G., Rubio M., Ramírez S., Johansson L.E.B., Thaddeus P., 1993, *A&A* **274**, 743.
Gardiner L.T., Hatzidimitriou D., 1992, *MNRAS* **257**, 195.
Gardiner L.T., Noguchi M., 1995, *MNRAS*, preprint.
Gardiner L.T., Sawa T., Fujimoto M., 1994, *MNRAS* **266**, 567.
Gardner F.F., Whiteoak J.B., 1985, *MNRAS* **215**, 103.
Garmany C.D., Conti P.S., Massey Ph., 1987, *AJ* **93**, 1070.
Garmany C.G., Massey Ph,. Parker J.W., 1994, *AJ* **108**, 1256.
Gascoigne S.C.B., 1966, *MNRAS* **134**, 59.
Gatley I., Hyland A.R., Jones T.J., 1982, *MNRAS* **200**, 521.
Gatley I., Becklin E.E., Hyland A.R., Jones T.J., 1981, *MNRAS* **197**, 17P.
Geisler D.,1987, *AJ* **93**, 1081.
Gerola H., Seiden P.L., 1978, *ApJ* **223**, 129.
Giaconi R., Murray S., Gursky H., Kellogg E., Schreier E., Tananbaum H., 1972, *ApJ* **178**, 281.
Gieren W.P., Barnes Th. G., Moffett T.J., 1993, *ApJ* **418**, 135.
Gieren W.P., Richtler T., Hilker M., 1994, *ApJ* **433**, L73.
Gilmore G., Wyse R.F.G., 1991, *ApJ* **367**, L55.
Gilmozzi R., Murdin P., Clark D.H., 1984, *A&A* **140**, 390.
Gilmozzi R., Murdin P., Clark D.H., Malin D., 1983, *MNRAS* **202**, 927.
Gilmozzi R., Kinney E.K., Ewald S.P., Panagia N., Romaniello M., 1994, *ApJ* **435**, L43.
Goldsmith P.F., Langer W.D., Ellder J., Irvine W.M., Kollberg E., 1981, *ApJ* **249**, 524.
Goldwurm A., Caraveo P.A., Bignami G.F., 1987, *ApJ* **322**, 349.
Gorenstein P., Hughes J.P., Tucker W.H., 1993, *ApJ* **420**, L25.
Graham J.A., 1975, *PASP* **87**, 641.
Graham J.A., 1977, *PASP* **89**, 425.
Graham J.A., 1984, *IAU Symp. No. 108* (eds. S. van den Bergh, K.S. de Boer), p. 207.
Gratton R.G., Ortolani S., 1987, *A&AS* **67**, 373.
Grebel E.K., Richtler T., 1992, *A&A* **253**, 359.
Green E.M., Demarque P., King C.R., 1987, *The Revised Yale Isochrones and Luminosity Functions* (Yale University Observatory, New Haven).
Greiner J., Hasinger G., Thomas H.-C., 1994, *A&A* **281**, L61.
Greve A., Castles J., McKeith C.D., 1991, *A&A* **251**, 575.
Greve A., van Genderen A.M., Laval A., 1990, *A&AS* **85**, 895.
Greve A., van Genderen A.M., Laval A., van Driel W., Prein J.J., 1988, *A&AS* **74**, 167.
Grondin L., Demers S., Kunkel W.E., 1992, *AJ* **103**, 1234.

Grondin L., Demers S., Kunkel W.E., Irwin M.J., 1990, *AJ* **100**, 663.
Grondin L., Demers S., Kunkel W.E., Irwin M.J., 1991, *IAU Symp. No.148* (eds. R. Haynes, D. Milne; Kluwer, Dordrecht), p. 478.
Gummersbach C.A., Zickgraf F.-J., Wolf B., 1995, *A&A* **302**, 409.
Han C., Ryden B.S., 1994, *ApJ* **433**, 80.
Hanuschik R.W., Schmidt-Kaler Th., 1991, *A&A* **249**, 36.
Hardy E., Buonanno R., Corsi C.E., Janes K.A., Schommer R.A., 1984, *ApJ* **278**, 592.
Hardy E., Suntzeff N.B., Azzopardi M., 1989, *ApJ* **344**, 210.
Harris H.C., 1981, *AJ* **86**, 1192.
Harris H.C., 1983, *AJ* **88**, 507.
Hartwick F.D.A., Cowley A.P., 1985, *AJ* **90**, 2244.
Hartwick F.D.A., Cowley A.P., 1988, *ApJ* **334**, 135.
Hartwick F.D.A., Cowley A.P., 1991, *The Magellanic Clouds, IAU Symp. No. 148*, p. 77 (eds. R. Haynes and D. Milne, Kluwer, Dordrecht).
Hatzidimitriou D., 1991, *MNRAS* **251**, 545.
Hatzidimitriou D., Cannon R.D., 1993, *New Aspects of Magellanic Cloud Research*, p. 17.
Hatzidimitriou D., Hawkins M.R.S., 1989, *MNRAS* **241**, 667.
Hawkins M.R.S., Brück M.T., 1982, *MNRAS* **198**, 935.
Haynes R.F., Caswell J.L., 1981, *MNRAS* **197**, 23P.
Haynes R.F., Klein U., Wielebinski R., Murra, J.D., 1986, *A&A* **159**, 22.
Haynes R.F., Klein U., Wayte S.R., Wielebinski R., Murray J.D., Bajaja E.E., Meinert D., Buczilowski U.R., Harnett J.I., Hunt A.J., Wark R., Sciacca L., 1991, *A&A* **252**, 475.
Hazen M.L., Nemec J.M., 1992, *AJ* **104**, 111.
Heap S., 1992, *Photo Release, STSCI-PRC92-12*.
Heap S.R., Altner B., Ebbets D. et al., 1991, *ApJ* **377**, L29.
Heap S.R., Ebbets D., Malumuth E.M., 1992, *Science with the HST* (eds. P.Benvenuti, E. Schreier; ESO, Garching), p. 347.
Heap S.R., Ebbets D., Malumuth E.M., Matan S.P., De Koter A., Hubeny I., 1994, *ApJ*, L39.
Helfand D.J., 1984, *IAU Symp. 108* (eds. S.van den Bergh, K.S. de Boer; Reidel, Dordrecht), p. 293.
Heller P., Rohlfs K., 1994, *A&A* **291**, 743.
Henize K.G., 1956, *ApJS* **2**, 315.
Herschel J., 1847, *Results of Astronomical Observations at the Cape of Good Hope*, p. 143 (London).
Hertzsprung E.,1920, *MNRAS* **80**, 782.
Heydari-Malayeri M., Hutsemékers D., 1991a, *A&A* **243**, 401.
Heydari-Malayeri M., Hutsemékers D., 1991b, *A&A* **244**, 64.
Heydari-Malayeri M., Testor G., 1982, *A&A* **111**, L11.
Heydari-Malayeri M., Testor G., 1983, *A&A* **118**, 116.
Heydari-Malayeri M., Testor G., 1985, *A&A* **144**, 98.
Heydari-Malayeri M., Magain P., Remy M., 1988, *A&A* **201**, 98.
Heydari-Malayeri M., Magain P., Remy M., 1989, *A&A* **222**, 41.
Heydari-Malayeri M., Niemela V.S., Testor G., 1987, *A&A* **184**, 300.
Heydari-Malayeri M., Grebel E.K., Melnick J., Jorda L., 1993, *A&A* **278**, 11.
Hilker M., Richtler T., Gieren W., 1995, *A&A* **294**, 648.
Hill V., Andrievsky S., Spite M., 1995, *A&A* **293**, 347.
Hill R.J., Madore B.F., Freedman W.L., 1994a, *ApJS* **91**, 583.
Hill R.J., Madore B.F., Freedman W.L., 1994b, *ApJ* **429**, 192.
Hill R.J., Madore B.F., Freedman W.L., 1994c, *ApJ* **429**, 204.
Hill J.K., Bohlin R.C., Cheng K.-P. et al., 1993, *ApJ* **413**, 604.
Hindman J.V., 1964, *IAU/URSI Symp. No. 20*, (eds. F.J .Kerr, A.W. Rodgers), p.255.
Hindman J.V., 1967, *Aust. J. Phys.* **20**, 147.
Hindman J.V., Kerr F.J., McGee R.X., 1963, *Aust. J. Phys.* **16**, 570.
Hodge P.W., 1960, *PASP* **72**, 308.
Hodge P.W., 1961, *ApJ* **133**, 413.
Hodge P.W., 1972, *PASP* **84**, 365.
Hodge P.W., 1974a, *PASP* **86**, 263.
Hodge P.W., 1974b, *ApJ* **192**, 21.
Hodge P.W., 1974c, *AJ* **79**, 860.
Hodge P.W., 1980, *AJ* **85**, 423.
Hodge P., 1985, *PASP* **97**, 530.

Hodge P., 1986, *PASP* **98**, 1113.
Hodge P., 1987, *PASP* **99**, 724.
Hodge P., 1988a, *PASP* **100**, 346.
Hodge P., 1988b, *PASP* **100**, 576.
Hodge P., 1988c, *PASP* **100**, 1051.
Hodge P.W., Flower Ph., 1987, *PASP* **99**, 734.
Hodge P.W., Lucke P.B., 1970, *AJ* **75**, 933.
Hodge P.W.,Sexton J.A., 1966, *AJ* **71**, 363.
Hodge P.W.,Wright F.W., 1967, *The Large Magellanic Cloud* (Smithsonian Inst., Washington, D.C.).
Houziaux L., Nandy K., Morgan D.H., 1980, *A&A* **84**, 377.
Houziaux L., Nandy K., Morgan D.H., 1985, *MNRAS* **215**, 5P.
Howarth I.D., Prinja R.K., Roche P.F., Willis A.J., 1984, *MNRAS* **207**, 287.
Huggins P.J., Gillespie A.R., Phillips T.G., Gardner F., Knowles S., 1975, *MNRAS* **173**, 69P.
Hughes J.P., 1987, *ApJ* **314**, 103.
Hughes J.P., 1994a, *ApJ* **427**, L25.
Hughes J.P., 1994b, *ROSAT Sci. Symp., Harvard-Smithsonian preprint No. 3863*
Hughes J.P., Smith R. Ch., 1994, *AJ* **107**, 1363.
Hughes J.P., Bronfman L., Nyman L.,1989, *Supernovae* (ed. S.E. Woosley; Springer-Verlag), p.679.
Hughes J.P., Hayachi I., Helfand D., Hwang U. *et al.*, 1995, *ApJ* **444**, L81.
Hughes S.M.G., 1989, *AJ* **97**, 1634.
Hughes S.M.G., Wood P.R., 1990, *AJ* **99**, 784.
Hughes S.M.G., Wood P.R., Reid N., 1991, *AJ* **101**, 1304.
Humphreys R. M., 1979, *ApJ* **231**, 384.
Humphreys R. M., Kudritzski R.P., Groth H.G., 1991, *A&A* **245**, 593.
Hunter D.A., 1994, *AJ* **107**, 565.
Hunter D.A., Gallagher III J.S., 1990, *ApJ* **362**, 480.
Hunter D.A., Shaya E.J., Scowen P., Hester J.J., Groth E.J., Lynds R., O'Neil Jr. E.J., 1995, *ApJ* **444**, 758.
Hutchings J.B., 1966, *MNRAS* **131**, 299.
Hutchings J.B., 1980, *PASP* **92**, 592.
Hutchings J.B., Cowley A.P., 1991, *IAU Symp. No. 148* (eds. R. Haynes, D. Milne; Kluwer, Dordrecht), p.285.
Hutchings J.B.,Thomson I.B., 1988, *ApJ* **331**, 294.
Hutchings J.B., Crampton D., Cowley A.P., Olszewski E., Thomson I.B., Suntzeff N., 1985, *PASP* **97**, 418.
Hutchings J.B.,Thomson I.B., Cartledge S., Pazder J.,1991, *AJ* **101**, 933.
Hyland A.R., Jones T.J., 1991, *The Magellanic Clouds, IAU Symp. No. 148* (eds. R. Haynes, D. Milne, Kluwer, Dordrecht), p. 202.
Hyland A.R., Thomas J.A., Robinson G., 1978, *AJ* **83**, 20.
Hyland A.R., Straw S., Jones T.L., Gatley I., 1992, *MNRAS* **257**, 391.
Ibata R.A., Gilmore G., Irwin M.J., 1994, *Nature* **370**, 194.
Inoue H., Koyama K., Tanaka Y., 1983, *IAU Symp. No. 101* (eds. J. Danziger, P. Gorenstein; Reidel, Dordrecht), p. 535.
IRAS Point Source Catalog., 1985, *Joint IRAS Science Working Group* (Washington D.C.: US GPO).
Irwin M.J., 1991, *IAU Symp. No.148* (eds. R. Haynes, D. Milne; Kluwer, Dordrecht), p. 453.
Irwin M.J., Demers S., Kunkel W.E., 1990, *AJ* **99**, 191, IDK.
Irwin M.J., Kunkel W.E., Demers S., 1985, *Nature* **318**, 160.
Israel F.P., 1980, *A&A* **90**, 246.
Israel F.P., 1984, *IAU Symp. No. 108* (eds. S. van den Bergh, K.S. de Boer; Reidel, Dordrecht) p. 319.
Israel F.P., de Graauw Th., 1991, *IAU Symp. No. 148* (eds. R.F Haynes, D. Milne; Kluwer, Dordrecht), p. 45.
Israel F.P., De Graauw Th., Lidholm S., Van de Stadt H., De Vries C.P., 1982, *ApJ* **262**, 100.
Israel F.P., De Graauw Th., Van De Stadt H., De Vries C.P., 1986, *ApJ* **303**, 186.
Israel F.P., Koornneef J., 1988, *A&A* **190**, 21.
Israel F.P., Maloney P.R., 1993, *New Aspects of Magellanic Cloud Research* (eds. B. Baschek, G. Klare, J. Lequeux; Springer-Verlag), p. 44.
Israel F.P., Johansson L.E.B., Lequeux J., Booth R. S.,Nyman L.-A.,Crane P., Rubio M., de Graauw Th., Kutner M.L., Gredel R., Boulanger F., Garay G., Westerlund B.,1993, *A&A* **276**, 25.
Isserstedt J., 1975, *A&A* **41**, 21.

Bibliography

Isserstedt J., 1984, *A&A* **131**, 347.
Jacobsen P., Jedrzejwski R., Macchetto F., Panagia N., 1994, *ApJ* **435**, L47.
Jacoby G.H., 1980, *ApJS* **42**, 1
Jacoby G.H., 1989, *ApJ* **339**, 39
Jacoby G.H., Walker A.R., Ciardullo R., 1990, *ApJ* **365**, 471.
Jasniewicz G., Thévenin F., 1994, *A&A* **282**, 717.
Jenniskens P., Ehrenfreund P., Foing B., 1994, *A&A* **281**, 517.
Jensen J., Mould J., Reid N., 1988, *ApJS* **67**, 77.
Johansson L.E.B., Booth R.S., 1989, *Recent developments of Magellanic Cloud Research* (eds. K.S. de Boer, F. Spite, G. Stasińska), p. 149.
Johansson L.E.B., Andersson C., Ellder J., et al., 1984, *A&A* **130**, 227.
Johansson L.E.B., Olofsson H., Hjalmarsson Å., Gredel R., Black J.H., 1994, *A&A* **291**, 89.
Johnson H.M., 1959, *PASP* **71**, 301.
Johnson H.M., 1961, *PASP* **73**, 20.
Johnson P.G., Meaburn J., Osman A.M.I., 1982, *MNRAS* **198**, 985.
Jones J.H., 1987, *AJ* **94**, 345.
Jones T.J., Hyland A.R., Straw S., Harvey P.M., Wilking B.A., Joy M., Gatley I., Thomas J.A., 1986, *MNRAS* **219**, 603.
Jones B.F., Klemola A.R., Lin D.N.C., 1994, *AJ* **107**, 1333.
Jüttner A., 1993, *New Aspects of Magellanic Cloud Research* (eds. B. Baschek, G. Klare, J. Lequeux; Springer-Verlag), p. 301.
Jüttner A., Reitermann A., Stahl O., Wolf B., 1989, *A&AS* **81**, 93.
Jüttner A., Stahl O., Wolf B., Baschek B., 1993, *New Aspects of Magellanic Cloud Research* (eds. B. Baschek, G. Klare, J. Lequeux; Springer-Verlag), p.337.
Kahabka P., Pietsch W., 1993, *New Aspects of Magellanic Cloud Research*, Lecture Notes in Physics 416 (eds. B. Baschek, G. Klare, J. Lequeux; Springer, Heidelberg), p.71.
Kahabka P., Pietsch W., Hasinger G., 1994, *A&A* **288**, 538.
Kaler J.B., Jacoby G.H., 1989, *ApJ* **345**, 871.
Kennicutt R., Chu Y.-H., 1988, *AJ* **95**, 720.
Kennicutt R.C., Hodge P.W., 1986, *ApJ* **306**, 130.
Kennicutt R.C.,Tamblyn P., Condon C.W., 1995b, *ApJ*, in press.
Kennicutt Jr. R.C., Bresolin F., Bomans D.J., Bothun G.D., Thomson I.B., 1995a, *AJ* **109**, 594.
Kerr F.J., 1971, *The Magellanic Clouds* (ed. A. Müller; Reidel, Dordrecht), p.50.
Kerr F.J., 1989, *The World of Galaxies* (eds. H.G. Corwin, L. Bottinelli; Springer, Berlin), p.160.
Kerr F.J., de Vaucouleurs G., 1955a, *Aust. J. Phys.* **8**, 508.
Kerr F.J., de Vaucouleurs G., 1955b, *Aust. J. Phys.* **9**, 90.
Kerr F.J., Hindman J.V., Robinson B.J., 1954, *Aust. J. Phys.* **7**, 297.
King I.R., 1962, *AJ* **67**, 471.
King I.R., 1966, *AJ* **71**, 64.
Kinman T.D., Stryker L.L., Hesser J.E., 1976, *PASP* **88**, 393.
Kinman T.D., Stryker L.L., Hesser J.E., Graham J.A., Walker A.R., Hazen M.L., Nemec J.M., 1991, *PASP* **103**, 1279.
Klein U., Wielebinski R., Haynes R.F., Malin D.F., 1989, *A&A* **211**, 280.
Klein U., Haynes R.F., Meinert D., Wielebinski R., 1993, *New Aspects of Magellanic Cloud Research* (eds. B. Baschek, G. Klare, J. Lequeux; Springer-Verlag), p.123.
Kontizas M., 1984, *A&A* **131**, 58.
Kontizas E., Kontizas M., Michalitsianos A., 1993a, *A&A* **267**, 59.
Kontizas M., Kontizas E., Michalitsianos A.G., 1993b, *A&A* **269**, 107.
Kontizas E., Kontizas M., Xiradaki E., 1989a, *Ap&SS* **156**, 81.
Kontizas E., Metaxa M., Kontizas M., 1988, *AJ* **96**, 1625.
Kontizas M., Chrysovergis M., Kontizas E., Hatzidimitriou D., 1986, *IAU Symp. No.116* (eds. C.W.H. de Loore, A.J. Willis, P. Laskarides; Reidel, Dordrecht), p. 407.
Kontizas M., Kontizas E., Dapergolas A., Argyropoulos S., Bellas-Velidis Y., 1994, *A&AS* **107**, 77.
Kontizas E., Kontizas M., Sedmak G., Smareglia R., Dapergolas A., 1990a, *AJ* **100**, 425.
Kontizas M., Morgan D.H., Hatzidimitriou D., Kontizas E., 1990b, *A&AS* **84**, 527.
Koornneef J., 1984, *IAU Symp. No. 108* (eds. S. van den Bergh, K.S. de Boer; Reidel, Dordrecht), p.333.
Koornneef J., Code A.D., 1981, *ApJ* **247**, 860.
Koornneef J., Israel F.P., 1985, *ApJ* **291**, 156.

Kron G.E., 1956, *PASP* **68**, 125.
Kroupa P., Röser S., Bastian U., 1994, *MNRAS* **266**, 412.
Kudritzki R.P., 1988, *Max-Planck-Institut für Physik and Astrophysik- MPA 380*.
Kudritzki R.P., Pauldrach A., Puls J., 1988, *O Stars and Wolf-Rayet Stars* **NASA SP-497** (eds. P.S. Conti, A.B. Underhill), p. 184.
Kudritski R.P., Cabanne M.L., Husfeld D., Niemela V.S., Groth H.G., Puls J., Herrero A., 1989, *A&A* **226**, 235.
Kudritzki R.P., Groth H.G., Butler K., Husfeld D., Becker S., Eber F., Fitzpatrick E., 1987, *ESO Workshop on the SN 1987 A*, (ed. I.J. Danziger), *ESO Conf. & Workshop Proc.* **26**, 39.
Kuiper L., van Paradijs J., van der Klis M., 1988, *A&A* **203**, 79.
Kumai Y., Basu B., Fujimoto M., 1993, *ApJ* **404**, 144.
Kunkel W.E., 1979, *ApJ* **228**, 718.
Kunkel W.E., 1980, *IAU Symp. No. 85*, (ed. J.E. Hesser; Reidel, Dordrecht), p.353.
Kunkel W.E., Demers S., Irwin M.J., 1995, *Proc. CTIO-ESO Workshop: The Local Group, La Serena*, preprint.
Laney C.D., Stobie R.S., 1986, *MNRAS* **222**, 449.
Lang F.L., Levine A.M., Bautz M., Hauskins S., Howe S., Primini F.A., Lewin W.H.G., Baity W.A., Knight F.K., Rotschild R.E., Petterson J.A., 1981, *ApJ* **246**, L21.
Langer N., Maeder A., 1995, *A&A* **295**, 685.
Langer N., Hamann W.-R., Lennon M., Najarro F., Pauldrach A.W.A., Puls J., 1994, *A&A* **290**, 819.
Lasker B.M., 1980, *ApJ* **237**, 765.
Lattanzio J.C., 1986, *ApJ* **311**, 708.
Lattanzio J.C., Vallenari A., Bertelli G., Chiosi C., 1991, *A&A* **250**, 340.
Lattanzi M.G., Hershey J.L., Burg R. *et al.* 1994, *ApJ* **427**, L 21.
Laval A., Greve A., van Genderen A.M., 1986, *A&A* **164**, 26.
Laval A., Gry C., Rosado M., Marcelin M., Greve A., 1994, *A&A* **288**, 572.
Laval A., Gry C., Rosado M. *et al.*, 1992a, *A&A* **253**, 213.
Laval A., Rosado M., Boulesteix J. *et al.*, 1989, *A&A* **208**, 230.
Laval A., Rosado M., Boulesteix J. *et al.*, 1992b, *A&A* **253**, 230.
Lawson, W.A., Cottrell P.L., Pollard K.R., 1991, *IAU Symp. 148* (eds. R. Haynes, D. Milne; Kluwer, Dordrecht), p.351.
Le Coarer E., Rosado M., Georgelin Y., Viale A., Goldes G., 1993, *A&A* **280**, 365.
Lee Y.-W., 1989, *PhD Thesis, Yale University*
Lee Y.-W., 1992, *PASP* **104**, 798.
Le Marne A.E., 1968a, *Proc. ASA* **1**, 97.
Le Marne A.E., 1968b, *MNRAS* **139**, 461.
Leong C., Kellogg E., Gursky H., Tananbaum H., Giaconi R., 1971, *ApJ* **170**, L67.
Lequeux J., 1989, *Recent Developments of Magellanic Cloud Research* (eds. K.S. de Boer, F. Spite, G. Stasińska), p. 119.
Lequeux J., 1994 *A&A* **287**, 368.
Lequeux J., Le Bourlot J., Pineau de Forèts G., Boulanger F., Rubio M., 1994, *A&A* **292**, 371.
Lequeux J., Maurice E., Prévot-Burnichon M.-L., Prévot L., Rocca-Volmerange B., 1982, *A&A* **113**, L15.
Lequeux J., Peimbert M., Rayo J.F., Serrano A., Torres-Peimbert S., 1979, *A&A* **80**, 155.
Li F., Rappaport S., Epstein A., 1978, *Nature* **271**, 37.
Liller W., 1990, *IAU Circ. 4964*.
Liller W., 1991, *IAU Circ. 5244*.
Lin D.N.C., 1993, *BAAS* **25**, 783.
Lin D.N.C., Lynden-Bell, D., 1977, *MNRAS* **181**, 59.
Lin D.N.C., Lynden-Bell, D., 1982, *MNRAS* **198**, 707.
Lin D.N.C., Jones B.F., Klenmola A., 1995, *ApJ* **439**, 652.
Linde P., Lyngå G., Westerlund B.E., 1995, *A&AS* **110**, 533.
Lindsay E.M., 1958, *MNRAS* **118**, 172.
Liu Y.-z, 1992, *A&A* **257**, 505.
Lloyd-Evans T., 1971, *The Magellanic Clouds* (ed. A. Muller), p. 74.
Loiseau N., Bajaja E., 1981, *Rev. Mex. Astron. Astrofis.* **6**, 55.
Loiseau N., Klein U., Greybe A., Wielebinski R., Haynes R.F., 1987, *A&A* **178**, 62.
Long K.S., Helfand D.J., 1979, *ApJ* **237**, L77.
Long K.S., Helfand D.J., Grabelsky D.A., 1981, *ApJ* **248**, 925.

Lortet M.C., Testor G., 1984, *A&A* **139**, 330.
Lortet M.-C., Testor G., 1988, *A&A* **194**, 11.
Lortet M.C., Testor G., 1991, *A&AS* **89**, 185.
Lu L., Savage B.D., Sembach K.R., 1994, *ApJ* **437**, L119.
Luck R.E., Lambert D.,1992, *ApJS* **79**, 303.
Lucke P.B., 1974, *ApJS* **28**, 73.
Lucke P.B., Hodge P.W., 1970, *AJ* **75**, 171.
Luks Th., Rohlfs K., 1992, *A&A* **263**, 41.
Lundgren K., 1988, *A&A* **200**, 85.
Lupton R.H., Fall S.M., Freeman K.C., Elson R.A.W., 1989, *ApJ* **347**, 201.
Lyngå G., Westerlund B.E., 1963, *MNRAS* **127**, 31.
McCall M.L., 1993, *ApJ* **417**, L 75.
McCarthy F.D., 1956, *Australia's AboriginesTheir Life and Culture* (Colourgravure Publications, Melbourne, Australia).
McClure R.D., 1984, *ApJ* **280**, L31.
McClure R.D., 1989, *Evolution of Peculiar Red Giant Stars, IAU Coll. No. 106*, p. 196 (eds. H.R. Johnson and B. Zuckerman, Cambridge Univ. Press, Cambridge).
McGee R.X., 1964, *Aust. J. Phys.* **17**, 515.
McGee R.X., Milton J.A., 1966, *Austr. J. Phys.* **19**, 343 and *Austr. J. Phys. Suppl. No.2*.
McGee R.X., Newton L.M., 1981, *Proc. ASA* **4**, 189.
McGee R.X., Newton L.M., 1982, *Proc. ASA* **4**, 305.
McGee R.X., Newton L.M., 1986, *Proc. ASA* **6**, 471.
McGee R.X., Brooks J.W., Batchelor R.A., 1972, *Austr. J. Phys.* **25**, 581.
McGee R.X., Newton L.M., Brooks J.W., 1974, *Austr. J. Phys.* **27**, 729.
McGregor P.J., Hyland A.R., 1981, *ApJ* **250**, 116.
McGregor P.J., Hillier D.J., Hyland A.R., 1988, *ApJ* **334**, 639.
McKee J.D., Fritz G., Cruddace R.G., Shulman S., Friedman H., 1980, *ApJ* **238**, 93.
Madore B.F., Freedman W.L., 1991, *PASP* **103**, 933.
Maeder A., 1990, *A&AS* **84**, 139.
Maeder A., Conti P.S., 1994, *ARA&A* **32**, 227.
Maeder A., Meynet G., 1994, *A&A* **287**, 803.
Magalhaes A.M., 1993, *New Aspects of Magellanic Cloud Research* (eds. B. Baschek, G. Klare, J. Lequeux; Springer-Verlag, Berlin), p. 276.
Malumuth E.M., Heap S.R., 1994, *AJ* **107**, 1054.
Mark H., Pice R., Rodriguez R., Seward F.D., Swift C.D., 1969, *ApJ* **155**, L143.
Markert T.H., Clark G.W., 1975, *ApJ* **196**, L55.
Martin N., Maurice E., Lequeux J., 1989, *A&A* **215**, 219.
Martin N., Prévot L., Rebeirot E., Rousseau J., 1976, *A&A* **51**, 31.
Masai K., Nomoto K., 1994, *ApJ* **424**, 924.
Massey Ph., Johnson J., 1993, *AJ* **105**, 980.
Massey Ph., Thomson A.B. 1991, *AJ* **101**, 1408.
Massey Ph., Parker J.W., Garmany C.D., 1989b, *AJ* **98**, 1305.
Massey Ph., Garmany C.D., Silkey M., DeGiola-Eastwood K., 1989a, *AJ* **97**, 107.
Massey Ph., Lang C.C., DeGioia-Eastwood K., Garmany C.D., 1995, *ApJ* **438**, 188.
Mateo M., 1988a, *ApJ* **331**, 261.
Mateo M., 1988b, *IAU Symp. No. 126* (eds. A.G.D. Philip, J.E. Grindley; Kluwer, Dordrecht), p. 557.
Mateo M., 1992, *PASP* **104**, 824.
Mateo M., 1993, *PASP* **104**, 824.
Mateo M., Hodge P.W., 1985, *PASP* **97**, 753.
Mateo M., Hodge P., 1986, *ApJS* **60**, 893.
Mateo M., Hodge P., 1987, *ApJ* **320**, 626.
Mateo M., Hodge P., Schommer R.A., 1986, *ApJ* **311**, 113.
Mathews W.G., 1989, *AJ* **97**, 42.
Mathewson D.S., Ford V.L., 1984, *IAU Symp. No. 108* (eds. S. van den Bergh, K.S. de Boer; Reidel, Dordrecht), p.125.
Mathewson D.S., Healey J.R., 1964a, *IAU/URSI Symp. No. 20* (eds. F.J. Kerr, A.W. Rodgers, Australian Acad. Sci., Canberra), p. 245.
Mathewson D.S., Healey J.R., 1964b, *IAU/URSI Symp. No. 20* (eds. F.J. Kerr, A.W. Rodgers, Australian Acad. Sci., Canberra), p.283.

Mathewson D.S., Cleary M.N., Murray J.D., 1974, *ApJ* **190**, 291.
Mathewson D.S., Healey J.R., Westerlund B.E., 1963, *Nature* **199**, 681.
Mathewson D.S., Ford V.L., Visvanathan N., 1986, *ApJ* **301**, 664.
Mathewson D.S., Ford V.L., Visvanathan N., 1988, *ApJ* **333**, 617.
Mathewson D.S., Schwarz M.P., Murray J.D., 1977, *ApJ* **217**, L5.
Mathewson D.S., Ford V.L., Dopita M.A., Tuohy I.R., Long K.S., Helfand D.J., 1983a, *ApJS* **51**, 345.
Mathewson D.S., Ford V.L., Dopita M.A., Tuohy I.R., Long K.S., Helfand D.J., 1983b, *Supernova Remnants and Their X-Ray Emission, IAU Symp. No. 101* (eds. P. Gorenstein and I.J. Danziger, Reidel, Dordrecht), p. 541.
Mathewson D.S., Ford V.L., Dopita M.A., Tuohy I.R., Mills B.Y., Turtle A.J., 1984, *ApJS* **55**, 189.
Mathewson D.S., Ford V.L., Schwarz M.P., Murray J.D., 1979, *IAU Symp. No. 84*, (ed. W. Burton; Reidel, Dordrecht), p.547.
Mathewson D.S., Ford V.L., Tuohy I.R., Mills B.Y., Turtle A.J., Helfand D.J., 1985, *ApJS* **58**, 197.
Mathewson D.S., Wayte S.R., Ford V.L., Ruan K., 1987, *Proc. Astron. Soc. Aust.* **7**, 19.
Maurice E., Bouchet P., Martin N., 1989, *A&AS* **78**, 445.
Mavridis L.N., 1967, *Coll. on Late-Type Stars* (ed. M. Hack), p. 420.
Meaburn J., 1979, *A&A* **75**, 127.
Meaburn J., 1980, *MNRAS* **192**, 365.
Meaburn J., 1981, *Investigating the Universe*, (ed. F.D. Kahn), p.61.
Meaburn J., 1984, *MNRAS* **211**, 521
Meaburn J., 1986, *MNRAS* **223**, 317.
Meaburn J., Laspias V.N., 1991, *A&A* **245**, 635.
Meaburn J., Bryce M., Holloway A.J., 1995, *A&A* **299**, L1.
Meaburn J., Marston A.P., McGee R.X., Newton L.M., 1987, *MNRAS* **225**, 591.
Meaburn J., Solomos N., Laspias V.N., Goudis C., 1989, *A&A* **225**, 497.
Meatheringham S.J., Dopita M.A., 1991a, *ApJS* **75**, 407.
Meatheringham S.J., Dopita M.A., 1991b, *ApJS* **76**, 1085.
Meatheringham S.J., Dopita M.A., Ford H.C., Webster B.L., 1988, *ApJ* **327**, 651.
Meatheringham S. J., Maran S.P., Stecher Th.P. *et al.*, 1990, *ApJ* **361**, 101.
Melcher N., Richtler T., 1989, *Recent developments of Magellanic Cloud research* (eds. K.S. de Boer, F. Spite, G. Stasinska), p.87.
Meliani M.T., Barbuy B., Richtler T., 1994, *A&A* **290**, 753.
Melnick J., 1985, *A&A* **153**, 235.
Meylan G., Maeder A., 1982, *A&A* **108**, 148.
Meynet G., Mermilliod J.-C., Maeder A., 1993, *A&AS* **98**, 477.
Mills B.Y., 1954, *Interim Report U.R.S.I.*.
Mills B.Y., Turtle A.J., 1984, *IAU Symp. No. 108* (eds. S. van den Bergh, K.S. de Boer; Reidel, Dordrecht), p. 283.
Mills B.Y., Turtle A.J., Watkinson A., 1978, *MNRAS* **185** 263.
Mills B.Y., Turtle A.J., Little A.G., Durdin J.M., 1984, *Aust. J. Phys.* **37**, 321.
Milne D.K., Caswell J.L., Haynes R.F., 1980, *MNRAS* **191**, 469.
Mishra R., 1985, *MNRAS* **212**, 163.
Mochizuki K., Nakagawa T., Doi Y., Yui Y.Y., Okuda H., Shibai H., Yui M., Nishimura T., Low F.J., 1994, *ApJ* **430**, L37.
Moffat A.F.J., Drissen L., Shara M.M., 1994, *ApJ* **436**, 183.
Moffat A.F.J., Shara M.M., Potter M., 1991, *AJ* **102**, 642.
Moffat A.F.J., Niemela V.S., Phillips M.M., Chu Y.-H., Seggewiss W., 1987, *ApJ* **312**, 612.
Molaro P., Vladilo G., Monai S., D'Odorico S., Ferlet R., Vidal-Madjar A., Dennefeld M., 1993, *A&A* **274**, 505.
Moore B., Davis M., 1994, *MNRAS* **270** , 209.
Morgan D.H., 1994, *A&AS* **103**, 235.
Morgan D,H., Good A.R., 1985, *MNRAS* **213**, 491.
Morgan D.H., Good A.R., 1987, *MNRAS* **224**, 435.
Morgan D.H., Hatzidimitriou D., 1995, *A&AS* **113**, 539.
Morgan D.H., Vassiliadis E., Dopita M.A., 1991, *MNRAS* **251**, 51P.
Morras R., 1983, *AJ* **88**, 62.
Morras R., 1985, *AJ* **90**, 1801.
Morse J.A., Winkler P.F., Kirshner R.P., 1995, *AJ* **109**, 2104.

Mould J.R., Aaronson M., 1986, *ApJ* **303**, 10.
Mould J.R., Da Costa G.S., 1988, *Progress and Opportunities in Southern Hemisphere Optical Astronomy, ASP Conf. Ser., no 1* (eds. V.M. Blanco and M.M. Phillips), p.197.
Mould J., Reid N., 1987, *ApJ* **321**, 156.
Mould J.R., Da Costa G.S., Crawford M.D., 1984, *ApJ* **280**, 595.
Mould J., Xystus D.A., Da Costa G.S., 1993, *ApJ* **408**, 108.
Mould J., Kristian J., Nemec J., Aaronson M., Jensen J., 1989, *ApJ* **339**, 84.
Murai T., Fujimoto M., 1980, *PASJ* **32**, 581.
Nandy K., Morgan D.H., 1978, *Nature* **276**, 478.
Nandy K., Morgan D.H., Willis A.J., Wilson R., Gondhalekar P.M., 1981, *MNRAS* **196**, 955.
Nandy K., Morgan D.H., Willis A.J., Wilson R., Gondhalekar P.M., Houziaux L., 1980, *Nature* **283**, 725.
Nemec J.M., Hesser J.E., Ugarte P. P., 1985, *ApJS* **57**, 287.
Olszewski E.W., 1988, *The Harlow-Shapley Symp. on Globular Cluster Systems in Galaxies* (eds. J.E. Grindlay and A.G.D. Philips), p. 159.
Olszewski E.W., Schommer R.A., Aaronson M., 1987, *AJ* **93**, 565.
Olszewski E.W., Harris H.C., Schommer R.A., Canterna R.W., 1988, *AJ* **95**, 84.
Olszewiski E.W., Schommer R.A., Suntzeff N.B., Harris H.C., 1991, *AJ* **101**, 515.
Orio M., Della Valle M., Massone G., Ögelman H., 1994, *A&A* **289**, L11.
Orio M., Ogelman H., 1993, *A&A* **273**, L56.
Ostriker J.P., Thuan T.X., 1975, *ApJ* **202**, 353.
Page Th., Carruthers G.R., 1981, *ApJ* **248**, 906.
Pagel B.E.J., 1993, *New Aspects of Magellanic Cloud Research* (eds. B. Baschek, G. Klare, J. Lequeux; Springer-Verlag), p.330.
Pagel B.E.J., Edmunds M.G., Fosbury R.A.E., Webster B.L., 1978, *MNRAS* **184**, 569.
Pakull M.V., Angebault L.P., 1986, *Nature* **322**, 511.
Pakull M.V., Swings J.P., 1979, *IAU Circular* **3318**.
Pakull M.V., Beuermann K., Angebault L.P., Bianchi L., 1987, *Ap&SS* **131**, 689.
Pakull M.V., Ilovaisky S.A., Chevalier C., 1985, *Space Sci. Rev.* **40**, 229.
Pakull M.V., Motch C., Bianchi I., Thomas H.-C., Guibert J., Beaulieu J.-P., Grison P., Schaeidt S., 1993, *A&A* **278**, L39.
Panagia N., Gilmozzi R., Macchetto F., Adorf H.-M., Kirshner R.P., 1991, *ApJ* **380**, L23.
Parker J.Wm., 1993, *AJ* **106**, 560.
Parker J. Wm., Garmany C.D., 1993, *AJ* **106**, 1471.
Parker J. Wm., Clayton G.C., Winge C., Conti P.S., 1993, *ApJ* **409**, 370.
Parker J. Wm., Garmany C.D., Massey P., Walborn N.R., 1992, *AJ* **103**, 1205.
Payne-Gaposchkin C., 1971, *Smithsonian Contr. Ap.* **13**, 1.
Payne-Gaposchkin C., Gaposchkin S., 1966, *Smithsonian Contr. Ap.* **9**, 1.
Pei Y.C., 1992, *ApJ* **395**, 130.
Penston M.V., 1982, *Observatory* **102**, 174.
Persson S.E., Aaronson M., Cohen J.G., Frogel J.A., Matthews K., 1983, *ApJ* **266**, 105.
Pierre M., Viton M., Sivan J.P., Courtès G., 1986, *A&A* **154**, 249.
Pietsch W., Kahabka P., 1993, *New Aspects of Magellanic Cloud Research*, Lecture Notes in Physics 416 (eds. B. Baschek, G. Klare, J. Lequeux; Springer, Heidelberg), p.59.
Pigafetta A., 1524, *Magellan's Voyage A Narrative Account of the First Navigation* (translated by R.A. Skelton, The Folio Society, London, 1975).
Plait Ph.C., Lundquist P., Chevalier R.A., Kirshner R.P., 1995, *ApJ* **439**, 730.
Prévot L., Rousseau J., Martin N., 1989, *A&A* **225**, 303.
Price R., Groves D., Rodriguez R.M., Seward F.D., Swift C.D., Toor A., 1971, *ApJ* **168**, L7.
Ramana Murthy P.V., Wolfendale A.W., 1993, *Gamma-Ray Astronomy*, Cambridge Astrophys. Ser. 22; Cambridge Univ. Press.
Rapley C.G., Tuohy I.R., 1974, *ApJ* **191**, L113.
Rappaport S., Levine A., Doxsey R., Bradt H.V., 1975, *ApJ* **196** , L15.
Rebeirot E., Azzopardi M., Westerlund B.E., 1993, *A&AS* **97**, 603.
Rebeirot E., Martin N., Mianes P., Prévot L., Robin A., Rousseau J., Peyrin Y., 1983, *A&AS* **51**, 277.
Reid N., 1991, *ApJ* **382**, 143.
Reid N., Freedman W., 1994, *MNRAS* **267**, 821.
Reid N., Mould J., 1984, *ApJ* **284**, 98.
Reid N., Mould J., 1985, *ApJ* **299**, 236.

Reid N., Mould J., 1990, *ApJ* **360**, 490.
Reid I.N., Strugnell P.R., 1986, *MNRAS* **221**, 887.
Reid N., Glass I.S., Catchpole R.M., 1988, *MNRAS* **232**, 53.
Reid N., Mould J., Thomson I., 1987, *ApJ* **323**, 433.
Reid N., Tinney C., Mould J., 1990, *ApJ* **348**, 98.
Reitermann A., Baschek B., Stahl O., Wolf B., 1990, *A&A* **234**, 109.
Renzini A., Voli M., 1981, *A&A* **94**, 175.
Rich R.M., Da Costa G.S., Mould J.R., 1984, *ApJ* **286**, 517.
Richer H.B., Frogel J.A., 1980, *ApJ* **242**, L9.
Richtler T., Nelles B., 1983, *A&A* **119**, 75.
Richtler T., Spite M., Spite F., 1989, *A&A* **225**, 351.
Robertson J.W., 1974, *A&AS* **15**, 261.
Roche P.F., Aitken D.K., Smith C.H., 1987, *MNRAS* **228**, 269.
Roche P.F., Aitken D.K., Smith C.H., 1993, *MNRAS* **262**, 30.
Rohlfs K., Kreitschmann J., Siegman B.C., Feitzinger J.V., 1984, *A&A* **137**, 343.
Rosado M., Le Coarer E., Laval A., Georgelin Y., 1993a, *Rev. Mex. Astron. Astrofis.* **27**, 41.
Rosado M., Le Coarer E., Georgelin Y., Viale A., 1993b, *New Aspects of Magellanic Cloud Research* (eds. B. Baschek, G. Klare, J. Lequeux; Springer-Verlag), p. 226.
Röser S., Bastian U., 1993, *Bull. Inf. CDS* **42**, 11.
Rosseau J., Martin N., Prévot L., Reiberot E., Robin A., 1978, *A&AS* **31**, 243.
Rubio M., Lequeux J., Boulanger F., 1993b, *A&A* **271**, 9.
Rubio M., Roth M., Garcia J., 1992, *A&A* **261**, L29.
Rubio M., Garay G., Montani J., Thaddeus P., 1991, *ApJ* **368**, 173.
Rubio M., Lequeux J., Boulanger F. et al.., 1993a, *A&A* **271**, 1.
Russell S.C., Bessell M.S., 1989, *ApJS* **70**, 865.
Russell S.C., Dopita M.A., 1992, *ApJ* **384**, 508.
Sagar R., Pandey A.K., 1989, *A&AS* **79**, 407.
Sagar R., Richtler T., 1991, *A&A* **250**, 324.
Sagar R., Richtler T., de Boer K.S., 1991a, *A&A* **249**, L5.
Sagar R., Richtler T., de Boer K.S., 1991b, *A&AS* **90**, 387.
Salpeter E.E., 1955, *ApJ* **121**, 161.
Sandage A., 1982, *ApJ* **252**, 553.
Sanduleak, N., 1984, *IAU Symp. No. 108* (eds. S. van den Bergh, K.S. de Boer; Reidel, Dordrecht), p. 231.
Sanduleak, N., 1989, *AJ* **98**, 825.
Sanduleak N., Philip A.G.D., 1977a, *IAU Circular* **3023**.
Sanduleak N., Philip A.G.D., 1977b, *IAU Circular* **3127**.
Sanduleak N., Philip A.G.D., 1977c, *IAU Circular* **3134**.
Sanduleak N., Philip A.G.D., 1977d, *Publ. Warner & Swasey Obs.* **2**, No. 5.
Sanduleak N., MacConnell D.J., Philip A.G.D., 1978, *PASP* **90**, 621 (SMP).
Savage B.D., de Boer K.S., 1979, *ApJ* **230**, L77.
Savage B.D., de Boer K.S., 1981, *ApJ* **243**, 460.
Scalise E., Braz M.A., 1981, *Nature* **290**, 36.
Scalise E., Braz M.A., 1982, *AJ* **87**, 528.
Scalo J.M., 1976, *ApJ* **206**, 474.
Scalo J.M., 1987, *Starbursts and galaxy evolution* (eds. T.X. Thuan, T. Montmarle, J.T.T.Van), p. 445.
Schaeidt S., Hasinger G., Trümper J., 1993, *A&A* **270**, L9.
Schertl D., Hofmann K.-H., Seggewiss W., Weigelt G., 1995, *A&A* **302**, 327.
Schild H., Testor G., 1992, *A&AS* **92**, 729.
Schlegel E.M., Marshall F.E., Mushotzky R.F. et al., 1994, *ApJ* **422**, 243.
Schmidtke P.C., Cowley A.P., Frattare L.M. et al., 1995, *PASP* **106**, 843.
Schmidtke P.C., McGrath T.K., Cowley A.P., Frattare L.M., 1993, *PASP* **105**, 866.
Schmidt Th., 1972, *A&A* **16**, 95.
Schmidt-Kaler Th., 1992, *ASP Conf. Ser.* **30**, 195.
Schmidt-Kaler Th., Feitzinger J.V., 1976, *Astroph. Space Sci.* **41**, 357.
Schmidt-Kaler Th., Gochermann J., 1992, *ASP Conf. Ser.* **30**, 203.
Schommer R.A., Olszewski E.W., Aaronson M., 1984, *ApJ* **285**, L53.
Schommer R.A., Olszewski E.W., Aaronson M., 1986, *AJ* **92**, 1334.

Schommer R.A., Olszewski E.W., Suntzeff N.B., Harris H.C., 1992, *AJ* **103**, 447.
Schwering P.B.W., 1988, *Starburst and galaxy evolution, XXIInd Recontres de Moriond* (eds. T.X, Thuan, T. Montmerle, J. Tran Thank Van; Editions Frontières, Gif sur Yvette) p.85.
Schwering P.B.W., 1989a, An Infrared Study of the Magellanic Clouds, Thesis, Leiden.
Schwering P.B.W., 1989b, *A&AS* **79**, 105.
Schwering P.B.W., Israel F.P., 1989, *A&AS* **79**, 79.
Schwering P.B.W., Israel F.P., 1990, *Atlas and Catalogue of Infrared Sources in the Magellanic Clouds* (Kluwer, Dordrecht).
Schwering P.B.W., Israel F.P., 1991, *A&A* **246**, 231.
Seal P., Hyland A.R., 1983, *Proc. ASA* **5**, 238.
Searle L., Wilkinson A., Bagnuolo W.G., 1980, *ApJ* **239**, 803 (SWB).
Seaton M.J., 1979, *MNRAS* **187**, 73.
Sebo K.M., Wood P.R., 1994, *AJ* **108**, 932.
Sedov L.I.,1959, *Similarity and Dimensional Methods in Mechanics* (transl. M. Friedman; Academic Press, New York).
Seggewiss W., Richtler T., 1989, *Recent Development of Magellanic Cloud Research* (eds.K.S. de Boer, F. Spite, G. Stasińska), p. 45.
Sekiguchi K., Kilkenny D., Winkler H., Doyle J.G., 1989, *MNRAS* **241**, 827.
Seward F.D., Harnden F.R., 1994, *ApJ* **421**, 581.
Seward F.D., Mitchell M., 1981, *ApJ* **243**, 736.
Shapley H., 1955, *Proc. Nat. Acad. Sci.* **41**, 824.
Shapley H., 1956, *American Scientist* **44**, 73.
Shapley H., Lindsay E.M., 1963, *Irish AJ* **6**, 74.
Shapley H., Nail V. McK., 1953, *PNAS* **39**, 358.
Shapley H., Paraskevopoulos J., 1940, *Harvard Bull.* **914**, 6.
Shore S.N., Sonneborn G., Starfield S.G. et al., 1991, *ApJ* **370**, 193.
Shull P., 1983a, *ApJ* **275**, 592.
Shull P., 1983b, *ApJ* **275**, 611.
Shuter W.L., 1992, *ApJ* **386**, 101.
Sievers J., 1970, *Inf. Bull. Var. Stars No.448*.
Sinclair M.W., Carrad G.J., Caswell J.L., Norris R.P., Whiteoak J.B., 1992, *MNRAS* **256**, 33P.
Singh K.P., Nousek J.A., Burrows D.N., Garmire G.P., 1987, *ApJ* **313**, 185.
Skinner G.K., Shulman S., Share G. et al., 1980, *ApJ* **240**, 619.
Smale A.P., Charles P.A., 1989, *MNRAS* **238**, 595.
Smale A.P., Corbet R.H.D., Charles P.A. et al.,1988, *MNRAS* **233**, 51.
Smith A.M., Cornett R.H., Hill R.S., 1987, *ApJ* **320**, 609.
Smith A., Cornett R.H., Hill R.S., 1990, *ApJ* **355**, 746.
Smith R.Ch., Raymond J.C., Laming J.M., 1994a, *ApJ* **420**, 286.
Smith R.Ch., Chu Y.-H., Mac Low M.-M., Oey M.S., Klein U., 1994b, *AJ* **108**, 1266.
Smith V.V., Lambert D.L., 1989, *ApJ* **345**, L75.
Smith M.G., Weedman D.W., 1971, *ApJ* **169**, 271.
Smith M.G., Weedman D.W., 1972, *ApJ* **172**, 307.
Smith M.G., Weedman D.W., 1973, *ApJ* **179**, 461.
Sofou Y., 1994, *PASJ* **46**, 431.
Songaila A., 1981, *ApJ* **243**, L19.
Songaila A., Blades J.C., Hu E.M., Cowie L.L., 1986, *ApJ* **303**, 198.
Spite F., 1989, *Recent Developments of Magellanic Cloud Research* (eds. K.S. de Boer, F. Spite, G. Stasinska), p.37.
Spite M., Spite F., 1990, *A&A* **234**, 67.
Spite M., Barbuy B., Spite F., 1989b, *A&A* **222**, 35.
Spite M., Barbuy B., Spite F., 1993, *A&A* **272**, 116.
Spite F., Richtler T., Spite M., 1991, *A&A* **252**, 557.
Spite F., Spite M., Françaois P., 1989a, *A&A* **210**, 25.
Spite M., Cayrel R., Françaois P., Richtler T., Spite F., 1986, *A&A* **168**, 197.
Spyromilio J., Malin D.F., Allen D.A., Steer C.J., Couch W.J., 1995, *MNRAS* **274**, 256.
Stacey G.J., Geis N., Genzel R. et al., 1991, *ApJ* **373**, 423.
Stahl O., 1989, *Recent developments of Magellanic Cloud Research* (eds. K.S. de Boer, F. Spite, G. Stasińska), p.13.

Stahl O., 1993, *New Aspects of Magellanic Cloud Research* (eds. B. Baschek, G. Klare, J. Lequeux; Springer-Verlag, Berlin), p. 263.
Staveley-Smith L., Sault R.J., McConnell D., Kesteven M.J., Hatzidimitriou D., Freeman K.C., Dopita M.A., 1995, *Proc. ASA* **12**, 13.
Storm J., Andersen J., Blecha A., Walker M.R., 1988, *A&A* **190**, L18.
Stothers R.B., 1983, *ApJ* **274**, 20.
Stothers R.B., 1988, *ApJ* **329**, 712.
Stothers R.B., Chin C.W., 1992a, *ApJ* **390**, L35.
Stothers R.B., Chin C.W., 1992b, *ApJ* **390**, 136.
Stryker L.L., 1983, *ApJ* **266**, 82.
Stryker L.L., 1984, *ApJS* **55**, 127.
Stryker L.L., Da Costa G.S., Mould J.R., 1985, *ApJ* **298**, 544.
Suntzeff N.B., Friel E., Klemola A., Kraft R.P., Graham J.A., 1986, *AJ* **91**, 275.
Suntzeff N.B., Phillips M.M., Elias J.H., Cowley A.P., Hartwick F.D.A., Bouchet P., 1993, *PASP* **105**, 350.
Suntzeff N.B., Schommer R.A., Olszewski E.W., Walker A.R., 1992, *AJ* **104**, 1743.
Sutherland R.S., Dopita M.A., 1995, *ApJ* **439**, 365.
Szeifert Th., Stahl O., Wolf B., Zickgraf F.J., Bouchet P., Klare G., 1993, *A&A* **280**, 508.
Tanaka K.I., 1981, *PASJ* **33**, 247.
Testa V., Ferraro F.R., Brocato E., Castellani V., 1995, *MNRAS* **275**, 454.
Testor G., Llebaria A., Debray B., 1988, *ESO Messenger No. 54*, p. 43.
Testor G., Lortet M.-C., 1987, *A&A* **178**, 25.
Testor G., Schild H., Lortet M.-C., 1993, *A&A* **280**, 426.
Thackeray A.D., 1951, *Observatory* **71**, 219.
Thackeray A.D., 1963, *Advances in Astronomy and Astrophysics* (ed. Z. Kopal) **3**, 264.
Thackeray A.D., Wesselink A.J., 1953, *Nature* **171**, 693.
Thackeray A.D., Wesselink A.J., 1955, *Observatory* **75**, 33.
Thévenin F., Jasniewicz G., 1992, *A&A* **266**, 85.
Torres G., Carranza G.J., 1987, *MNRAS* **226**, 513.
Treves A., Belloni T., Bouchet P. et al., 1988, *ApJ* **335**, 142.
Trümper J., Hasinger G., Aschenbach B. et al., 1991, *Nature* **349**, 579.
Tucholke H.-J., Hiesgen M., 1991, *Proc IAU Symp. 148* (eds. R.Haynes, D.Milne; Kluwer, Dordrecht), p.491
Tuohy I.R., Dopita M.A., 1983, *ApJ* **268**, L11.
Vallenari A., Aparicio A., Fagotto F., Chiosi C., Ortolani S., Meylan G., 1994a, *A&A* **284**, 447.
Vallenari A., Aparicio A., Fagotto F., Chiosi C., 1994b, *A&A* **284**, 424.
Vallenari A., Aparicio A., Fagotto F., Chiosi C., Ortolani S., 1995, *A&A* in press, .
Vallenari A., Bertelli G., Chiosi C., Ortolani S., 1994c, *The Messenger* **No.76**, 30.
Vallenari A., Chiosi C., Bertelli G., Meylan G., Ortolani S., 1992, *AJ* **104**, 1100.
Valtonen M. J., Innanen K.A., Tähtinen L., 1984, *Aph. Space Sci.* **107**, 209.
Van den Bergh S., 1974, *ApJ* **193**, 63.
Van den Bergh S., 1981, *A&AS* **46**, 79.
Van den Bergh S., 1988, *PASP* **100**, 1486.
Van den Bergh S., 1989, *A&AR* **1**, 111.
Van den Bergh S., 1991, *ApJ* **369**, 1.
Van den Bergh S., 1993, *ASP Conf. Ser.* **48**, 346.
Van der Klis M., Hammerschlag-Hensberge G., Bonnet-Bidaud J.M. et al., 1982, *A&A* **106**, 339.
van Paradijs J., Zuiderwijk F., 1977, *A&A* **61**, L19.
van Paradijs J., van der Klis M., Augustein T. et al., 1987, *A&A* **184**, 201.
Vassiliadis E., Meatheringham S.J., Dopita M.A., 1992, *ApJ* **394**, 489.
Vidal-Madjar A., Andreani P., Cristiani S., Ferlet R., Lanz T., Vladilo G., 1987, *A&A* **177**, L17.
Visvanathan N., 1985, *ApJ* **288**, 182.
Visvanathan N., 1986, *MNRAS* **219**, 495.
Visvanathan N., 1989, *ApJ* **346**, 629.
Vladilo G., Crivellari L., Molaro P., Beckman J.E., 1987, *A&A* **182**, L59.
Vladilo G., Molaro P., Monai S. et al., 1993, *A&A* **274**, 37.
Walborn N.R., 1984, *Structure and Evolution of the Magellanic Clouds, IAU Symp. No. 108* (eds. S. van den Bergh and K. S. de Boer; Reidel, Dordrecht), p. 243.

Walborn N.R., 1986, *IAU Symp. No. 116* (eds. C.W.H. de Loore, A.J. Willis, P. Laskarides; Reidel, Dordrecht), p. 185.
Walborn N.R., 1989, *IAU Coll. No.113*, (eds.K. Davidson, A.F.I. Moffat, H.J.K.L.M. Lamers; Kluwer), p.27.
Walborn N.R., 1991, *The Magellanic Clouds, IAU Symp. No. 148* (eds. R. Haynes, D. Milne, Kluwer, Dordrecht), p. 145.
Walborn N.R., 1993, *The MK Process at 50 Years: A Powerful Tool for Astrophysical Research, ASP Conf. Ser. No. 60* (eds. C.J. Corbally, R.O. Gray, R.F. Garrison), p. 84.
Walborn N.R., Blades J.C., 1986, *ApJ* **304**, L17.
Walborn N.R., Blades J.C., 1987, *ApJ* **323**, L65.
Walborn N.R., Parker J.Wm., 1992, *ApJ* **399**, L 87.
Walborn N.R., Lasker B.M., Laidler V.G., Chu Y.-H., 1987, *ApJ* **321**, L41.
Walborn N.R., Ebbets D.C., Parker J. Wm., Nichols-Bohlin J., White R.L., 1992, *ApJ* **393**, L13.
Walborn N.R., MacKenty J.W., Saha A., White R.L., Parker J.Wm., 1994, *ApJ* **439**, L47.
Walborn N.R., MacKenty J.W., Parker J. Wm., Saha A., White R.L., 1995, *The interplay between massive star formation, the ISM and Galaxy Evolution* (eds. D. Kunth, B. Guiderdoni, M. Heydari-Malayeri, T.X. Thuan; Editions Frontires), preprint.
Walker A.R., 1985, *MNRAS* **217**, 13P.
Walker A.R., 1987, *MNRAS* **225**, 627.
Walker A.R., 1988, *The Extragalactic Distance Scale, ASP. Conf. Ser. 4* (eds. S. van den Bergh and C.J. Prichet), p.89.
Walker A.R., 1989, *AJ* **98**, 2086.
Walker A.R., 1990, *AJ* **100**, 1532.
Walker A.R., 1991, *The Magellanic Clouds, IAU Symp. No. 148* (eds. R. Haynes, D. Milne; Kluwer, Dordrecht), p. 307.
Walker A.R., 1992a, *AJ* **103**, 1166.
Walker A.R., 1992b, *ApJ* **390**, L91.
Walker A.R., 1993, *AJ* **106**, 999.
Walker A.R., Mack P., 1988a, *AJ* **96**, 872.
Walker A.R., Mack P., 1988b, *AJ* **96**, 1362.
Walker A.R., Suntzeff N.B., 1990. *PASP* **102**, 131.
Wang L., Dyson J.E., Kahn F.D., 1993, *MNRAS* **261**, 391.
Wang Q., 1991a, *MNRAS* **252**, 47P.
Wang Q., 1991b, *ApJ* **377**, L85.
Wang Q., Helfand D.J., 1991a, *ApJ* **370**, 541.
Wang Q., Helfand D.J., 1991b, *ApJ* **379**, 327.
Wang Q., Wu X.: 1992, *ApJS* **78**, 391.
Wang Q., Hamilton T., Helfand D.J., Wu X.: 1991, *ApJ* **374**, 475.
Wannier P., Wrixon G.T., Wilson R.W., 1972, *A&A* **18**, 224.
Wayte S.R., 1989, *Proc. ASA* **8**, 195.
Wayte S.R., 1990, *ApJ* **355**, 473.
Weaver R., McCray R., Castor J., Shapiro P., Moore R., 1977, *ApJ* **218**, 377.
Webster B.L., 1978, *MNRAS* **185**, 45P.
Weigelt G., Albrecht R., Barbieri C., Blades J.C., et al., 1991, *ApJ* **378**, L 21.
Welch D.L., 1992, *Pulsating variable stars*
Welch D.L., McLaren R.A., Madore B.F., McAlary Ch.W., 1987, *ApJ* **321**, 162.
Welch D.L., Mateo M., Coté P., Fischer P., Madore B.F., 1991, *AJ* **101**, 490.
Werner M.W., Becklin E.E., Gatley I. et al., 1978, *MNRAS* **184**, 365.
West S.R.D., Tobin W., Gilmore A.C., 1992, *MNRAS* **254**, 419.
Westerlund B., 1960, *Uppsala Astron. Obs. Ann* **IV**, no.7.
Westerlund B., 1961a, *Uppsala Astron. Obs. Ann* **V**, no.1.
Westerlund B., 1961b, *Uppsala Astron. Obs. Ann* **V**, no.2.
Westerlund B., 1964a, *IAU/URSI Symp. No.20* (eds. F.J. Kerr and A.W. Rodgers, Austr. Acad. Sci., Canberra), p.239.
Westerlund B., 1964b, *IAU/URSI Symp. No.20* (eds. F.J. Kerr and A.W. Rodgers, Austr. Acad. Sci., Canberra), p. 316.
Westerlund B.E., 1964c, *Observatory* **84**, 253.
Westerlund B.E., 1964d, *MNRAS* **127**, 429.
Westerlund B.E., 1982, *C.R. Journées Strasbourg* **4**, p.20.

Westerlund B.E., 1987, *Stellar Evolution and Dynamics in the Outer Halo of the Galaxy, ESO Conf. Workshop Proc.* (Eds. M. Azzopardi, F. Matteucci), p. 207.
Westerlund B.E., 1989, *Recent Developments of Magellanic Cloud Research* (eds. K.S. de Boer, F. Spite, G. Stasinska), p. 159.
Westerlund B.E., 1990, *A&AR* **2**, 29.
Westerlund B.E., Glaspey J., 1971, *A&A* **10**, 1.
Westerlund B.E., Krelowski J., 1989, *A&A* **218**, 216.
Westerlund B.E., Mathewson D.S., 1966, *MNRAS* **131**, 371.
Westerlund B.E., Smith L.F., 1964, *MNRAS* **128**, 311.
Westerlund B.E., Linde P., Lyngå G., 1995a, *A&A* **298**, 39.
Westerlund B.E., Olander N., Hedin B., 1981, *A&AS* **43**, 267.
Westerlund B.E., Azzopardi M., Breysacher J., Rebeirot E., 1991, *A&AS* **91**, 425.
Westerlund B.E., Azzopardi M., Breysacher J., Rebeirot E., 1995b, *A&A* **303**, 107.
Westerlund B.E., Olander N., Richer H.B., Crabtree D.R., 1978, *A&AS* **31**, 61.
White N.E., 1978, *Nature* **271**, 38.
White N.E., Marshall F.E., 1984, *ApJ* **281**, 354.
White N.J., 1981, *Aph & Space Sci.* **78**, 443.
Whitelock P.A., Feast M.W., 1993, *IAU Symp. No. 155* (eds. R.Weinberger, A. Acker; Kluwer, Dordrecht), p.251.
Whitelock P.A., Feast M.W., Menzies J.W., Catchpole R.M., 1989, *MNRAS* **238**, 769.
Whiteoak J.B., Gardner F.F., 1976a, *MNRAS* **174**, 51P.
Whiteoak J.B., Gardner F.F., 1976b, *MNRAS* **176**, 25P.
Whiteoak J.B., Gardner F.F., 1986, *MNRAS* **222**, 513.
Whiteoak J.B., Gardner F.F., Höglund B., 1980, *MNRAS* **190**, 17P.
Whiteoak J.B., Wellington K.J., Jauncey D.L. et al.,1983, *MNRAS* **205**, 275.
Wielebinski R., 1989, *Recent developments of Magellanic Cloud Research* (eds. K.S. de Boer, F. Spite, G. Stasinska), p. 129.
Will J.-M., Bomans D.J., de Boer K.S., 1995, *A&A* **295**, 54.
Will J.-M., Vásquez R.A., Feinstein A., Seggewiss W.,1995, *A&A* **301**, 396.
Wolf B., 1989, *A&A* **217**, 87.
Wolf B., Reitermann A., 1989, *Recent Developments of Magellanic Cloud Research* (eds. K.S. de Boer, F. Spite, G. Stasinska), p.23.
Wood P.R., 1990, *From Miras to Planetary Nebulae* (eds. M.O. Mennessier, A. Omont; Editions Frontiers), p. 67.
Wood P.R., Faulkner D.J., 1987, *Proc. Astron. Soc. Aust.* **7**, 75.
Wood P.R., Bessell M.S., Dopita M.A., 1986, *ApJ* **311**, 632.
Wood P.R., Bessell M.S., Fox, M.W., 1983, *ApJ* **272**, 99.
Wood P.R., Bessell M.S., Paltoglou G., 1985, *ApJ* **290**, 477.
Wood P.R., Whiteoak J.B., Hughes S.M.G., Bessell M.S., Gardner F.F., Hyland A.R., 1992, *ApJ* **397**, 552.
Xu C., Klein U., Meinert D., Wielebinski R., Haynes R.F., 1992, *A&A* **257**, 47.
Ye T., Turtke A.J., Kennicutt R.C., 1991, *MNRAS* **249**, 722.
Ye T., Amy S.W., Wang D., Ball L., Dickel J.,1995, *MNRAS* **275**, 1218.
Zickgraf F.-J., 1993, *New Aspects of Magellanic Cloud Research* (eds. B. Baschek, G. Klare, J. Lequeux; Springer-Verlag, Berlin), p. 257.
Zickgraf F.-J., Wolf B., Stahl O., Leitherer C., Appenzeller L., 1986, *A&A* **163**, 119.
Zinn R., West M.J., 1984, *ApJS* **55**, 45.

Object index

Clusters
NGC 121 9, 14, 40, 55, 57, 58, 59, 61, 62, 63, 65, 112, 113, 116, 117, 223, 230, 233, 248
NGC 152 64, 112, 113, 223
NGC 269 188
NGC 299 68
NGC 330 67, 69, 75, 77, 78, 79, 86, 93, 112, 113, 223, 224, 226, 227, 229, 233
NGC 339 56, 61, 62, 63, 64
NGC 346 86, 93
NGC 376 68
NGC 411 62, 64, 112, 113, 119, 223
NGC 416 223
NGC 419 113
NGC 458 69, 77, 78, 113
NGC 1466 23, 39, 40, 55, 57, 58, 62, 117
NGC 1711 16, 69
NGC 1754 57
NGC 1755 70
NGC 1756 63, 65, 67
NGC 1777 54
NGC 1783 12, 63, 64, 65, 111, 117, 118
NGC 1786 13, 40, 57, 58, 230
NGC 1787 111
NGC 1806 63, 65
NGC 1818 67, 68, 69, 223, 224, 226, 227, 229
NGC 1831 16, 54, 64, 65, 67
NGC 1835 40, 57, 58, 66, 241
NGC 1841 7, 13, 16, 23, 40, 55, 57, 58, 60, 62, 230, 231, 242
NGC 1849 67
NGC 1850 63, 68, 69, 70, 74, 75, 226, 227
NGC 1854 70, 227
NGC 1856 67
NGC 1858 69
NGC 1866 9, 10, 12, 16, 64, 65, 66, 67, 68, 69, 70, 71, 72, 76, 111, 224
NGC 1868 65, 66, 67
NGC 1872 67
NGC 1885 67
NGC 1895 67
NGC 1898 57
NGC 1913 69
NGC 1922 69
NGC 1916 57
NGC 1978 61, 63, 64, 65, 66, 72, 117, 223
NGC 1984 223
NGC 1987 63, 65, 66, 67

NGC 2002 67
NGC 2004 46, 67, 68, 226, 227
NGC 2005 57, 241
NGC 2010 16, 70
NGC 2019 57, 66
NGC 2030 195
NGC 2031 9, 10, 12, 68, 69, 70
NGC 2044 216
NGC 2058 69
NGC 2060 193, 202, 216, see also N157B (HII regions)
NGC 2065 69
NGC 2070, see 30 Doradus
NGC 2100 46, 67, 68, 75, 106, 77, 226, 227
NGC 2102 106
NGC 2107 63, 65, 67
NGC 2108 65, 67
NGC 2134 64, 68, 70
NGC 2136 70
NGC 2155 223
NGC 2157 9, 10, 12, 70
NGC 2162 15, 16, 65
NGC 2164 69, 70, 71, 72
NGC 2173 63, 65, 223
NGC 2190 15, 16
NGC 2209 63, 65, 66, 67, 223
NGC 2210 13, 16, 40, 57, 58, 60, 66, 118, 223, 231
NGC 2213 16
NGC 2214 63, 69, 70, 71, 75
NGC 2241 16, 64
NGC 2249 64, 66, 67
NGC 2257 7, 9, 12, 13, 23, 26, 40, 55, 56, 58, 59, 60, 62, 117, 118, 119, 223, 230
E2 62, 64
ESO121-SC03 52, 53, 61, 62, 64, 223
GLC 0435−59 (Reticulum) 7, 13, 40, 56, 57, 58, 60, 62, 230, 231
H 4 16, 54
H7 56
H11 16, 40, 56, 57, 60
H88-159 74
HS96 111
K3 48, 61, 62, 63, 64, 112, 113, 223
L1 61, 62, 64, 112, 116, 223
L56 90
L101 109
L103 109

271

Object index

L104 109
L113 61, 62, 64, 112, 113, 116, 223
LW55 111, 117
LW 79 ≡ SL61 16, 54
LW127 54
LW134 54
NC 1 54
SL 4, 54
SL196 111
SL258 54
SL639 7
Clusters, binary
  NGC 1850/NGC 1850A/H88-159 74, 75
  NGC 1938/NGC 1939 73
  NGC 2006/SL538 73, 74
  NGC 2011a/b 73, 74
  NGC 2042a/b 73, 74
  NGC 2058/NGC 2059 73, 74
  NGC 2214 75
Clusters, galactic
  47 Tucanae 116, 128
  Hyades 13, 16
  M 3 16, 116, 117
  M 71 119
  M 92 16, 119, 124, 230
  NGC 362 62
  NGC 752 16
  NGC 2420 16
  NGC 2506 16
  NGC 3603 219–220
  NGC 7789 16
  ω Centauri 125
  Pleiades 13, 16
  Sher 1 219
  Westerlund 2 219
Dark nebulae
  SMC H1 159
  SMC H 12 159
  SMC H 13 159
  SMC H 14 159
Galaxies
  Draco dwarf galaxy 126
  Fairall 9 35
  M 31 18, 21, 22, 45, 68, 70, 71, 117, 180
  M32 45
  Milky Way/ Galaxy 21–25, 27, 31, 36, 38, 41, 43, 44, 45, 47, 48, 49, 51, 53, 55, 56, 60, 62, 68, 70, 76, 77, 81, 84, 85, 86, 88, 89, 92, 94, 99, 109, 116, 118, 120, 123, 125, 128, 129, 130, 131, 143, 144, 150, 157, 159, 160, 162, 163, 165, 166, 173, 174, 175, 177, 180, 189, 196, 205, 210, 211, 219, 222, 234, 235, 236, 240, 242, 244
  NGC4472 18
  UMi dwarf galaxy 128
HII regions, LMC
  30 Doradus ≡ NGC 2070 7, 9, 18, 19, 21, 24, 29, 30, 45, 50, 80, 84, 85, 90, 96, 99, 101, 103, 105, 106, 109, 144, 145, 146, 147, 149, 150, 151, 155, 157, 158, 160, 161, 162, 164, 169, 173, 174, 176, 186, 187, 188, 193, 194, 202–220, 222, 236, 237, 240
  N4A 191
  N9 191
  N11 80, 95, 100, 150, 152, 153, 157, 161, 164

N11A 90
N11A 'Blob' 90
N11B 150
N11C 90, 150
N11E 150
N44 ≡ NGC 1929-34-35-36-37, IC 2126-27-28
  ≡ DEM 150, 151, 152 ≡ Shapley's
  Constellation I 151, 153, 157, 186
N44A 175
N51D 186
N57, N58 ≡ LH75, LH76, LH78 103
N57A 186
N63 195
N66 91
N70 186
N103A 75
N103B 185
N105A 163, 164
N113 153, 164
N114 153
N119 153
N120 ≡ DEM134 90, 153, 154
N120A 89, 153, 154
N120B, C, D 153, 154
N130 90
N132A–J 196
N135 202
N144 ≡ DEM199 96
N154 186
N157 ≡ DEM263 186, 202, *see also* 30 Doradus
N157A 163, 194, 202
N157B ≡ NGC 2060 185, 193, 202, 207, *see also* SNRs
N157C 201
N158 91, 146, 186
N158C 157
N159 89, 90, 106, 107, 146, 157, 158, 160, 161, 162, 163, 165
N159A 89, 161
N159 'Blob' 90
N160 106, 146, 161
N160A 106, 157, 161, 163
N180 80
N186E 197
DEM177 148
DEM186 196
DEM203 148
DEM224 148
DEM232 148
DEM309 100
DEM310 100, 148
DEM328 100
HII regions, SMC
  N12 155, 157
  N27 155, 157, 161
  N66 ≡ NGC 346 161
  N81 159
  N88 109, 155
  N88A 173
  N89 110
  DEM16 159
  DEM37 159
  DEM38 159
  DEM167 148

Object index

IRAS sources
  00350–7436 88
  00483–7347 88
  00554–7351 88
  01074–7140 88
  01432–7455 ≡ R50 88
  04553–6825 88, 130
  05346–6949 88
  05389–6922 130
Novae
  LMC 1988 No.2 141
  LMC 1990 No.2 142
  LMC 1991 18
Planetary nebulae
  SMC N2 138, 139
  SMC N5 138, 139
  SMC N67 140
  SMC SMP28 140, 231
  LMC N66 ≡ SMP83 ≡ WS35 138, 139, 140, 141
  LMC N201 138, 139
  LMC SMP64 141
  LMC SMP85 140
  SMC 2 138
  SMC 15 138
  LMC 1 138
  LMC 52 138
  LMC 62 138
  LMC 83 138
  LMC 97 138
Planetary nebulae, galactic
  NGC 2440 139
  NGC 6302 139
  NGC 6357 139
  H2-111 139
Protostars and 'Knots'
  N159-P1 92, 93
  N159-P2 93
  N76B-K2 93
  N160A 93
  N105A 93
  30Dor-P1 209
  30Dor-P2 209
  30Dor-P3 209
  30Dor-P4 209
'Knots'
  30Dor-Knot 1 208
  30Dor-Knot 2 209
  30Dor-Knot 3 209
Stars
  AV304 228
  BI141 89, 153, 154
  HDE 269551 96
  NGC 330/A7 225
  NGC 330/3 224, 225
  NGC 1818/B12 225
  NGC 1818/D12 224, 225
  NGC 2004/B30 225
  R71 88
  R126 88
  R136 18, 151, 202, 203, 205, 206, 210, 211, 212, 213, 214–219
  R136a, b, c 214
  R136a 205, 213, 215, 216, 217, 219

R136a1 216, 219
R136a2 216, 219
R136b 215
R139 18, 211
R145 18, 205, 211
R147 205
R150 88
Sk67-108 173
Sk69-108 173, 174, 177
Sk86-67 151
Sk98-69 153
Sk103-69 153
Sk105-69 153
Sk106-69 153
Sk107-69 153
Sk−69°202 199
Sk−69°212 91
Sk35 177
Sk143 173
Sk151 177
Sk 155 177
Stars, galactic
  Canopus 229
  HR 2618 224, 226
  HR 3663 224
  ιHer 224
Stars, multiple
  HD 32228 ≡ R64 96, 151, 152
  HDE 269676 91
  HDE 269828 69, 91
  DD13 89, 90
  N11A 'blob' 90
  N159 'blob' 90
  R136 ≡ HD 38268, see R136 above
  Sk157 91
  Sk−66°41 91
  Sk−69°212 91
  Sk−69°253 ≡ HDE269936 91
Stars, variables
  HV829 14
  HV5497 228
  SDor 69, 83
  R40 85
  R71 ≡ HDE269006 83
  R110 ≡ HDE269662 ≡ S116 84
  R127 ≡ HDE269858 83
  R143 84
  Sk−69°142a ≡ HDE269582 84
  S111 ≡ HDE269599 84
Stars, Wolf–Rayet
  Brey9 96
  Brey32 ≡ HD36521 96
  Brey33 96
  Brey34 ≡HDE269546 96
  Brey65 91
  Brey73 194, 216
  Brey77 ≡ Mk42 218
  Brey83 215
  Brey88 215
  NGC 2044 Heydari-Malayeri 5 216
  R136a, c 214, 217, see also stars
  R136a5 219
Stellar associations
  NGC 346 91, 94, 95, 173, see clusters

NGC 456-460-465 93, 95, 99, 109, 165, 173, 177
NGC 465 91
NGC 602 95
LH9 95, 96, 97, 150
LH10 95, 96, 97, 150
LH13 96, 150
LH14 150
LH39 187
LH42 ≡ NGC 1918 153, 186
LH47 151
LH48 151
LH49 151
LH53 186
LH58 ≡ NGC 1962-65-66-70 95, 96
LH65 187
LH72 ≡ N55 103
LH75 104, 186, 187
LH76 95, 103, 187
LH77 103
LH78 103
LH80 187
LH83 ≡ NGC 2030 186, 195
LH88 186
LH90 91, 95, 186, 199
LH96 186
LH99 186, 194
LH105 95
LH111 ≡ N70 186
LH112 100
LH116 100
LH117 94, 95
LH118 94, 95
LH119–121 100
LH122 100

Supergiant shells
SGSLMC 1 – 9 102, 108
SGSLMC 1
SGSLMC 2 101, 104–109, 144, 146, 161, 176, 180, 202, 203, 207, 237
SGSLMC 3 101, 105, 146, 202, 203, 207
SGSLMC 4 100–105, 122, 144, 156, 185, 187, 193, 203, 210, 227, 236
SGS LMC 5 103, 104
SGS LMC 9, 102
SGS SMC 1 93, 101, 103, 147, 165, 173
SGS SMC 2 109
SGS SMC 3 110

Supernovae (SN)
1987A 5, 18–20, 79, 88, 145, 156, 175–178, 199–201, 210, 211, 222

Supernova remnants (SNR)
Crab nebula 193, 194, 197
30DorB 193–195, 216
30DorC 216
0101–72.4 9
0101–7236 198
0505–67.9 ≡ DEM71 192
0509–67.5 197
0518–696: 154
0519–69.0 197
0519–696 154
0532–67.5 186
0534-69.9 187
0540–693 185, 197
N49 156, 185, 195, 197
N49B 192
N63A 185, 191, 192, 195, 196
N103B 185, 197
N132D 179, 185, 191, 196, 197
N157B, see HII regions
N186D 197

X-ray sources
LMC X-1 ≡ CCH 32 107, 179, 180, 181, 188
LMC X-2 179, 180, 181
LMC X-3 179, 180, 182
LMC X-4 179, 180, 182
LMC X-5 179
LMC X-6 179
A0538-66 179, 180, 183
X0544-665 179
CAL39 ≡ DEM204 180
CAL83 181, 182, 183, 184
CAL85 180
CAL87 181, 183, 184
SMC(N67) 184
SMC X-1 179, 180, 188, 189
SMC X-2 179, 1801, 188, 189
SMC X-3 179, 180, 188, 189
RX J0048.4–7332 184, 188
RX J0058.6–7146 184, 188
RX J0059.2–7138 184, 188
RX J0122.9–7521 184, 188
RX J0439.8–6809 184
RX J0513.9–6951 181, 183, 184
RX J0527.8–6954 183
RX J0537.7–7034 183
RX J0534.6–7056 183
1E 0035.4–7230 184, 188 189
1E 0056.8–7154 183, 184, 188, 189
1E 0102.2–7219 188, 197, 198

Subject index

See the Object list for individual clusters, nebulae, stars, X-ray sources, etc.

abundances 9–17, 36, 41, 45, 48, 49, 52, 53, 55–62, 64, 68, 69, 76, 77, 78, 81, 83, 85, 86, 88, 89, 93, 94, 98, 112, 118, 119, 122, 125, 129, 131, 135, 136, 139, 140, 155, 157, 162, 165, 173, 196, 197, 218, 221–234
active galactic nuclei (AGN) 180, 188
ADH telescope 170
Anglo-Australian telescope (AAT) 200, 201, 211
Ariel V satellite 179
Astro 1 mission 181 205
asymptotic giant branch (AGB) 15, 16, 32, 40, 41, 43, 52, 58, 63–66, 68, 76, 87, 114, 115, 119–124, 127, 128, 129, 130, 131, 135, 136, 139, 140, 244, 247, 248
    E-AGB 65, 114, 119, 121
    TP-AGB 66, 119, 120
Australia telescope 104, 195, 198, 245

Baade's method 18
Balloon-borne Infrared Carbon Explorer (*BICE*) 162
Balmer decrement 174
Bok region 68
Brackett-γ 83, 92
bremsstrahlung 181
Bridge region, *see* Magellanic System.
Broad Band X-Ray Telescope (*BBXRT*) 181

Case (classification) system 86, 115
circumstellar dust 81, 123
circumstellar environment 177
circumstellar material/medium 82, 196, 200, 201
circumstellar shell 87–89
clusters 15, 16, 27, 28, 30–34, 39, 40, 41, 43–78, 79, 80, 86, 91, 94, 96, 100, 103, 106, 109, 110, 113, 115, 120, 126, 132, 150, 173, 221, 222–227, 231, 233, 241, 242, 244
    age distribution 51–70
    binary 72–75
    classification 48–50
        s parameter 49, 52, 63, 65
        SWB type 49, 50, 52, 56, 57, 61, 64–71, 73, 74, 77, 120, 121, 241
    extra-tidal 23, 56, 60
    formation-rate (CFR) 54
    galactic 24, 49, 53, 55, 56, 57, 58, 66, 71, 77, 119, 127, 219, 230
    globular 2, 23, 24, 30, 32, 37, 40, 44, 45, 48, 51, 55–60, 66, 70, 72, 73, 76, 113, 116, 117, 118, 119, 230
    intermediate-age 23, 45, 52, 54, 60–67, 109, 124, 241
    isochrones 4, 9, 13–16, 39, 43, 44, 52, 76, 97, 223
    radial velocities of 241
    relaxation time 72
    rotation solutions for 241, 242
    surface brightness profile 71, 218
    surface distribution of 49–51
    surface mass density profile 217
    tidal capture 72
    tidal radius 48, 72
    tidal truncation 71
    velocity dispersion in 72
    youngest 67–70
colour excess, *see* interstellar extinction
colour–magnitude diagram (CMD) 4, 9, 15, 16, 43, 44, 52–54, 56, 59, 61, 63, 68, 69, 74, 75, 76, 77, 78, 86, 112–114, 124, 218
Copernicus satellite 179
core-helium burning 43, 44, 66, 76, 122, 128, 130
CTIO 4 m telescope 48, 170
Curtis Schmidt telescope 170

diffuse interstellar bands (DIB) 177, 178
    galactic DIBs 178
distance/modulus, $m - M$ 4, 6–20, 23, 28, 38, 39, 56, 60, 61, 62, 64, 69, 76, 91, 94, 102, 121, 128, 134, 177, 203
    mass-equivalency (ME) method 11
    maximum magnitude vs. rate of decline (MMRD) 17, 18
    visual surface brightness technique (VSB) 11
distance between the Clouds 23
distance from centre of LMC 51, 54, 58, 167
distance scale 117, 127
dust 5, 6, 9, 21, 28, 32, 35, 41, 48, 80, 81, 82, 87, 88, 89, 92, 96, 101, 104, 105, 108, 131, 140, 160, 162, 169, 170, 177, 194, 202, 205, 206, 209, 218, 235, 239
    to gas ratio 155, 160, 174, 207, 240
dwarf galaxy 2, 21, 22, 126

Einstein satellite 3, 106, 179, 180, 183, 184, 186

275

Subject index

ESO 3.6 m telescope, La Silla 155
ESO/SERC Southern Sky Atlas 47

Fabry–Perot interferometry 107, 148, 194, 196, 213
Faraday rotation 196
far-infrared (FIR) 5, 87, 92, 104, 105, 122, 153, 157, 158, 161, 166, 168, 170, 171, 172, 174, 209
 FIR/6.3 cm brightness ratio 87, 240
Fleurs Synthesis Telescope 207

galactic associations 93, 94, 150
galactic bulge 122, 126
galactic centre 27, 35, 37, 130, 244, 248
galactic disk, 60, 109, 122
galactic halo 36, 38, 55, 57, 59, 60, 108, 118, 125, 173, 181, 211, 230
galactic mass 243
galactic OB-forming regions 162
galactic plane 23, 35, 36
generations in the Clouds 40–42
 galactic 165
Ginga data 182

Hα filaments 107, 139, 147, 148, 149, 207; froth 148
Hα emission 68, 92, 147, 148, 154, 166, 167, 186, 192, 218, 246
Hα flux 101, 147, 148, 149, 174, 191, 203
Hα/Hβ ratio 90, 146, 148, 154
Hα imagery/photometry 38, 83, 146, 148, 152, 154, 187, 202, 203, 207, 208, 214, 215
[SII]/Hα ratio 186, 191
Hα luminosity 102, 149, 150, 202
Hα surface brightness 148
Hα survey 109, 148
Harvard College Observatory 2
Hβ emission/flux 132, 138, 139, 148, 174, 194
HEAO satellite 179, 182 185
helium flash 66, 119, 125, 128
helium shell burning 114; helium exhaustion 114
Hertzsprung–Russell (HR) diagram 52, 65, 68, 70, 78, 79, 81, 83, 84, 90, 97, 114, 118, 119, 125, 126, 135, 137, 215, 230
high-velocity clouds (HVC) 36
Hodge–Wright Atlas 47
Hubble flow 22; time 22; type 133
Hubble Space Telescope, HST 3, 18, 74, 132, 137, 138, 139, 140, 141, 152, 200, 208, 216, 218, 219
hydrogen
 neutral, HI, 21 cm line 2, 6, 24, 25, 27, 28, 29, 30, 31, 32, 33, 34, 35, 36, 37, 38, 41, 50, 54, 58, 99, 100, 103, 104, 105, 106, 107, 109, 113, 123, 143–146, 150, 151, 154, 156, 157, 161, 166, 169, 170, 174, 185, 187, 195, 202, 207, 210, 211, 222, 235–240, 241, 242, 243, 248
 molecular, H$_2$ 28, 155, 159, 160
 HII regions, emission nebulae 6, 27, 29, 30, 31, 32, 34, 38, 41, 45, 74, 87, 88–92, 94, 95, 100, 103, 105, 106, 107, 108, 130, 132, 136, 143, 146–154, 155, 156, 157, 159, 160, 161, 163–167, 168, 170, 173, 174, 184, 185, 186, 187, 188, 193, 194, 195, 196, 197, 202, 204, 207, 213, 219, 221, 222, 225, 227, 228, 230, 231, 234, 235, 237, 239, 241, 248
hypergalaxy 22

initial mass function (IMF) 39, 75, 81, 94, 95, 98, 215, 216
interaction 194, 197, 199, 200, 235, 240
InterCloud region (IC), *see* Magellanic System.
interstellar absorption
 $A_{H\alpha}$ 149
 A_V 89, 90, 91, 92, 170
 $A_{H\beta}$ 90
interstellar extinction, colour excess
 E_{B-V} 6–9, 13, 16, 17, 18, 58, 64, 69, 74, 77, 78, 84, 89, 95, 96, 102, 134, 153, 173, 178, 199, 204, 205, 207, 218, 224
 E_{Bump} 205
 E_{J-K} 92
interstellar extinction curve 173, 177, 205, 206
interstellar lines 33, 35, 175–177, 210–212, 236, 244, 247
interstellar medium (ISM) 4, 28, 55, 79, 120, 143–178, 191, 210, 222, 223, 229, 231, 233, 234, 236
 non-thermal 28, 33, 167, 168, 208
 thermal 29, 37, 166, 167, 207, 208, 240
interstellar reddening, *see* colour excess.
interstellar reddening law 5, 102
ionized carbon (C$^+$, [CII])160–162
IRAS 3, 5, 28, 35, 87, 88, 104, 106, 122, 123, 130, 131, 150, 153, 155, 157, 158, 161, 166, 170, 173, 237
isophotes
 far-UV 38, 39
 HI 28, 143
 45 MHz 30, 31, 168
 408 MHz 30, 31
 in clusters 70
 optical 30, 31, 146
 3.5 m 143
 yellow light 33
IUE 3, 18, 33, 36, 49, 105, 108, 142, 173, 182, 183, 191, 196, 224

Lick expedition 2, 143
LMC
 Bar 6, 7, 21, 27, 29, 30, 31, 32, 33, 47, 48, 50, 68, 69, 73, 94, 109, 115, 116, 120, 121, 122, 128, 132, 144, 151, 153, 169, 170, 179, 185, 187, 193, 196, 236, 237, 242
 disk, plane 28, 32, 107, 109, 144, 145, 165, 167, 169, 176, 202, 211, 222, 232, 236, 240, 241
 tilt of 28
 East–West asymmetry 21, 240
 halo 13, 32, 53, 104, 118, 175, 191, 229, 232, 244
 HI rotation curve 32, 236
 kinematics 235–244
 mass 28
 population centroids 33
 radio spectral index 166
 surface photometry 30
 structure 235–244
 tidal field 71, 73
 tidal radius 23, 62
Local Group 21, 126, 202
Lyman α absorption 90; continuum, photons 90, 138, 147, 153, 160

Magellanic System 4, 21–26, 235–249

Subject index

Bridge/InterCloud region (IC) 7, 9, 21, 25, 35, 37, 38–41, 47, 115, 248
gaseous coronae 33
Magellanic Stream 4, 21, 22, 24, 248
 origin of 35–38
  diffuse ram pressure 36, 37
  discrete ram pressure 36
  primordial 35
  tidal 36
  turbulent wake 36
 tidal effects:
  acceleration 22
  bridge 25
  disruption 22
  disturbance 45
  interaction 22, 23, 202, 242
  perturbation 26
  tail 25

magnetic field 166, 169
 pan-Magellanic 169
magnitudes, absolute
 M_B 46, 83, 218
 M_I 116
 M_V 50, 76, 80, 89, 91, 94, 103, 111, 116, 118, 119, 142, 182, 183, 210, 219, 220, 221, 224, 226, 229
 bolometric M_{bol} 77, 78, 83, 85, 86, 91, 114, 115, 120, 122, 123, 124, 125, 126, 127–132
 bolometric correction (BC) 91, 94
 bolometric luminosity L_{bol} 142, 218, 219

masers
 H_2O 92, 106, 163–165, 209
 methanol (CH_3OH) 164
 OH 92, 106, 130, 163–165
 polarized emission 163
mass-to-light M/L ratio 78
metallicity, [Fe/H], see abundances.
molecular clouds 92, 105, 153, 155, 157, 160, 162, 165, 173, 195, 196, 209, 235, 239
 giant molecular clouds (GMC) 73
molecules 5, 88, 119, 124, 155–165, 170
 CO 5, 82, 88, 89, 92, 106, 120, 124, 150, 155–159, 174, 195, 202, 209, 210, 244
 CO/H_2 ratio 155, 159
 C/O 88, 119
motion of LMC, SMC 23
 non-circular 45
 proper 26–27
 translational 166
 transverse 26–27, 37
 see also velocity.
Mount Stromlo Observatory 115

N-body simulation 25
NLTE 218, 221, 227

Observatoire de Marseille 154
OSO 7 satellite 179

Paczynski relation 114
Parkes 64 m telescope 130
perigalactic(on) 22, 23, 25, 103, 243
photometry (pe, pg, CCD and/or imagery)

BV 46, 76
BVI 10
$BVRI$ 229
$H\beta$ 15, 146
$H\gamma$ 15
IV 124
JHK 87, 123, 125, 127, 128, 209
$JHKL$ 130, 131
$RIJHK$ 127
Strömgren system 213, 224, 229
UBV 49, 91, 96, 215
Walraven 155, 174
planetary nebulae (PNe) 30, 33, 34, 87, 130, 132–141, 173, 199, 234, 242, 244, 246, 248
 element abundances 231
 dimensions 138–141
 evolution 135–137
 helium burners 137, 140
 luminosity 139–140
 luminosity function (PNLF) 133, 134
 nuclei 136
 Peimbert type 138
polarization 6, 82, 168, 169
population centroids 31–34
PPM catalogue 26
pulsars 184, 189, 197, 198

Q-method 89, 215, 216

radio continuum 165–169, 191
radio sources 106, 150, 156, 161, 163, 194, 237
ROSAT 3, 37, 104, 181, 183, 187, 188, 190, 191, 197, 198, 200
r-process elements 229, 230, 234

SAS 3 satellite 179, 189
Seyfert galaxy 184
Shapley constellations 17, 50, 84, 93, 96, 99–103, 109, 116, 121, 122, 128, 144, 151, 185, 187, 237
SMC
 Bar 25, 29, 32, 34, 41, 91, 94, 110, 113, 124, 133, 146, 149, 155, 156, 158, 161, 167, 170, 173, 177, 191, 246, 247, 248
 depth 245–247
 disk 165
 halo 13, 25, 190, 230, 232, 248
 mass 28
 Mini-Magellanic Cloud (MMC) 37, 245, 246
 population centroids 34
 Remnant (SMCR) 37, 245, 246
 surface photometry 31
 tidal radius 61, 247
 wing 3, 21, 34, 38, 39, 41, 50, 99, 109, 113, 146, 147, 149, 165, 170, 177, 179, 249
solar neighbourhood/vicinity 188, 221, 232
Spacelab I 38
Space Shuttle, FAUST telescope 39
speckle interferometry 132, 138
s-process elements 120, 122, 123, 127, 128, 229, 230, 234
star formation 39, 52, 62, 95, 96, 100, 103, 105, 108, 109, 180, 190, 196, 209, 210, 215, 235–240
 bursts of 4, 23, 31, 54, 63, 72, 99, 109, 115, 215, 229, 240
 deterministic self-propagating (DSPSF) 239

star formation (cont.)
  rate (SFR) 121, 147, 148
  stochastic self-propagating (SSPSF) 79, 80, 99, 100, 103, 105, 150, 235–240
stars
  B-type 182, 212, 216, 221, 224, 225, 227, 239
  Be 77, 183, 198
  B[e] 81, 82, 88
  binary 87, 125, 127, 184, 213, 244
    Be/X-ray 9, 201, see also X-ray binaries
  blue stragglers 60
  carbon (C) 25, 34, 40, 61, 63–65, 115, 119, 120, 122, 124, 125, 126, 129, 135, 174, 247, 248
  CH 125, 126, 242, 244
  early-type 8, 15–17, 88, 161, 175
  field 81, 100, 111–131, 221, 223, 226, 227–230
  galactic 126, 177, 180, 189
  horizontal-branch (HB) 5, 25, 56, 59, 60, 114–118, 119, 247
    morphology index 56, 58, 59
    RHB/BHB ratio 116
  J-type 124
  M-type 115, 120, 125–126, 232
  main-sequence (MS) 7, 43, 49, 52, 61, 75, 77, 81, 85, 86, 90, 103, 111–113, 125, 140, 200, 205, 219, 221, 228, 231
  MS-type 120, 124
  multiple 80, 91, 94, 96, 153
  neutron 183
  O-type 79–81, 88, 89, 90, 91, 96, 154, 181, 202, 208, 211, 212, 214–216, 218–220, 239
  OB 6, 7, 17, 20, 32, 34, 75, 79, 80, 81, 92, 93–98, 100, 101, 145, 151, 162, 174, 185, 186, 215, 240
    OB/HII systems 186
  Of/WN 81, 82, 84, 202
  OH/IR 88, 122, 123, 130, 131
    galactic 87, 131
  P Cyg profile/type 81, 82, 83, 84, 96, 141
  post-main-sequence (post-MS) 77, 82
  pre-main-sequence (PMS) 75, 90, 95
  proto- 92, 93, 151, 209
  red-giant- 5, 17, 34, 85, 86, 92, 111, 113–115, 127
    branch (RGB) 5, 63–65, 114, 116, 119
    clump (RGC) 25, 61, 116, 118–119, 247
  S-type 116, 120, 124
  sub-giant branch (SGB) 114
  supergiants 2, 3, 8, 17, 27, 30, 31, 33, 34, 50, 68, 76, 77, 78, 79–81, 82, 85–89, 92, 94, 96, 99, 100, 103, 108, 109, 115, 121, 123, 128, 130, 131, 175, 192, 199, 215, 218, 221, 224, 225, 226–230, 233, 235, 240, 244, 245, 246
    galactic 92, 228, 229, 230
    blue/red (B/R) 86, 87
  supermassive 80, 90, 216
  variables
    cepheids 2, 3, 6, 7, 9–12, 13, 14, 17, 20, 25, 30, 34, 62, 69, 70, 80, 115, 119, 122, 123, 124, 127, 128, 129, 193, 228, 232, 246
    eclipsing 129
    long-period (LPV) 87, 115, 122, 124, 127–129, 131, 243, 244
    luminous blue (LBV) 83–85, 219

Mira 6, 14, 15, 20, 56, 116, 120, 123, 129–130, 243
novae 17, 18, 20, 33, 141, 142, 184
RR Lyrae 6, 7, 12, 13, 14, 20, 32, 34, 56, 57, 59–61, 116–118, 230, 231
  period–amplitude relation 118, 230
  period–luminosity (PL) relation 2, 9, 128, 131
  period–luminosity–colour relation (PLC) 9, 129
S Dor ≡ Hubble–Sandage 81, 83, 84, 96
white dwarf 18, 188
Wolf–Rayet (WR) 5, 79, 82, 85–86, 91, 96, 100, 153, 194, 209, 212, 214–216, 218–220, 239
stellar associations (OB) 30, 32, 34, 39, 41, 44, 45, 50, 79, 93–98, 99, 105, 112, 145, 147, 151, 174, 180, 186, 188, 191, 195, 217, 240
stellar encounters 71
stellar evolution 5, 43, 99, 110, 111, 128
stellar luminosity (L) 82, 114, 128, 130, 131
  amplitude–luminosity relation 83
  function (LF) 76, 96, 103, 111, 112, 128, 217
  Salpeter luminosity function 112, 148
stellar mass 82, 94, 114, 115, 120, 126, 128, 129, 140, 148, 151, 216
  loss 71, 83, 89, 114, 123, 199, 219
  rate 79, 122, 129, 131
  models 43, 52, 75–78
stellar populations 23, 31–34, 42, 58, 59, 60, 71, 73, 86, 87, 92, 108, 109, 110, 112, 116, 243, 245
stellar winds 38, 63, 79, 81, 83, 89, 90, 105, 108, 114, 131, 144, 150, 182, 183, 184, 186, 187, 199, 205, 209, 211, 212, 218
superassociations 4, 9, 21, 31, 41, 45, 79, 99, 108, 109, 110, 144, 151, 193, 222, 236, 240, 242
superbubbles 79, 106, 108, 186, 187, 191, 198
supergiant shells (SGS) 4, 21, 41, 45, 99–110, 149, 150, 151, 185, 187, 201, 213, 237, 240, 245
supernovae (SNe) 79, 105, 135, 190, 239
  explosions 38, 108, 150, 153, 207, 210, 213
  remnants (SNR) 103, 153, 156, 170, 180, 185, 186, 187, 190–199, 213, 220, 232
Swedish/ESO/Submillimetre Telescope, SEST 3, 157, 158, 159, 164
synchrotron radiation 168, 194, 197

temperature, effective T_{eff} 78, 83, 89, 90, 94, 130, 137, 140, 189, 215, 218, 219, 221, 224
temperature, black-body T_{BB} 184, 190

Uhuru catalogue 179
UK 1.2 m Schmidt telescope 132
Uppsala Southern Station 115
ultraviolet (UV)
  absorption features 246
  brightness distribution 237
  colour–magnitude diagram 218
  emission/radiation 28, 82, 100, 140, 144, 155, 157, 159, 160, 162, 165, 173, 177, 189, 195, 202, 239
  extinction 6, 173, 174, 204
  far- 5, 38, 81, 173, 174, 177
  imaging telescope (*UIT*) 205
  light curve 182, 189
  lines 18, 89, 182, 189, 211
  spectrophotometry 139, 140

Subject index

spectroscopy 91
two-colour diagram 57, 66, 68

velocity
  circular 242
  expansion 89, 102, 195, 196, 210, 213
  field 107, 213, 235, 236
  galactocentric 26, 27, 242, 243
  heliocentric 27, 210, 213
  isovelocity lines 236
  Local Standard of Rest (LSR) 36, 102, 105, 154, 156, 159
  microturbulent 224
  radial (RV) 2, 11, 26, 27, 35, 36, 38, 56, 71, 72, 75, 77, 109, 145, 197, 211, 212, 242–244, 245, 246
  shock 209
  system(ic) 105, 197, 241, 244
  tangential 26, 237
  terminal 89
  transverse 27, 241, 242

warps 33, 34

X-ray
  Broad Band Telescope ($BBXRT$) 181
  emission 4, 29, 34, 87, 103, 104, 105, 106, 108, 140, 153, 156, 169, 180–201, 240
  extreme ultrasoft spectra (EUS) 190
  binaries 179–183, 184, 187, 188–190, 191, 198
    low-mass (LMXRB) 180–183, 188
  sources 4, 107, 108, 180–184, 189–190, 213, 220
  spectral region 3, 5
    supersoft $\equiv$ ultrasoft (SSS $\equiv$ SUSO) 181, 184, 188
  superbubble 103, 185
  surface brightness 103, 104
  variability
    non-periodic 179
    quasiperiodic 181

Zeeman doublet 163
Zero-Age-Main-Sequence ($ZAMS$) 15, 17, 39, 81, 84, 88, 90, 96, 218